Earth History and Plate Tectonics
An Introduction to Historical Geology

Carl K. Seyfert
STATE UNIVERSITY OF NEW YORK
COLLEGE AT BUFFALO

Leslie A. Sirkin
ADELPHI UNIVERSITY

HARPER & ROW, PUBLISHERS
New York, Evanston, San Francisco, London

EARTH HISTORY AND PLATE TECTONICS: An
Introduction to Historical Geology

Standard Book Number: 06-045919-0

Library of Congress Catalog Card Number: 72-8262

Contents

Preface

In 1915, Alfred Wegener proposed that the continents were once part of a single large landmass and that the geographic poles had moved relative to this landmass. He based his conclusions largely on the study of the ancient climates of the continents. His ideas were not widely accepted at the time, but during the 1960s, studies of the ancient magnetic field of the earth and the magnetics of the ocean floor stimulated renewed interest in Wegener's ideas. Paleomagnetic evidence provided strong support not only for Wegener's concept of continental drift but also for polar wandering as well. Marine magnetics furnished almost conclusive proof that large blocks of sea floor were moving relative to each other. At the present writing, most geologists accept the ideas of continental drift, polar wandering, and sea-floor spreading, although they may disagree about how and when these movements have taken place.

Acceptance of these concepts requires a great many revisions in the account of the physical history of the earth. We have undertaken the task of incorporating the idea of large-scale crustal movements and plate tectonics into the content of historical geology. The evidence for and against large-scale crustal instability is presented, and then, using the necessary global approach, we attempt to trace the history of continental movements. Finally, we explore fully the consequences of large-scale crustal movements on the structural development of the earth and the development of both life and climates.

Some geologists may object to this type of approach, maintaining that the theories on which we are basing this text are not sufficiently well-established. However, we contend that the concepts of continental drift, polar wandering, and sea-floor spreading are more reasonable than the concept of a static crust. When writing a history of the earth, it is not possible to reserve judgment on whether or not the continents, poles, and sea floor have moved, for in doing so, we automatically accept the concept of a static crust. Nor can such a project be accomplished by considering a single continent. Whatever continent is studied, there is likely to be geologic evidence, such as sources of sedimentation or glaciation which may be traced beyond the present limits of the continental margins. Similarly, regional deformations cannot be fully understood without considering crustal movements, that is, plate tectonics, on a global scale.

It is a major purpose of this text to integrate modern and classical concepts of geology into a workable earth history. We make use of some recently published data in an attempt to come closer to the solution of some old geological problems. We are aware that our use of the latest publications as material for a geological text introduces the possibility that some of the data or interpretations may be proven inaccurate with time. In order to render this book as useful as possible, we have tried to separate the basic data from the interpretions drawn from those data.

The text may be divided into two parts. The first seven chapters introduce the student to the fundamental concepts of historical geology and to the means by which these concepts are formulated. The remaining chapters are devoted to a discussion of the geologic history of the earth. Included are such topics as the location and shape of ancient landmasses, ocean basins, and mountain ranges, and the evolution of plants and animals.

We wish to thank the many people who helped in the preparation of this book. Parts have been read critically by Parker E. Calkin, Anthony E. Cok, G. Gordon Connally, Rhodes W. Fairbridge, Robert M. Finks, Richard Foster, Dennis S. Hodge, James R. Orgren, Allison R. Palmer, Paul H. Reitan, Harold F. Roelling, William D. Romey, Holmes A. Semken, James W. Skehan, Robert Stein, Irving H. Tesmer, Fred R. West, and E-an Zen. Rotations of coastlines and paleomagnetic poles were preformed using the HYPERMAP program of Robert L. Parker. Karen Seyfert, the wife of Carl Seyfert, provided valuable assistance in editing the manuscript and drafting many of the illustrations. Thanks are also due the secretarial and library staff of Buffalo State University College and Adelphi University, especially Mrs. Myra Ball and Miss Bonnie Fletcher.

Carl K. Seyfert
Leslie A. Sirkin

1
Introduction

*T*he geological sciences are presently experiencing a revolution comparable to the Copernican revolution in astronomy, the Darwinian revolution in biology, and the Einsteinian revolution in physics. Those profound changes in scientific thought were all focused on single concepts: in astronomy, heliocentricism—in biology, evolution—and in physics, relativity. For geology the unifying concept has been continental drift, first developed into an integrated theory by the German meteorologist, Alfred Wegener. Consequently, the revolution in geology could be called the Wegenerian revolution.

Attention to geological problems by scientists from other fields has given new directions and emphases in the earth sciences, particularly through studies of the ocean basins and of the moon. Chemists, physicists, astronomers, and oceanographers have been attracted by the exciting prospect of achieving a greater understanding of the origin and development of the earth. Newly created fields in marine geochemistry, marine geophysics, and astrophysics reflect these interdisciplinary interests.

More specifically, the marine sciences have radically changed our view of the stability of the earth's crust. For years, geologists believed the earth to be a relatively rigid body in which only minor crustal movements could occur. In the 1960s, oceanographic and seismic studies brought forth impressive evidence of the creation of new sea-floor at mid-ocean ridges and of the movement of tremendous blocks of crust away from these ridges.

As a result of these studies, the concept of sea-floor spreading has received widespread acceptance. Paleontologic, paleomagnetic, paleoclimatic, and geochronological studies have lent additional support to Wegener's contention that the continents were once part of a single landmass. The same lines of evidence also suggest that this landmass itself had been formed by an earlier joining of several smaller continents. Moreover, the presence of extremely ancient mountain belts on the margins of these ancient continental blocks indicates that there has been an underflow of the sea-floor at the margins of one continental block or another at various times for several billion years.

THE HISTORY OF GEOLOGIC STUDY

Early investigations of the earth were confined primarily to isolated observations of geological features; only minimal speculation was focused on their origin. There was little of scientific value in these studies, since a framework did not exist into which new ideas and observations might be placed. It was not until the seventeenth and eighteenth centuries that the fundamental geological principles were established. These principles provided the basis for assembling a reliable account of the formation and development of the earth and the life on it.

The ancient Greek and Roman philosophers believed that the earth was at the mercy of capricious gods, a supposition which all but discouraged the study of geologic features. How-

Figure 1.1 Fault scarp produced during a violent earthquake near West Yellowstone, Montana, in August, 1959. The uplift of the block in the background occurred in less than a minute. (Photo by J. B. Hadley, courtesy of the U. S. Geological Survey.)

ever, Xenophanes in the fifth century B.C. studied the fossil fish and sea shells of Italy and concluded that the sea had once covered the areas where these fossils were found.

The importance of water in shaping the land, both by erosion and deposition, was perceived in the fourth century B.C. by Aristotle who pointed out that the Nile River Delta had been built up by the slow deposition of sediments from the river. He also suggested that deposition of these sediments displaced seawater and thus caused a rise in sea level. Strabo, a Greek who lived in Asia Minor during the first century A.D., stated that land was often uplifted following earthquakes (Fig. 1.1), and that this type of uplift had been important in shaping the earth. He also recognized that land could be uplifted quite slowly without perceptible disturbance (Fig. 1.2).

With the fall of the Roman Empire, the people of Europe were more concerned with defense against foreign invasions than with the cause of science. Although the center of learning shifted to the Middle East, little scientific progress was made beyond that of the Greeks. During the eleventh century, the philosopher Avicenna noted that soft mud hardened into stone with time and that water dripping from the roof of a cave also turned into stone. Avicenna concluded that the cause of the petrifaction was a mystic "congealing virtue." His concepts of erosional processes were more sound. For example, he stated that mountain peaks were shaped by wind and streams over a long period of time and that softer rocks were eroded to form valleys between ridges of more resistant rocks.

Four hundred years after Avicenna, Leonardo da Vinci (1452 - 1519) observed and speculated on the origin of a number of geologic features. He recognized the organic nature of fossils and postulated that marine forms which were found in rocks

Figure 1.2 Raised marine terraces on the north coast of the island of Islay, western Scotland. Such terraces provide evidence of slow uplift of the land relative to the sea. (Crown copyright Geological Survey photo. Reproduced by permission of the Controller, Her Britannic Majesty's Stationery Office.)

well above sea level indicated either that the land had been uplifted or that the sea level had been lowered.

STENO'S PRINCIPLES

It was not until the seventeenth century that detailed investigations were made of rocks and the fossils they contain. In 1669 Nicolas Steno pub-

lished a study of the mountainous region around Tuscany, Italy, and proposed in this work several basic geologic principles. Of lasting importance were his observations that sedimentary strata are deposited layer upon layer, that younger beds are laid down on top of older beds, and that strata are initially deposited in horizontal layers (Fig. 1.3).

Steno's observations, today known respectively as the *principle of superposition* and the

Figure 1.3   The horizontal strata of the Grand Canyon record 300 million years of geologic history. (Santa Fe Railway photo.)

principle of original horizontality, have enabled geologists to determine relative ages of strata, to recognize rock deformation, and to compare rock units in structurally complex areas with sequences in areas that have not been significantly deformed. The principles are basic to the reconstruction of regional structural geology and the history of the earth itself. Steno was far ahead of his time however, and most of his ideas were rejected by his contemporaries. It was not until the nineteenth century that these principles finally received general acceptance.

CATASTROPHISM

Until the nineteenth century, clergymen and scientists alike believed that the earth had been created only a few thousand years ago. With so little time available for the formation and development of the earth, it is not surprising that many scientists believed that sudden, violent forces or catastrophes had shaped the earth's surface. This philosophy has become known as "catastrophism." Some catastrophists were convinced that canyons, such as the Grand Canyon, were simply giant cracks in the ground formed during a series of violent earthquakes and that tidal waves and floods were caused by huge meteorites or comets that struck the earth.

Other catastrophists postulated that the thick sequences of sedimentary rocks in the earth's crust had been deposited during the worldwide flood depicted in the Bible, that is, Noah's flood. They thought that fossils found in these rocks were the remains of organisms killed by the deluge. Still others realized the difficulty of accounting for the

Figure 1.4 Anchorage, Alaska, following the "Good Friday" earthquake of March 27, 1964. Such displacements are produced by the slumping of unconsolidated sediments as a result of the passage of seismic waves. (Courtesy of the U. S. Geological Survey.)

Figure 1.5   The Missouri River near Plattsmouth, Nebraska, during a flood in April 1952. (U.S. Army, Corps of Engineers.)

apparent changes in fossils from older to younger beds with only one flood. William Buckland, a theologian who became the popular professor of geology at Oxford University during the early nineteenth century, accommodated this discrepancy by adding two more worldwide floods to the original deluge. This led to the belief that practically all life had been sucessively wiped out by such floods and replaced by new forms, which resulted from separate, divine creations.

Georges Cuvier (1769 - 1832), a leading vertebrate paleontologist of his day, proposed that all the organisms that had existed from time to time in Europe had perished in a series of such floods. Cuvier postulated that these organisms had subsequently been replaced by migrations from other localities. He believed that the floods were caused by "instantaneous rather than gradual" episodes of crustal subsidence, which alternated with sudden crustal uplifts. Cuvier was an influential man whose catastrophist teachings were widely accepted in both Europe and North America.

Although most of the catastrophists' proposals were related to observation of natural events, they were generally distorted by gross exaggerations in scale. For example, uplift and subsidence of the crust do occur, but generally at a very slow rate. Earthquakes do cause cracks in the

earth, but the width of such cracks is generally not more than a few feet (Fig. 1.4). Floods are common, but always limited in extent, never engulfing an entire continent, much less the entire earth (Fig. 1.5).

NEPTUNISTS

The neptunist school was established in the late eighteenth century by Abraham Werner, a professor at the Freiberg Mining Academy. He adopted a theory, originally proposed by Johann Lehmann, which stated that the majority of the earth's rocks had been chemically precipitated in a universal sea. Although he had observed only the rocks in the vicinity of Freiberg, Werner assumed that the rock formations were the same everywhere. A persuasive and popular teacher, Werner presented his theories with authority and tolerated no criticism. Students came from all over Europe to study under this man who inspired so much confidence in his ideas.

Werner called the oldest formation of the earth the "Primitive" or "Primary." It consisted of granite and presumably had been precipitated out of a murky, universal sea onto the nuclei of the continents. It was assumed to have become subsequently mixed with gneiss, schist, porphyry, basalt, and marble. The overlying "Transitional" strata also consisted mainly of precipitates but included some sediments weathered from the primary islands that were exposed as the sea retreated. Rocks belonging to the Transitional strata included sparsely fossiliferous slate, graywacke, and limestone.

As the waters of the sea further receded, according to the neptunists, sandstone, conglomerate, limestone, chalk, and coal became more abundant. This division, the "*Flötz*," was characterized by steeply inclined bedding planes, due presumably to precipitation on the slopes of undersea mountains and to slumping of soft sediments from inclined slopes. The youngest deposits, the "alluvial," were deposited by streams following the retreat of the sea from the continents. This category included clay, peat, sand, gravel, and some volcanic materials. Werner believed that volcanic eruptions were of minor importance and were caused by the burning of coal beds in the "*Flötz*."

UNIFORMITARIANISM

In opposition to the catastrophists and the neptunists, there grew up a school of geologists who believed in the uniformity of nature. They reasoned that the development of the earth's crust could be best understood by observing geologic processes at work today. In 1788 James Hutton, a Scottish geologist, published a paper entitled *Theory of the Earth*; this paper laid the groundwork for the modern science of geology and led to the eventual abandonment of neptunism and catastrophism.

The formation of ancient rocks, according to Hutton, could be explained without invoking any processes other than those which could be directly observed. Hutton believed that most geologic phenomena could be understood through careful observations of modern processes, and that all geologic processes that operated in the past also operate at the present time. Faith in the constancy of these processes became the essence of the *principle of uniformitarianism*. The history of the

earth could then be interpreted without positing unpredictable catastrophes, divine intervention, or processes that could not be observed or tested. The phrase "*the present is the key to the past*" sums up the concept of uniformitarianism.

When Hutton published his ideas, Werner had been teaching at Freiberg for 20 years, and the neptunist philosophy was widely accepted. In opposing neptunism, Hutton insisted that granites had crystallized from hot molten rock reservoirs or "magmas" situated deep in the earth and that basalts had flowed from volcanoes whose superficial parts had, in many cases, long since been obliterated by erosion. He further observed that sand and mud were transported to the sea by streams and rivers and that these sediments were deposited in the ocean. Ultimately, sandstone and shale were formed by the hardening or "lithification" of these sediments, whereas the cementation of small fragments of calcareous shells produced limestone. Hutton's ideas seemed impossibly overdramatic to the neptunists, who persisted in the belief that all important rock types were chemical precipitates from cold water.

Even though Hutton's ideas were backed by careful field observations, his paper was written in such a difficult style that it was not widely read. An associate, John Playfair, realized this fault and in 1802 published what has become a scientific classic, *Illustrations of the Huttonian Theory of the Earth* (1) in which Hutton's ideas were clarified.

Uniformitarianism received increasing support from a group of geologists who were known as the "plutonists" because of their belief in the importance of heat as a factor in the formation of certain rocks. (Pluto was king of the underworld in Roman mythology.) The debate which ensued between the neptunists and plutonists has been one of the most bitter and drawn out clashes in the discipline of geology. The neptunists persisted in claiming that basalts were chemically precipitated from the sea, and the plutonists believed them to be crystallized from a magma.

The work of Nicholas Desmarest (1725 -

1815) on the basalts of Auvergne, in France, greatly strengthened the plutonists' case. Desmarest traced basalt to extinct volcanic cones, which were identical in form to active cones such as Mount Vesuvius, and he concluded that the basalt had crystallized from lava. Such empirical observations, combined with Werner's eventual retirement as a teacher, led to the ultimate abandonment of neptunism (Fig. 1.6).

Gradually catastrophism also lost considerable support in favor of uniformitarianism. In fact, Charles Lyell used the principle of uniformitarianism as the unifying theme in his highly successful textbook, *Principles of Geology,* the first volume of which appeared in 1830. Lyell became interested in geology when, as a law student at Oxford, he took a course in geology from William Buckland. Vivid descriptions by the Reverend Buckland of the conflict between the neptunists and the plutonists sparked Lyell's interest.

Lyell used one of his vacations to study the

Figure 1.6 Vertical granitic dike cutting Triassic sandstones and diabase. It was observations such as this that led Hutton to conclude that granites were intruded in the molten state rather than precipitated on the sea floor. (Crown Copyright Geological Survey photo. Reproduced by permission of the Controller, Her Britannic Majesty's Stationery Office.)

problem in the field and became convinced of the value of direct observation. Becoming a staunch supporter of uniformitarianism, he postulated that geologic processes occurred at about the same rate in the past as they do at present. Uniformity in the rate of change became known as *gradualism,* a concept often confused with uniformitarianism, but now known to be an oversimplification. There is much in the geologic record to indicate that geologic processes do *not* always operate at the same rate today as they did in the past. Volcanic activity, glaciation, and deposition rates, among other factors, have varied considerably from time to time throughout geologic history.

GEOLOGIC MAPS

Before the eighteenth century, the geologic literature consisted of descriptions of the rocks at different localities. No attempt was made to relate the rocks of one locality to those of another. In the latter part of the eighteenth century, William Smith, an English surveyor, observed that definite relationships existed between rock units in different localities. While supervising the building of a canal in the west of England, Smith traced certain rock units over hundreds of square miles and always found them in the same distinctive sequence.

In his travels throughout the country, Smith also observed that each stratigraphic level likewise had its own distinctive fossil content. He noted that "the same strata were found always in the same order and contained the same peculiar fossils." These observations gave rise to the *principle of faunal and floral succession,* which emphasizes the succession of fossil assemblages through geologic time and the correlation of rock sequences based on such fossil assemblages.

Using lithologic character and fossil content as a means of tracing strata, William "Strata" Smith, as he came to be known, published in 1815, a hand-colored geologic map of England, Wales, and part of Scotland. The map showed the boundaries between different rock units (Fig. 1.7). The accompanying detailed descriptions of the rock

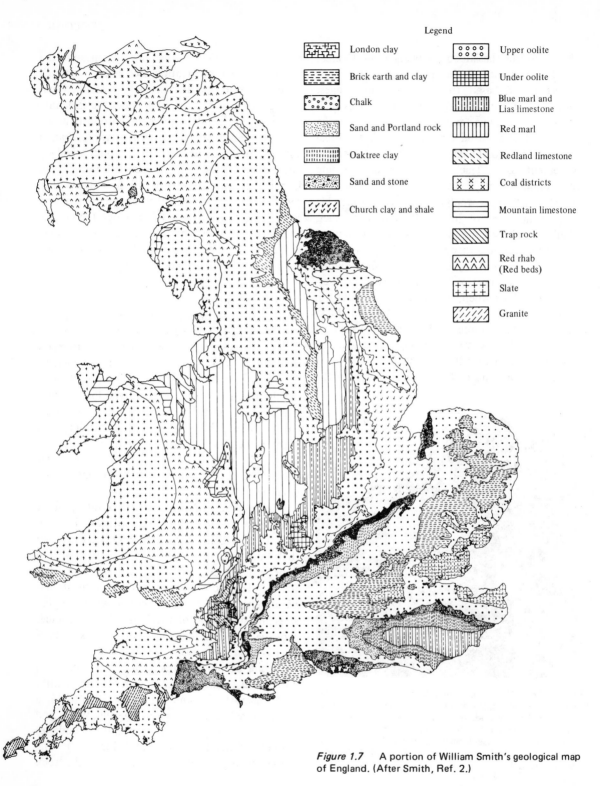

Legend

London clay		Upper oolite	
Brick earth and clay		Under oolite	
Chalk		Blue marl and Lias limestone	
Sand and Portland rock		Red marl	
Oaktree clay		Redland limestone	
Sand and stone		Coal districts	
Church clay and shale		Mountain limestone	
		Trap rock	
		Red rhab (Red beds)	
		Slate	
		Granite	

Figure 1.7   A portion of William Smith's geological map of England. (After Smith, Ref. 2.)

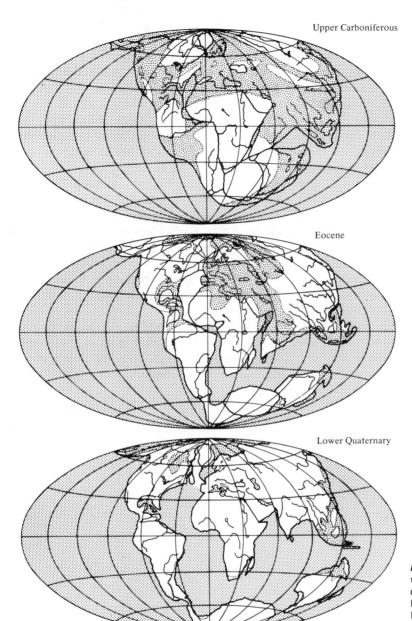

Upper Carboniferous

Eocene

Lower Quaternary

Figure 1.8 Wegener's reconstruction of the continents during three different intervals. (After Wegener, Ref. 4, printed by permission of Dover Publications, Inc., New York.)

units, which included thicknesses, lithologies, and fossils, were the forerunners of modern *columnar sections.* These vertical sections graphically depicted the sequence and resistance to weathering of rock units from the oldest on the bottom to the youngest on the top of the column. Smith also made *geological cross sections* or profiles, showing the rock strata exposed in hypothetical vertical slices in the earth's crust. Through the use of geologic maps, columnar sections, and cross sections, Smith illustrated the stratigraphy and outcrop pattern of the rock units of England and provided the science of geology with some of its most valuable tools.

CONTINENTAL DRIFT

Early maps of the world clearly illustrated a "jigsaw puzzle fit" of the shorelines of Africa and South America. As early as 1801, Alexander von Humboldt pointed out that not only were the coastlines parallel, but the rocks of the opposite coasts were similar as well. However, von Humboldt postulated that these phenomena had resulted from erosion of a single large continent by marine currents (3). Antonio Snider in 1858 suggested that the similarity of the shorelines on opposite sides of the Atlantic was due to a catastrophic splitting apart of the continents. He proposed that America might even be the ancient continent of Atlantis (3). At that time, however, the scientific world did not seriously consider the possibility of continental drift.

In the early part of the twentieth century, Alfred Wegener, the German meteorologist, also became impressed by the similarity of opposing coastlines. Wegener found the problem intriguing and for several years gathered data on ancient climates, paleontology, and the structural history of the continents. In 1915, Wegener put forth his observations and interpretations in a book entitled *Die Enstehung der Kontinente und Ozeane (The Origin of Continents and Oceans).* He proposed that the present continents once comprised one large landmass which he named Pangaea (Fig. 1.8).

According to Wegener, Pangaea began to break apart and the individual continents started to move toward their present positions during the Mesozoic Era. Wegener's work also provided evidence of movement of the earth's rotational poles relative to Pangaea. It is now believed that there is both a shift of the crust of the earth relative to the poles, as well as displacement of individual continents.

Wegener's book, which was translated into English in 1924, created a storm of controversy among geologists and geophysicists. Most of them rejected Wegener's conclusions and the evidence

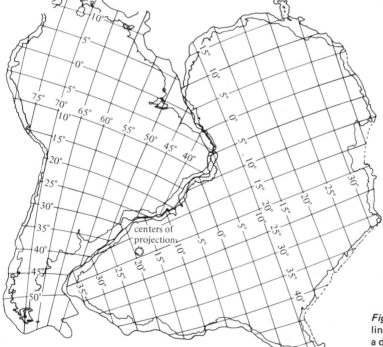

Figure 1.9 Matching of the outlines of South America and Africa at a depth of 2000 m below sea level. (After Carey, Ref. 5.)

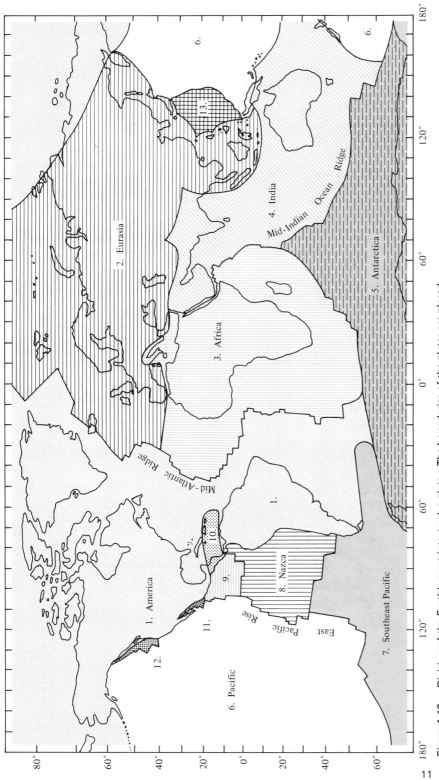

Figure 1.10 Division of the Earth's crust into twelve plates. The boundaries of the plates are placed at mid-oceanic ridges (e.g., the Mid-Atlantic Ridge), trenches (e.g., Peru-Chile Trench), great faults (e.g., the San Andreas Fault), and active fold belts (such as the Himalaya Mountains). (Adapted from Morgan, Ref. 6.)

11

supporting continental drift because the author had failed to offer a really convincing mechanism, and because of some minor errors and inconsistencies.

In the 1950s, the results of measurements of the earth's ancient magnetic field provided quantitative evidence of both continental drift and movement of the poles relative to the continents. At the same time, an Australian geologist, Samuel Warren Carey, showed that the outlines of South America and Africa matched almost exactly at a depth of 2000 m below sea level (Fig. 1.9). At this depth, approximately half-way down the continental slope, both erosion and deposition would have been minimal since the formation of the continental margins. Chester Longwell remarked that "if the fit between South America and Africa is not genetic, surely it is a device of Satan for our frustration" (5).

The majority of earth scientists did not support the concept of continental drift, however, until the publication of the studies on marine geomagnetic anomalies associated with mid-ocean ridges, such as the Mid-Atlantic Ridge and East Pacific Rise. These studies provided convincing evidence for the movement of large blocks of sea floor relative to one another.

In 1968, W. Jason Morgan (6) introduced the concept of plate tectonics in which the earth's crust is considered to be divided into a series of rigid plates bounded by mid-oceanic ridges, oceanic trenches, great faults, and active fold belts (Fig. 1.10). According to this theory the movements of the continents and the sea-floor are part of large-scale movements of plates. Subsequent chapters will develop the many facets of plate tectonics and integrate these concepts into the history of the earth.

REFERENCES CITED

1. J. Playfair, 1956, *Illustrations of the Huttonian Theory of the Earth:* Facsimile reprint by University of Illinois Press, Urbana.

2. W. Smith, 1815, *A Delineation of the Strata of England and Wales with Part of Scotland:* S. Gosnell, London.

3. A. V. Carozzi, 1970, New historical data on the origin of the theory of continental drift: *Geol. Soc. Amer. Bull.,* v. 81, p. 283.

4. A. Wegener, 1966, *The Origin of Continents and Oceans:* Dover, New York.

5. S. W. Carey, 1958, *Continental Drift: A Symposium:* Geology Department University of Tasmania, Hobart.

6 W. J. Morgan, 1968, Rises, trenches, great faults, and crustal blocks: *J. Geophys. Res.,* v. 73, p. 1959.

SUGGESTED READINGS

F. D. Adams, 1938, *The Birth and Development of the Geological Sciences:* Dover, N.Y.

E. W. Berry, 1925, Shall we return to cataclysmal geology?: *Amer. J. Sci.,* v. 17, p. 12.

C. L. Fenton and M. A. Fenton, 1952, *Giants of Geology:* Doubleday, Garden City, N.Y.

N. D. Newell, 1956, Catastrophism and the fossil record: *Evolution,* v. 10, p. 97.

N. Steno, 1958, *Canis Carchariae Dissectum Caput:* transl. A. Garboe, St. Martin's Press, New York.

K. Von Zittel, 1962, James Hutton, *in* J. F. White, ed., *Study of the Earth-Readings in Geological Science:* Prentice-Hall, Englewood Cliffs, N.J., p. 11.

2
Geologic Time

The development of a radiometrically dated geological time scale during recent decades has provided a standardized chronology with worldwide applicability. Thus the histories of different regions may be directly compared. Such comparisons of similar rock units on different continents have furnished evidence for continental drift. Moreover, with the study of meteorites and with lunar exploration a reality, radiometric dating has enabled geologists to relate the history of the Earth to that of the Moon and other planets in the solar system. This has yielded data for the formulation and testing of theories on the origin of the Earth-Moon system and planetary systems in general.

According to the principle of superposition, the relative ages of strata increase with depth, in a sequence of sediments that has not been disturbed by folding or faulting. However, in structurally complex regions marked by intense fracturing and crumpling, the relative ages of beds must be determined by other means.

In reconstructing the history of the earth, geologists have developed methods of dating and arranging geologic events in meaningful, chronologic sequences based on the relative or absolute ages of rock units. Relative ages are established through the use of primary structures, the sequence of strata, unconformities, cross-cutting structures, and fossils; absolute ages are measured by various "nuclear clocks."

RELATIVE TIME

Primary Structures

Primary structures, which are produced at the time of deposition of a rock, include layering, cross-bedding, graded bedding, ripple marks, and pillow lavas. Cross-beds are produced by wind or subaqueous currents and are inclined at angles up to 40° to the horizontal. Erosion may truncate the tops of the cross-beds, but the bottoms of the beds are generally parallel to the underlying beds (Fig. 2.1a). Thus the relative positions of the truncated and tangent portions of cross-beds may be used to establish the relative age of a sedimentary sequence.

In graded beds, the particle size varies from coarse to fine. Graded bedding is produced by intermittent submarine currents such as turbidity currents. As the current slows, coarser particles settle out first, followed by successively finer particles. In general, the particle size decreases in the direction of the younger beds (Fig. 2.1b).

Ripple marks are produced by both wind and water currents. Oscillating water currents produce nearly symmetrical ripples with pointed crests and rounded troughs (Fig. 2.1c). In a bedded sequence, the angular crests point toward the younger beds. Translational currents produce ripples having a steep face in the direction of current flow (Fig. 2.1d). The crests of current ripples are generally more pointed than the troughs, and therefore their

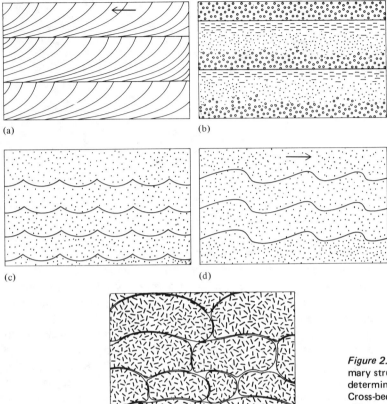

Figure 2.1 Diagrammatic sketches of primary structures that may be used in the determination of the "tops" of beds: (a) Cross-bedding, (b) graded bedding, (c) oscillatory ripple marks, (d) current ripple marks, (e) pillow lavas.

shape may be used to determine relative ages. Other bedding plane structures, such as tracks, raindrop prints, and erosion channels, have their concave surfaces on the original top of the bed.

Pillow lavas are masses of volcanic rock which were probably produced by submarine volcanic eruptions. Somewhat rounded in cross section, they have convex upper and concave basal surfaces. Projections on the bottoms of some pillows point toward the older beds (Fig. 2.1e).

Unconformities

In many sequences of sediments, not all the layers that were originally deposited are preserved. Uplift may have resulted in erosion surfaces which may have been subsequently covered by younger sediment. Buried erosion surfaces are termed uncon-

formities. Some varieties of unconformities are nonconformities, angular unconformities, and disconformities. Unconformities may be used in conjunction with metamorphic grade and intensity of deformation to determine the relative age of a sequence.

At a *nonconformity*, sedimentary rocks overlie older metamorphic or plutonic igneous rocks (Fig. 2.2a). An *angular unconformity* separates tilted or folded sedimentary rocks from younger, less deformed sedimentary rocks (Fig. 2.2b). A *disconformity* is a buried erosion surface between beds that are essentially parallel to each other (Fig. 2.2c). A break in sedimentation between two layers may be recognized if fossils indicate a significant difference in age between the two layers. By comparison, breaks in sedimentation,

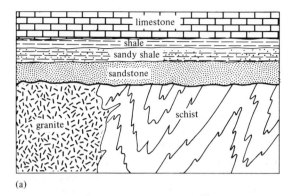

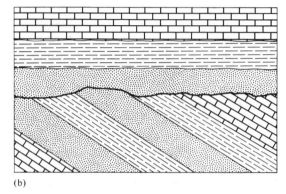

(a)

(b)

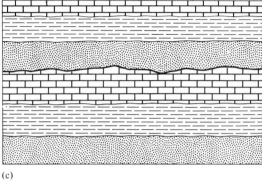

(c)

Figure 2.2 Diagrammatic sketches illustrating three types of unconformities: (a) Nonconformity, (b) angular unconformity, (c) disconformity.

which represent short time intervals, are termed *diastems*.

Cross-Cutting Structures

Cross-cutting igneous intrusions, such as dikes, stocks, and batholiths, are always younger than the youngest beds which they intrude and older than the oldest bed which unconformably overlies the intrusion (Fig. 2.3a). The age of one igneous intrusion relative to another may also be determined by cross-cutting structures. If two intrusions are in contact, dikes of the younger intrusion commonly intrude the older intrusion (Fig. 2.3b).

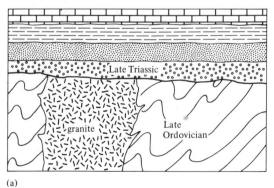

(a)

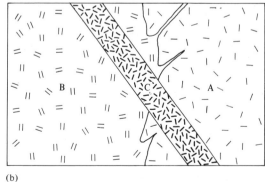

(b)

Figure 2.3 Relative age determination of igneous rocks: (a) Granite pictured was emplaced between the Late Ordovician and the Late Triassic, (b) multiple episodes of igneous activity—intrusion A was emplaced first, followed by B and then C.

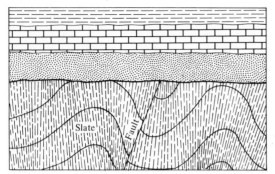

Figure 2.4   Fault and fracture cleavage terminating at an unconformity.

Using similar reasoning, it is possible to determine the relative time during which movement along a fault occurred. The final movement along a fault must have occurred after the deposition of the youngest beds that are displaced by the fault and before the deposition of the oldest bed that unconformably overlies the fault (Fig. 2.4)

THE GEOLOGICAL TIME SCALE

One of the first geological time scales was developed in 1756 by Johann Lehmann, who recognized three ages of rocks:

Primitive—all crystalline rocks, such as granite and gneiss

Secondary—consolidated sedimentary rocks containing fossils

Alluvial—soils and gravels

A subsequent time scale, proposed in 1760 by Giovanni Arduino, an Italian mining geologist, listed four ages:

Primitive—crystalline rocks in the cores of mountains

Secondary—sedimentary rocks

Tertiary—unconsolidated sediments

Volcanics—extrusive igneous rocks

Of these terms, "Tertiary" alone has survived in the modern system of chronologic nomenclature.

In 1778 Georges de Buffon, a French naturalist, proposed a new historical time scale which recognized six epochs. The epochs were cast in

biblical conformity and were accompanied by a precise chronology. Buffon, a pre-Wernerian neptunist, believed in the growth of continents and "the separation of the old and new world," which may be interpreted as an early consideration of continental drift.

The Modern Time Scale

The modern time scale was developed during the nineteenth century (Table 2.1). Following the lead of William Smith, geologic periods were established on the basis of relative ages and the fossil content of sequences of rock. The Paleozoic (from the Greek: ancient life) periods were established through the works of several English geologists. In 1831 Adam Sedgwick began a detailed study of a sequence of deformed sedimentary and volcanic rocks in northern Wales. After several years of work, he published a description of the rocks in this area and proposed that the sequence be named the "Cambrian System" after Cambria, the Roman name for Wales. At about the same time, Roderick Murchison published a description of the rock sequence in southern Wales, which he named the "Silurian System," after an ancient Welsh tribe, the Silures.

A study of the fossils in Sedgwick's Upper Cambrian System showed that they were almost identical to those of Murchison's Lower Silurian System. Evidently, these two rock units were overlapping, and a bitter feud developed between the two men over where to place the disputed rocks. The conflict was not resolved until 1879, when Charles Lapworth recognized that this intermediate sequence was in fact distinctive and proposed that it be placed in a new system, the "Ordovician System," named after another ancient Welsh tribe, the Ordovices.

The "Carboniferous System" was named in 1822 by two English geologists, William Coneybeare and William Phillips. The name was chosen because large quantities of coal occur in many sequences of this age. North American geologists felt that the Carboniferous should be subdivided into two systems, the Mississippian and the Pennsylvanian. The Mississippian System, defined by Alexander Winchell in 1870, corresponds approxi-

Table 2.1   The geologic time scale

	Era	Period	Age of beginning, millions of years ago		
			(Kulp [5])	(Holmes [6])	(Harland et al. [7])
Phanerozoic	Cenozoic	Quaternary	1	1	1.5 - 2
		Tertiary	63	70	65
	Mesozoic	Cretaceous	135	135	136
		Jurassic	180	180	190 - 195
		Triassic	230	225	225
	Paleozoic	Permian	280	270	280
		Pennsylvanian	310		325
		Mississippian	355	350	345
		Devonian	405	400	395
		Silurian	425	440	430 - 440
		Ordovician	500	500	500
		Cambrian	600	600	570
Precambrian	Proterozoic	Late	1000		
		Middle	1700		
		Early	2500		
	Archean		4600		

mately to the Early Carboniferous. The Mississippian System takes its name from the Mississippi River, along which strata of this age are exposed. The Upper Carboniferous is approximately equivalent to the Pennsylvanian System. Rocks of this age are widely exposed in the state of Pennsylvania, where they were mapped and named by Henry Shaler Williams in 1891.

Sedgwick and Murchison together studied a thick marine sequence in the County of Devon in southwestern England. At first they thought, based on the lithology of the beds, that this sequence was either Cambrian or what is now called Ordovician. However, the paleontologist William Lonsdale studied the fossils in this sequence and concluded that they were intermediate in development between those of the Silurian and Carboniferous systems. Ultimately, Sedgwick and Murchison accepted Lonsdale's conclusion and named these rocks the "Devonian System"

Murchison is also credited with defining the "Permian System." On an invitation from the Czar of Russia, Murchison went to the European part of Russia to study the strata of the region. Just west of the Ural Mountains, he found a very thick sequence of rocks which contained fossils younger in appearance than those of the Carboniferous System and yet older than those of a unit known as the "Triassic." He named this sequence after the Russian province of Perm.

The three subdivisions of the Mesozoic (Greek: middle life) were named after rock units or sites in western Europe. The type region for the "Triassic System" is in central Germany, where this unit consists typically of two continental rock sequences (lake and river deposits) separated by a marine sequence. The name Triassic refers to the tripartite nature of the unit, which was named by Von Alberti in 1834.

The "Jurassic System" is named for the Jura Mountains, which form the border between France and Switzerland. The rocks of the Jurassic were first studied by the German geologist von Humboldt in 1799, and have become famous for their richness in fossils.

The "Cretaceous System" includes the chalk units exposed in the Chalk Cliffs of Dover, and in other parts of the London and Paris Basins; the name is derived from the Latin word for chalk, *creta*. These deposits were studied extensively by the Belgian geologist d'Halloy, who named them in 1822. D'Halloy subdivided the system into seven stages based on their characteristic fossils.

Finally the names for the periods of the Cenozoic (Greek: recent life), the Tertiary and Quaternary, reflect the original time scale terminology of Lehmann and Arduino. The deposits of these two periods are largely unconsolidated sediments and contain a wide variety of modern-looking organisms, although many are now extinct.

EARLY ESTIMATES OF THE DURATION OF GEOLOGIC TIME

In 1658 Archbishop James Ussher, Primate of Ireland, set the year of the earth's creation as 4004 B.C. This date, based on biblical chronology, was no longer acceptable to the geologists of the late eighteenth and nineteenth centuries. Even neptunists, such as Werner and Buffon, felt that at the least tens of thousands of years were required for the deposition of the great thicknesses of sedimentary rocks in the earth's crust.

Rate of Deposition

As early as the fifth century B.C., the Nile River Delta had been studied by Herodotus, who reasoned that the delta must be thousands of years old, given the yearly rate of deposition of sediment. Similar attempts have been made to determine the age of strata deposited since the beginning of the Paleozoic Era. The maximum cumulative thickness of Phanerozoic (post-Precambrian) strata from various parts of the world has been estimated at 450,000 ft (1). A deposition rate of 1 ft of sediment per 1000 years for 450,000 ft of rock would yield an age for the beginning of the Cambrian of nearly half a billion years, which is remarkably close to that determined by radiometric dating. The age determined by this method is only a minimum for that of the earth, since Precambrian deposits are not included in the estimate of the total thickness of strata, and because of the difficulty of making accurate measurements in contorted and fractured formations.

Salt in the Oceans

Another early attempt to determine the age of the earth involved calculating the total amount of salt in the ocean compared with the amount of salt added yearly. In this scheme, it was assumed that

the oceans originated early in the earth's history as bodies of fresh water. It was estimated that the total amount of salt in the ocean was 16 quadrillion (16×10^{12}) tons and that 160 million tons of salt was added yearly, carried there in solution by rivers and derived from the weathering of rocks and soil. Using this line of reasoning, the age of the earth was placed at approximately 100 million years. However, these calculations ignore the vast quantities of salt which are present in sedimentary sequences in many parts of the world.

Rate of Cooling

Lord Kelvin, the nineteenth century English physicist, set the age of the earth at 70 million years based on the earth's present cooling rate. Kelvin assumed that the earth originated as a molten ball which had been pulled out of the sun and that the original temperature of the earth was that of the melting point of an average igneous rock. The fallacy in these calculations is that Kelvin also assumed that no internal source of heat existed within the earth. At that time, it was not known

Figure 2.5 Autophotograph of the uranium-bearing mineral uraninite from Grafton Center, New Hampshire. The photograph was made from a negative that had been enclosed in black paper while the specimen lay upon it. (Courtesy of Ward's Natural Science Establishment.)

that radioactive decay releases a significant amount of heat.

NUCLEAR CLOCKS

In 1896 Antoine Henri Becquerel observed that uranium-bearing minerals, such as uraninite, caused a covered photographic plate to darken—an effect previously associated only with X-rays (Fig. 2.5). Subsequently, it was demonstrated that uranium decays spontaneously and gives off energy in the form of particles and electromagnetic radiation (i.e. radioactivity). The elementary particles radiated are helium nuclei (alpha rays) and electrons (beta rays), and the electromagnetic radiation is in the form of gamma rays, which are similar to X-rays but have shorter wavelengths. For the geologist, radioactive decay has become important as a means of determining the age of the earth and its rock formations.

Although the atoms of a given element always have the same number of protons, many elements have atoms that differ in the number of neutrons they contain. These are known as isotopes of the element. For example, uranium has two naturally occurring isotopes, ^{235}U and ^{238}U. The numbers 235 and 238 represent the mass numbers of the two isotopes of uranium and are the sums of the protons and neutrons in the atoms. Both isotopes of uranium have 92 protons, but ^{235}U has 143 neutrons, whereas ^{238}U has 146 neutrons.

Some isotopes are unstable and with time decay into one or more isotopes with a different mass number, a different number of protons; or both. The process of radioactive decay proceeds at a constant rate which is unique for each isotope. The time required for one half of the original amount of an isotope to decay is termed the *half-life* of that isotope.

The ratio of the remaining amount of the original isotope to the accumulated decay products is used in determining the age of rocks containing radioactive minerals. The utilization of "nuclear clocks" for the determination of geologic age relies on the basic assumptions that the decay occurs at a constant rate and that the radioactive minerals have suffered no loss or addition of decay products. Laboratory studies of short-lived radioactive isotopes have confirmed the first assumption.

Radiometric dating has enabled geologists to assign absolute ages to episodes of metamorphism,

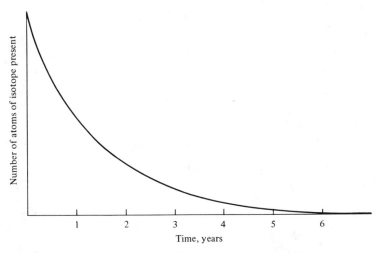

Figure 2.6   Decay curve of a hypothetical radioactive isotope with a half-life of one year.

igneous intrusion, and deposition of sediments. Radiometric dating is often the only reliable method of correlating metamorphic rock units within and between continents, and this method of dating has provided a means of assigning absolute ages to the geological time scale.

Uranium - Lead (U - Pb) Dating

Both ^{235}U and ^{238}U decay spontaneously with the loss of an alpha particle. The half-life of ^{238}U is 4.5

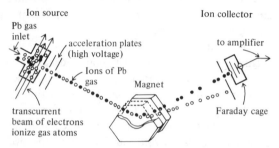

Figure 2.7 Diagrammatic sketch of a mass spectrometer. An ion of mass M_1 and electric charge e will be separated from another ion of mass M_2 and the same charge as it passes through a magnetic field. (After Hurley, Ref. 2. Copyright © 1959 by Educational Services, Incorporated.)

billion years. This means that starting with 1 gm of ^{238}U, 0.5 gm will remain after 4.5 billion years, 0.25 gm after 9 billion years, and so on (Fig. 2.6). ^{238}U breaks down to form ^{234}U, which in turn decays rather rapidly through a series of daughter products to form an isotope of lead, ^{206}Pb. Similarly, ^{235}U decays to ^{207}Pb, and thorium-232 (^{232}Th) decays to ^{208}Pb. As a result of radioactive decay, uranium-bearing minerals continuously accumulate lead.

In order to determine the age of a uranium-bearing mineral, the amounts of uranium and lead present in the material are carefully measured by chemical analysis. Next, the relative amount of each isotope of lead present is determined through the use of a *mass spectrometer*. Lead is vaporized and the lead ions are passed through a strong magnetic field (Figs. 2.7 and 2.8). The amount of deflection of the lead ions depends on the mass and the charge of the particles; for example, ions of ^{204}Pb are lighter than ions of ^{206}Pb, and therefore ^{204}Pb is deflected more than ^{206}Pb. The relative proportions of the isotopes of lead are

Figure 2.8 An Avco surface ionization mass spectrometer. Source on left, collector on right. (Courtesy of Avco Electronics Division.)

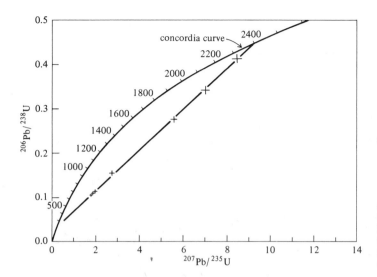

Figure 2.9 Concordia plot for the Baggot Rocks Granite, Medicine Bow Mountains, Wyoming, which is inferred to have crystallized approximately 2400 million years ago. (From Hills et al. Ref. 3.)

measured either photographically or with a photoelectric cell.

The apparent age of a rock may be calculated from the ratio of ^{238}U to ^{206}Pb, ^{235}U to ^{207}Pb, ^{232}Th to ^{208}Pb, or ^{206}Pb to ^{207}Pb. If the ages calculated for a single sample using each of the ratios agree approximately, they are said to be *concordant*; if the ages disagree significantly, they are said to be *discordant*. When the apparent ages are discordant for several samples of a suite of rocks from a limited area, the age of original crystallization may be determined through the use of a concordia diagram (Fig. 2.9). In the diagram, the concordia curve is the locus of concordant uranium - lead ages. If there has been no loss of uranium or lead from the minerals in a rock, the ratios of ^{238}U to ^{206}Pb and ^{235}U to ^{207}Pb will change. For rock having discordant ages, the ^{238}U to ^{206}Pb and ^{235}U to ^{207}Pb ratios plot as a straight line intersecting the concordia curve. The uppermost intersection of this line with the concordia curve gives the age of most intense metamorphism of metamorphic rock or the age of original crystallization of an igneous rock.

Thorium - Protactinium (^{230}Th - ^{231}Pa) Dating

Age determinations made by way of ^{230}Th - ^{231}Pa dating utilize decay products of ^{238}U and ^{235}U. The relatively short half-lives of ^{230}Th (75,000 years) and ^{231}Pa (32,480 years) permit dating of sediments and corals less than 250,000 years old. These isotopes are useful in establishing rates of deposition of sediment and in dating changes of sea level.

Potassium - Argon (K - Ar) Dating

Naturally occurring potassium contains three isotopes. The stable isotopes, ^{39}K and ^{41}K are far more abundant than the unstable isotope, ^{40}K. Through the capture of an electron by a proton, ^{40}K is converted into ^{40}Ar. The half-life of ^{40}K is 11.85 billion years. Thus the age of very ancient rocks may be measured using the potassium - argon (K - Ar) method.

Potassium-bearing minerals, such as biotite, muscovite, hornblende, sanidine, glauconite, and glaucophane, as well as whole rocks may be dated by the potassium - argon method. This method of dating is widely used owing to the abundance of potassium-bearing minerals in most rocks. The potassium content of a mineral or rock is determined by chemical analysis, and the percentage of argon is measured indirectly by a mass spectrometer. Since argon is a gas, it may be lost from the mineral or rock during metamorphism. For this reason, the potassium - argon method dates the

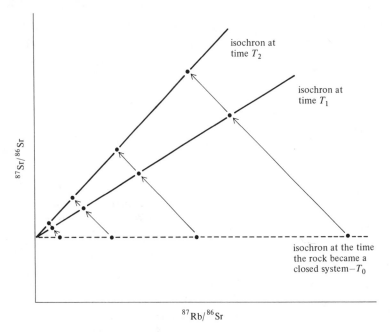

isochron at
time T_2

isochron at
time T_1

$^{87}Sr/^{86}Sr$

$^{87}Rb/^{86}Sr$

isochron at the time
the rock became a
closed system—T_0

Figure 2.10   Change in the rubidium-strontium ratio of four samples of igneous
rock with time on an isochron plot.

time of the last heating of a granitic rock or the
time of the last metamorphism of a metamorphic
rock.

Rubidium - Strontium (Rb - Sr) Dating

Most rock-forming minerals contain small quantities of rubidium. The unstable isotope, ^{87}Rb
decays with the emission of an electron to ^{87}Sr. It
is assumed that at the time of crystallization of a
rock, all its minerals contained ^{87}Sr and ^{86}Sr in the
same ratio. The amount of ^{87}Sr accumulated
within a rock is directly proportional to the
original amount of ^{87}Rb present. Therefore, the
ratio of ^{87}Sr to ^{86}Sr of a rock increases and the
ratio of ^{87}Rb to ^{86}Sr decreases with time. (Fig.
2.10). In a rubidium - strontium isochron diagram,
the ratio of ^{87}Sr to ^{86}Sr is plotted against the ratio
of ^{87}Rb to ^{86}Sr.

Since the half-life of ^{87}Rb is known, the age of
the sample may be calculated from the slope of
the isochron, which is a line connecting points on
the graph. Whole-rock rubidium - strontium dates

are determined through the use of a rubidium -
strontium isochron diagram (Fig. 2.11). The 50
billion year half-life of ^{87}Sr makes it useful for
dating Paleozoic and Precambrian events. Since
biotite and muscovite may lose strontium when
metamorphosed, rubidium - strontium dating of
these minerals provides the age of the last heating
of the rock. On the other hand, strontium is
usually not lost from the other constituents of the
rock during heating; this means that a rubidium -
strontium date on the whole rock may provide an
age of the original crystallization of an ingeous
rock or the most intense metamorphism of a
metamorphic rock.

Radiocarbon Dating

The isotope carbon-14 (^{14}C) is produced in the
upper atmosphere through collisions of neutrons
with nitrogen-14 (^{14}N). Entering the carbon cycle
as carbon dioxide, ^{14}C is fixed in the tissues of plants
and animals as well as in shells and bones. This
isotope of carbon is unstable and decays in several

steps to the stable isotope, ^{12}C. Willard Libby (4) realized that if the ratio of ^{14}C to ^{12}C remains constant and if ^{14}C is produced at a fixed rate, then the ratio of ^{14}C to ^{12}C could be used to date carbon-bearing materials. The relatively short half-life of 5570 years limits this method to materials less than about 50,000 years old. Dating by ^{14}C has greatly advanced our understanding of man's prehistory, the Pleistocene ice age, and of sea-level changes. The major drawbacks to radiocarbon dating are its time span limitation, the effects of atmospheric testing of nuclear bombs on ^{14}C production, and increase in ^{12}C in the atmosphere owing to the increasing use of fossil fuels.

RADIOMETRIC DATING OF THE GEOLOGICAL TIME SCALE

Radiometric dating has provided ages for the boundaries between geologic periods and epochs by the dating of suitable igneous and sedimentary rocks (Table 2.1 and Appendix A). For example, a date on a granite that intrudes Late Ordovician sedimentary rocks and is unconformably overlain by Early Silurian beds gives an age for the boundary between the Ordovician and Silurian periods.

A radiometric date for a volcanic flow or volcanic ash interbedded with fossiliferous sedimentary rocks of latest Silurian age will provide an approximate age for the Silurian - Devonian boundary. Most sedimentary rocks are not suitable for radiometric dating, since the uranium-, potassium-, or rubidium-bearing minerals which most of them contain were not formed when the rock was deposited. However, some black shales contain uranium that was precipitated from seawater during deposition of the shale. Such shales may be dated by the uranium - lead method using the whole rock. Glauconite, a potassium-bearing mineral, also forms during sedimentation in warm, marine environments. Potassium - argon dates on glauconite provide a minimum age for deposition of the enclosing rock, since some of the argon may

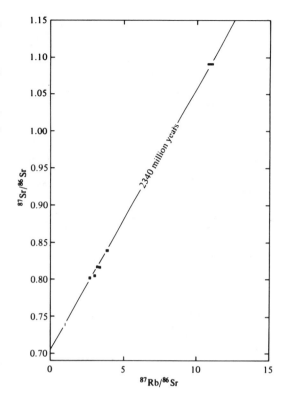

Figure 2.11   Rubidium-strontium isochron diagram for the Baggot Rocks Granite. (From Hills et al. Ref. 3.)

have been lost from the glauconite at only slightly elevated temperatures.

VARVES

Varve counts may assist in the determination of the ages of events during the retreat of the last ice sheet. *Varves* form on the bottom of lakes and consist of alternating light and dark layers of silt or clay (Fig. 2.12). The light colored layers are generally coarser than the dark layers, and they form during the spring, summer, and fall when streams carry coarse sediment along with dissolved oxygen into the lake. During the winter when the lakes are frozen, only fine silt and clay are deposited. The organic material in these sediments is not oxidized on the stagnant lake bottom. This

Figure 2.12 A varve set consisting of a dark layer over-lying a lighter layer.

REFERENCES CITED

1. L.R. Wager, 1964, The history of attempts to establish a quantitative time-scale, *in* W.B. Harland et al., Eds., *The Phanerozoic Time-Scale: A Symposium:* Geological Society of London, v. 120s, p. 13.

2. P. M. Hurley, 1959, *How Old is the Earth?:* Doubleday, Garden City, N.Y.

3. F. A. Hills et al., 1968, Precambrian geochron-ology of the Medicine Bow Mountains, south-eastern Wyoming: *Geol. Soc. Amer. Bull.,* v. 79, p. 1757.

4. E. C. Anderson et al., 1957, Radiocarbon from cosmic radiation: *Science,* v. 105, p. 576.

5. J. L. Kulp, 1961, Geologic time scale: *Science,* v. 133, p. 1105.

6. A. Holmes, 1959, A revised geological time scale: *Trans. Edinburgh Geol. Soc.,* v. 17, p. 183.

7. W. B. Harland, A. G. Smith, and B. Wilcock, eds., 1964, *The Phanerozoic Time-Scale: A Symposium:* Geolological Society of London, v. 120s.

SUGGESTED READINGS

W. B. N. Berry, 1968, *Growth of a Prehistoric Time Scale:* Freeman, San Francisco and London.

H. Faul, ed., 1954, *Nuclear Geology:* Wiley, New York.

H. Faul, 1966, *Ages of Rocks, Planets, and Stars:* McGraw-Hill, New York.

E. I. Hamilton, 1965, *Applied Geochronology:* Academic Press, New York.

K. Rankama, 1963, *Progress in Isotope Geol-ogy:* Wiley-Interscience, New York.

O. A. Schaffer and J. Zähringer, 1966, *Potassium - Argon Dating:* Springer-Verlag, New York.

material gives a dark color to the layers deposited during the winter. A varve set consists of a light colored, relatively coarse-grained layer and a darker fine-grained layer. Each varve set represents one year of deposition. Varved sediments are typically formed in lakes near glacial margins. Varve counts provide an estimate of the duration of such lakes. By radiocarbon dating of one of the varve sets, an absolute chronology of the lake history may be reconstructed.

3
Evolution and the Fossil Record

Fossils were once defined as any strange object dug out of the earth. This definition covered inorganic structures, such as crystals and concretions, as well as the remains of preexisting life. In modern usage, the term *fossil* denotes any evidence of prehistoric life. *Paleontology* is the study of ancient life based on fossil evidence. In the development of the geological sciences, paleontology has provided an indispensable means of determining relative age and correlating widely separated rock units. Fossils are an aid in geologic mapping and also in the reconstruction of ancient geography, environments, and climates. They also provide basic evidence in support of the concepts of evolution and continental drift.

THE NATURE OF FOSSILS
Early naturalists speculated about the nature of fossils, but few pursued any kind of systematic study or understood the organic nature of such specimens. In fact, many early naturalists had remarkably bizarre ideas. One sixteenth century writer thought that fossils were the remains of organisms that grew within a rock, after having been carried there by germ-laden air. Medieval scholars thought that fossils were the remains of poor, unfortunate individuals who had been trapped in Noah's flood. These men were not entirely wrong, because some mode of burial is necessary for the preservation of fossils. However, such early explanations of the origin of fossils were largely the result of attempts to fit the history of life into a 6000-year period, in accor-dance with contemporary notions of the age of the earth.

As late as the eighteenth century, Johann Beringer, a professor at the University of Würz-burg, maintained that fossils need not have an organic origin. His evidence consisted of replicas of the sun, moon, planets, and Hebrew letters, which he found along with true fossils. Beringer was so fascinated by the "fossils" he found that he published descriptions and illustrations of the specimens (Fig. 3.1). Some time later, Beringer found a "fossil" in his own likeness with his name on it. Only then did he realize that the specimens had been "planted" where he could find them. Eventually, the culprits turned out to be two envious colleagues (1).

The paleontological sciences have often been subjected to this type of hoax. Two other infam-ous frauds that nearly achieved scientific stature prior to their exposure were the Piltdown man, a faked "missing link," and the Cardiff giant, a solid stone replica of a "giant man fossil."

METHODS OF FOSSILIZATION
Fortunately, fossils are quite common in spite of the vast amount of organic debris that is destroyed and eventually reenters the organic cycle. The most important factors in fossilization are rapid burial and the presence of resistant structures in the original organism. Rapid burial prevents the remains of an organism from being broken and scattered or destroyed by weathering. Even after burial, the fleshy soft parts of an

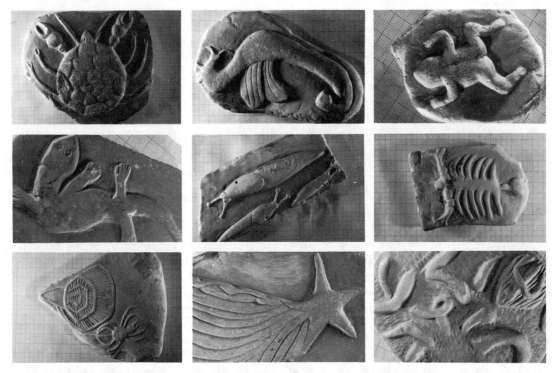

Figure 3.1 Johann Beringer's "fossils". (From Jahn, Ref. 1. Originally published by University of California Press; reprinted by permission of The Regents of the University of California.)

organism are decomposed by bacteria, so that generally only the hard parts are preserved. In rare instances, however, entire soft-bodied organisms have been preserved.

Original Remains

Essentially unaltered calcareous shells of clams and snails are common in relatively young sedimentary deposits (Fig. 3.2). Such shells may retain the original iridescent mother-of-pearl luster, the leathery connective tissues, and the chitinous outer layers. In older deposits, the original shell may have been recrystallized, but it generally has the same composition as the original shell.

Occasionally decay has been retarded to the extent that the soft parts of an organism retain their original form, although greatly reduced in volume as a result of dehydration. This type of preservation may occur in dry caves, bogs, and tar deposits. Freezing has preserved the entire remains of large Pleistocene mammals in the permafrost of the Arctic region. Some of these animals, such as woolly mammoths, appear to have been trapped in bogs or swamps. Most of the organism, including hair, muscles, blood, and even stomach contents, has been preserved.

The walls or shells of certain plants and animals are composed of nearly indestructible compounds of carbon, oxygen, hydrogen, and nitrogen; cellulose and chitin are examples of such materials. Pollen and spores, for example, are extremely resistant, and thus provide good evidence of ancient vegetation and climates (Fig. 3.2b).

Carbonization

Soft organic material may be preserved by carbonization or distillation, a process in which

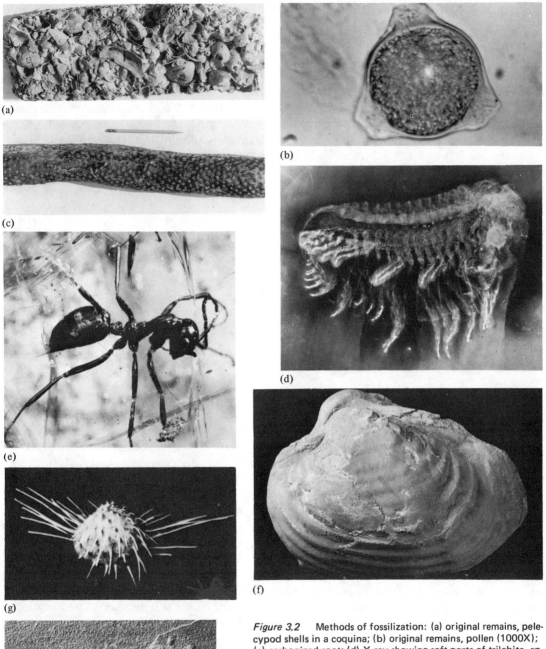

Figure 3.2 Methods of fossilization: (a) original remains, pelecypod shells in a coquina; (b) original remains, pollen (1000X); (c) carbonized root; (d) X-ray showing soft parts of trilobite, approximately natural size (Courtesy of Stuermer, Ref. 2, copyright © 1970 by the American Association for the Advancement of Science); (e) ant in amber (Courtesy of F. M. Carpenter, Ref. 3, copyright © by American Association for the Advancement of Science); (f) Internal mold of pelecypod shell (Courtesy of the Smithsonian Institution); (g) brachiopod shell replaced by silica (Courtesy of G. A. Cooper, photo by Smithsonian Institution); (h) worm trail.

nitrogen, oxygen, and hydrogen are lost and a carbon replica of the organism is produced. Jellyfish, fish, and parts of trees have been fossilized in this manner (Fig. 3.2c). Such fossils are generally found in black shales that were deposited under reducing conditions. X-Rays of black shales have revealed carbonized films of the soft parts of organisms which would not otherwise be visible (Fig. 3.2d). Insects in amber are actually distilled remains (Fig. 3.2e).

Casts and Molds

The original shell of an organism will leave an impression or *mold* of itself in the enclosing sediments. If the original shell is later dissolved, the mold may be filled with sediment or mineral deposits. The characteristics of the original shell, which have been preserved in the mold, are impressed on the sedimentary filling to form a *cast* of the original (Fig. 3.2f).

Replacement

The original remains of an organism may be partially or wholly replaced by other minerals, such as quartz, pyrite, or galena (Fig. 3.2g). In the related process of *permineralization,* minerals are precipitated inside the cell walls of wood or bone. Subsequently, the wall material may be removed and additional minerals precipitated in the open spaces. Thus a hard, compact petrification may be formed in which the original cell walls, growth rings, or other structures are preserved in detail. This kind of preservation occurs in Petrified Forest National Monument, Arizona. Permineralized dinosaur bone is common in the Colorado Plateau of eastern Utah and western Colorado.

Tracks and Trails

Bottom-dwelling organisms may leave tracks or trails in wet sediments. Such markings may be the only preserved record of soft-bodied organisms such as worms (Fig. 3.2h). Tracks made by reptiles, birds, and mammals may provide evidence of the ecology or habitats of these animals.

CLASSIFICATION OF FOSSILS

By the middle of the eighteenth century, the biologic origin of fossils had been accepted. Fossils could then be classified according to the system used by biologists for living organisms. This system was described by the Swedish botanist Carolus Linnaeus (Carl von Linné, 1707 - 1778). In his text *Systema Naturae*, Linnaeus formulated the general principles of biologic classification, which includes the ordering and naming of plant and animal groups.

As it is used today, the Linnaean system includes a hierarchy of seven basic categories:

Example:
Man (*Homo sapiens*)

Kingdom	Animalia
Phylum (pl. phyla)	Chordata
Class	Mammalia
Order	Primates
Family	Hominidae
Genus (pl. genera)	*Homo*
Species	*sapiens*

Additional groupings have been added by using the prefixes super-and sub-, to form such categories as superfamilies and subclasses. Linnaeus proposed that each organism be designated by two descriptive Latin names, in keeping with the custom of scholars of that day who Latinized their own names. In this scheme, which is known as *binomial nomenclature,* the first letter of the generic name, or *genus,* is capitalized; following is the specific name, or *species,* in lower case. Both genus and species are italicized or underlined.

The species is the basic unit of classification. A living species is comprised of an interbreeding population of individuals, which is capable of producing fertile offspring. In paleontology, particularly when there are no living representatives, species designation depends on comparative morphology of fossilized structures and, when available, physiological and ecological characteristics. Modern paleontologists establish the relationship of similar fossils by precisely measuring various dimensions such as shell length, width, thickness, shape, and numbers of surface ornaments of large numbers of fossil specimens. Computer analysis of these statistics provides more accurate species determinations. Closely related species are grouped into a single genus, related genera are incorporated into a family, and so on.

Fossil and living organisms may be assigned to one of three kingdoms, Protista, Plantae, and Animalia. The protists include both one-celled plants and one-celled animals. This kingdom has been established because some one-celled organisms exhibit characteristics of both plants and animals. For example, some protists are mobile like animals and yet manufacture their own food like plants. A brief description of the major fossil groups is given in Table 3.1 and Appendix B.

EVOLUTION

The eighteenth century conflict between the catastrophists and those who took a uniformitarian approach carried over into the study of fossils. Georges Cuvier, for example, attempted to show that the succession of different fossils in the strata of the Paris Basin resulted from a series of extinctions brought about by catastrophic floods. Each flood was followed by the creation or immigration of new life forms. Although the catastrophic approach was ultimately rejected, Cuvier and other catastrophists, such as Alexandre Brongniart, made significant contributions to descriptive paleontology.

According to the doctrine of evolution, new life forms develop over a long period of time from more primitive organisms. Although this idea had been suggested by some of the early Greek philosophers, the French biologist Jean de Lamarck developed the first fully consistent theory of evolution. In 1809 Lamarck proposed that characteristics acquired during the life of an organism could be inherited by its offspring. Accordingly, the modern giraffe's long neck would have resulted from the neck stretching of its browsing ancestors. Although this is an attractive idea and Lamarck had many followers, modern biologists do not believe that acquired characteristics can be inherited.

The modern version of the theory of evolution stems from the observations of Charles Darwin (4). Darwin collected considerable biological data during his five-year voyage aboard the H. M. S. *Beagle* in the southern hemisphere (1831 - 1836). His observations led him to conclude that evolution occurred by a process of natural selection among organisms in the struggle for food and space. Darwin proposed that the characteristics which enabled an organism to compete successfully were often passed on to their offspring, so that eventually a new species would develop. In 1859 Darwin published his theories in *The Origin of Species*, in which he documented evidence in support of evolution. This work has become the basis for modern concepts of evolution.

Darwin's theory of evolution has profoundly altered man's concept of the world by providing a dynamic rather than a static view of the history and development of life. In this sense, the theory of evolution complements the Copernican view of a dynamic earth in space.

Evidence For Evolution

Similarity in Body Chemistry. The chemical compositions of the blood of many animal groups are strikingly similar. Furthermore, many of the ions present in seawater are also present in blood in approximately the same abundances. These similarities have led to the suggestion that all animals had a common origin within the early oceans.

Another similarity in body chemistry is seen in the nucleic acids. The chromosomes in the nucleus of each living cell contain deoxyribonucleic acid and ribonucleic acid, two nucleic acids commonly known by their abbreviations, DNA and RNA. The molecules of DNA are enormously long and are shaped like two intertwined coiled springs, a double helix (5). Each of these molecules is incredibly complex and contains thousands of subunits lined up in a precise order like numbers on a tape. The pattern apparently forms a code of instructions which dictates what each cell

Table 3.1 Phyla that are important in the fossil record

Kingdom	Phylum	Class	Brief description
Protista			Most are one-celled organisms
	Cyanophyta		Blue-green algae — form stromatolites
	Chrysophyta		Diatoms, coccoliths
	Pyrrophyta		Dinoflagellates, etc.
	Protozoa		One-celled animals with calcareous shells (Foraminifera) or siliceous shells (Radiolaria)
Animalia			Multicellular animals
	Porifera		Sponges
	Coelenterata		Coelenterates
		Scyphozoa	Jellyfish
		Hydrozoa	Stromatoporoids
		Anthozoa	Corals — most have calcareous skeleton
	Bryozoa		Bryozoans — "moss-animals"
	Brachiopoda		Brachipods
	Mollusca		Mollusks
		Pelecypoda	Pelecypods — clams, oysters, etc.
		Gastropoda	Gastropods — snails
		Cephalopoda	Cephalopods — nautiloids, ammonoids, belemnities, squids
	Arthropoda		Jointed-legged organisms
		Trilobita	Trilobites
		Crustacea	Crustaceans — crabs, lobsters
		Insecta	Insects
	Echinodermata		Echinoderms
		Cystoidea	Cystoids
		Blastoidea	Blastoids
		Crinoidea	Crinoids
		Echinoidea	Echinoids — sea urchins, sand dollars
		Asteroidea	Starfishes
	Chordata		Chordates — Subphyla Hemichordata (graptolites, etc.) and Vertebrata (vertebrates)
		Graptolithina	Graptolites
		Agnatha	Primitive jawless fish
		Placodermi	Placoderms — primitive fish with jaws
		Chondrichthyes	Sharks
		Osteichthyes	Bony fish
		Amphibia	Amphibians
		Reptilia	Reptiles
		Aves	Birds
		Mammalia	Mammals
Plantae			Plants
	Bryophyta		Mosses, etc.
	Psilopsida		Primitive land plants
	Lycopsida		Club mosses and scale trees
	Sphenopsida		Horsetail rushes, etc.
	Pteropsida	Filicinae	Ferns and seed plants
		Gymnospermae	Gymnosperms — seed ferns, cycads, ginkgoes, conifers
		Angiospermae	Angiosperms — flowering plants

of an organism will be, how it will grow, what substances will be produced, and what raw materials will be required.

The basic building materials of all living organisms are proteins, which are composed of various combinations of about 20 amino acids. The DNA code specifies how the amino acids are to be joined in order to form the proteins. Studies of the DNA molecule have shown that, with few exceptions, a given code "word" in the DNA molecule always specifies the joining of the same amino acids. What is startling about this discovery is that this relationship holds true for *all* organisms, from the protists through the vertebrates. The similarity in the DNA code provides additional evidence for the idea that all life on earth had a common origin.

Similarity in Body Structures. The skeletal structures of terrestrial and marine vertebrates are remarkably similar. The bones in the arm of man closely correspond in number and distribution

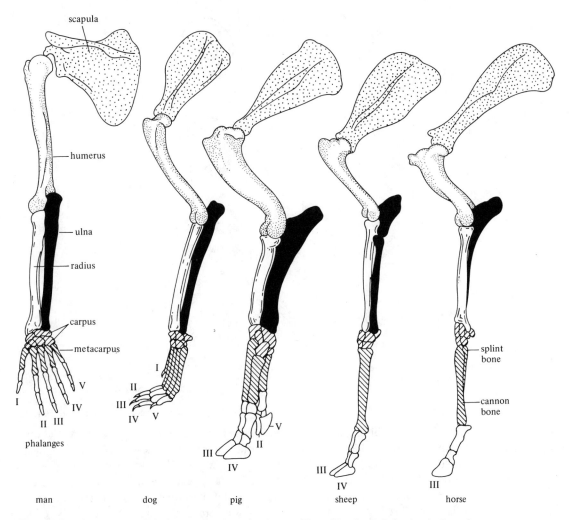

Figure 3.3   Comparison of the bones of various vertebrates. (From Moody, Ref. 6, after LeConte.)

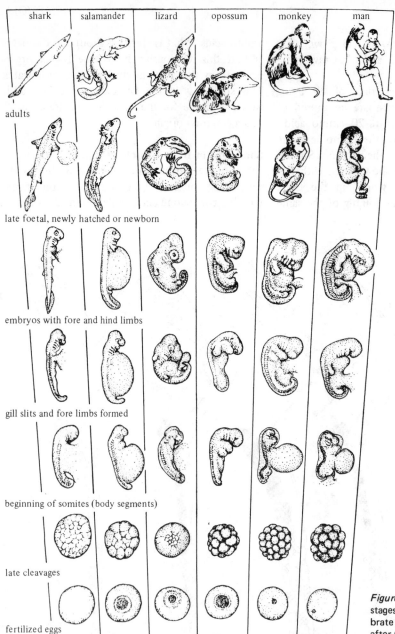

shark | salamander | lizard | opossum | monkey | man

adults

late foetal, newly hatched or newborn

embryos with fore and hind limbs

gill slits and fore limbs formed

beginning of somites (body segments)

late cleavages

fertilized eggs

Figure 3.4 Comparison of the stages of development of some vertebrate embryos. (From Moody, Ref. 6, after Gregory and Roigneau, Ref. 7.)

to the appendages of reptiles, birds, and other mammals, including marine forms (Fig. 3.3). The teeth of many vertebrates are also quite similar. For example, primitive placental mammals had 3 incisors, 1 canine, 4 premolars, and 3 molars on each side of each jaw; humans have 2 incisors, 1 canine, 2 premolars, and 3 molars.

Presence of Vestigial Structures. Vestigial structures are small, imperfectly developed parts of organs which were more fully developed in earlier generations. Man's appendix seems to be of little use to him now, but it corresponds to a digestive organ in some herbivorous mammals. The presence of the appendix serves as an indication that man may have evolved from a herbivorous ancestor. The coccyx, a small triangular bone forming the lower extremity of man's spinal column, may be part of a vestigial tail.

Stages in the Growth of an Organism. In its development from the embryonic to the mature stage, an organism may undergo changes similar to the evolutionary changes that have occurred during the development of the species. This process is technically referred to as the *biogenetic law* which states that generally "ontogeny recapitulates phylogeny." For example, the human embryo passes through stages in which it resembles, in turn, the embryos of fish, amphibians, and reptiles (Fig. 3.4). Thus the stages in the development of the human embryo suggest that man evolved from fish, through amphibian and reptile, to mammal.

The feet of the embryo of the modern horse develop in stages similar to those through which the ancestors of the horse evolved, as indicated by fossil evidence. At an early stage of development, the embryo has three toes, as did horses in Early Tertiary time (Fig. 3.5). The mature modern horse has one toe, with only vestiges of the other two.

Divergent and Convergent Evolution

Dispersal of a species into a new environment may result in the development of new species through the process of *divergent evolution* or *adaptive radiation*. Excellent examples of this were found by Charles Darwin in the Galapagos Islands, 600 miles west of Ecuador. Darwin observed that the fauna of these islands was in many ways significantly different from its apparent ancestral stock in mainland South America. Adaptation coupled with complete isolation from the continent gave rise to new species probably over a period of several million years. Apparently the species which initially occupied the islands diversified and occupied many of the available habitats.

The finches on the islands have had little competition and have produced 14 distinct species (Fig. 3.6). They developed a number of modifications that allowed them to better exploit the

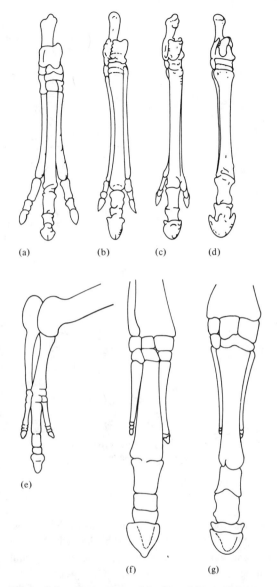

Figure 3.5 Comparison of the feet of Cenozoic horses with the feet of the embryo of a modern horse: (a) *Hyracotherium*, a horse which lived during the Eocene (From Moody, Ref. 6, after Cope); (b) *Miohippus*, a horse which lived during the Oligocene; (c) *Merychippus*, a horse which lived during the Miocene [(b) and (c) from Moody, Ref. 6, after Romer, Ref. 8] ; (d) *Equus*, the modern horse (From Moody, Ref. 6); (e) Limb of embryo of *Equus* at six weeks, showing three toes; (f) the same at eight weeks, showing the side toes much reduced relative to the middle toe; (g) the same at five months. [(e), (f) and (g) from Wells et al., Ref. 9, after Prof. Cossar Ewart. Reprinted by permission of Collins-Knowlton-Wing, Inc.]

small insectivorous
tree finch

large insectivorous
tree finch

warbler
finch

medium insectivorous
tree finch

woodpecker
finch

large ground finch
(seed-eater)

medium ground finch
(seed-eater)

small ground finch
(seed-eater)

large cactus
ground finch

Figure 3.6   Representatives of Darwin's finches. (From Volpe, Ref. 10.)

various food supplies in the islands. The most notable changes are in the beaks, which have become variously adapted for woodpecking and eating insects or cactus (11). One group, whose beaks are too short for probing rotten wood for grubs, developed the technique of poking small twigs or thorns into holes to retrieve the grubs (Fig. 3.7). This is one of the few examples of the use of a tool by one of the lower animals.

The fossil record also contains numerous examples of *convergent evolution*, resulting from the development of organisms with a similar external morphology by quite different evolutionary paths. For example, the shark (a fish), the ichthyosaur (a Mesozoic marine reptile), and the dolphin (an aquatic mammal) are all similar in appearance as a result of similar selective forces in the marine environment (Fig. 3.8). However, these animals evolved from different basic stocks. Similarities produced as a result of parallel evolution are also evident in birds, the pterosaur (a Mesozoic flying reptile), and the bat (a mammal). These all developed wings and other structures which permitted them to adapt to an aerial environment.

Evolution and Earth History

The fossil record provides extensive documentation of the evolutionary development of many

Figure 3.7 Galapagos Islands finch using a cactus spine to extract insects from a log. (Photograph by Irenaus Eibesfeldt, Max-Planck-Institut.)

groups of terrestrial and marine organisms. This evidence will be discussed in detail in Chapters 9 through 14. It is evident that a definite pattern exists in the first appearance of major groups of organisms. Only quite simple, primitive organisms are present in very ancient rocks, and the more advanced forms are found only in younger sequences. Such a pattern is the natural consequence of the evolution of the advanced organisms from simpler organisms.

As we hope to demonstrate in subsequent chapters, the assemblages of fossil plants and animals are distinctive not only for given rock sequences or geologic ages, but also for ancient seaways, geosynclines, and continents. Correlation of similar floras and faunas between certain landmasses has provided some of the evidence used

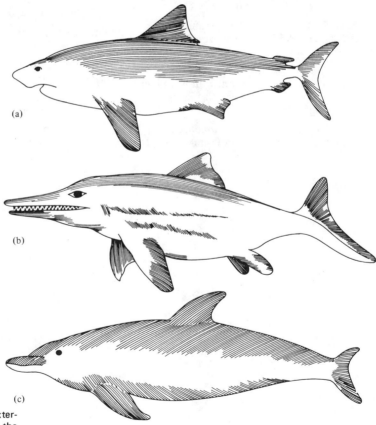

(a)

(b)

(c)

Figure 3.8 Comparison of the external morphology of (a) the shark, (b) the ichthyosaur, and (c) the dolphin.

in the reconstructions of the supercontinents for specific time intervals. Alternatively, the lack of paleontologic correlation between regions known to be of the same age, has assisted the reconstructions by locating physical barriers, such as deep oceans, which may have separated the continents. Paleontology is, therefore, vital in the interpretation of earth history.

REFERENCES CITED

1. M. E. Jahn and D. J. Woolf, 1963, *The Lying Stones of Dr. Johann Bartholomew Adam Beringer Being his Lithographiae Wirceburgensis:* University of California Press, Berkeley.

2. W. Stuermer, 1970 Soft parts of cephalopods and trilobites: Some surprising results of X-ray examinations of Devonian slates: *Science,* v. 170, p. 1300, Dec. 18.

3. E. O. Wilson, F. M. Carpenter and W. L. Brown, Jr., 1967, The first Mesozoic ants: *Science,* v. 157, p. 1038, Sept. 1.

4. C. Darwin, 1964, *The Origin of Species,* a facsimile of the 1st edition, with introduction by E. Mayr: Harvard University Press, Cambridge, Mass.

5. J. D. Watson, 1968, *The Double Helix:* paperback, New American Library, Inc., New York, hardcover, Atheneum, New York

6. P. A. Moody, 1970, *Introduction to Evolution,* 3rd ed.: Harper & Row, New York.

7. W. K. Gregory and M. Roigneau, 1934, *Introduction to Human Anatomy:* American Museum of Natural History, New York.

8. A. S. Romer, 1966, *Vertebrate Paleontology,* 3rd. ed: University of Chicago Press.

9. H. G. Wells, J. S. Huxley, and G. P. Wells, 1929, *The Science of Life:* Doubleday, Garden City, New York.

10. E. Peter Volpe, 1970, *Understanding Evolution:* Wm. C. Brown, Dubuque, Iowa.

11. D. Lack, 1947, *Darwin's Finches:* Cambridge University Press, Cambridge, England.

SUGGESTED READINGS

D. L. Clark, 1968, *Fossils, Paleontology and Evolution:* Wm. C. Brown, Dubuque, Iowa.

G. Himmelfarb, 1959, *Darwin and the Darwinian Revolution:* Doubleday, Garden City, N. Y.

A. S. Romer, 1966, *Vertebrate Paleontology,* 3rd ed.: University of Chicago Press, Chicago.

G. G. Simpson, 1949, *The Meaning of Evolution:* Yale University Press, New Haven, Conn.

4
Methods of Correlation

*I*n most parts of the world, exposures of rock are not continuous over great distances. For this reason, the relations between rocks exposed in different areas are not always evident. In mapping the geology and compiling a geologic history of a region, it is necessary to determine the continuity and chronologic relationship of rocks from one exposure to the next.

Rock sequences are said to correlate if they are equivalent to one another. Rocks that are lithologically similar and continuous are equivalent rock (lithostratigraphic) units. Rocks that were deposited during the same time interval are said to be time-stratigraphic equivalents, and rocks containing the same fossil assemblages are biostratigraphic equivalents. The comparison and correlation of sequences of rocks requires the determina- of many physical, chemical, and paleontological characteristics of the rocks.

ROCK UNITS
Rock units are observable stratigraphic bodies having a distinctive lithology and recognizable physical boundaries. *Formations,* which are the fundamental rock units, must be thick enough and sufficiently distinctive to be represented on a geologic map. A formation may consist of one or several lithologies—for example, a thick deposit of limestone or an interbedded sequence of sandstone and shales. In the latter case, the individual beds would be too thin to be mapped at the usual scale of a quadrangle map (1:24,000). The boundaries between formations are generally located at breaks in sedimentation (bedding planes or unconformities) or changes in rock type. A lithologically distinct portion of a formation may be designated as a *member* of that formation. Related formations may be combined to form a *group*.

The site at which a rock unit was first described is the *type locality*. An exposure of that rock unit at the type locality is the *type section*. This exposure is the standard to which similar sequences are compared. The name of the rock unit generally comes from a geographical feature in the vicinity of the type section. For example, the thick dolostone, a rock containing the mineral dolomite, which crops out near the town of Lockport, New York, is named the Lockport Formation. Similarly, the interbedded shales and sandstones that are exposed near Martinsburg, West Virginia, are known as the Martinsburg Formation.

William Smith was one of the first to trace and correlate rock units over large areas (see Chapter 1). His geologic map of England (Fig. 1.7) shows the extent of these rock units. Smith and his French contemporaries, Cuvier and Brongniart, mapped parts of the London and Paris basins. Their detailed maps and cross sections have provided a basis for correlation of rocks in the London Basin with those of the Paris Basin (Fig. 4.1).

Correlation by Lateral Continuity
The most reliable method of correlating rock units is to establish their physical continuity. Where the strata are well exposed, as in areas of little

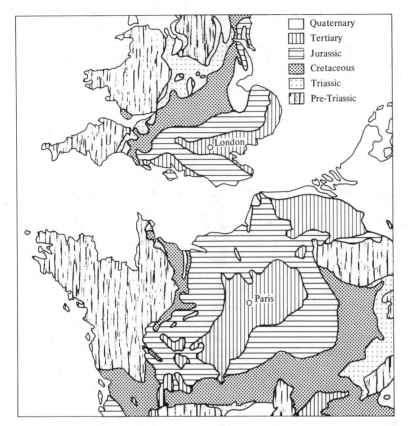

Figure 4.1   Geologic map of part of Europe showing the correlation of London and Paris basins.

vegetation, rock units can be followed on foot or traced visually on aerial photographs (Fig. 4.2). However, where exposures are rare, the rock units must be traced by other techniques. Small rock fragments or distinctive soils may serve to locate the unit. In structurally uncomplicated regions, rock units may be traced by projecting their contacts laterally for short distances (Fig. 4.3). Rock units may also be correlated on the basis of the continuity of land forms, such as valleys cut in less resistant rock or ridges supported by a resistant unit.

Correlation by Lithology

Correlation by lithology alone is reliable only on a local basis, owing to the repetition of rock types in the geologic record. Lithologic similarities are generally used in conjunction with other lines of evidence, such as the fossil content and character of the adjacent strata. In the absence of other more reliable criteria, correlations may be made between lithologically similar sequences of strata. Rock units may also be correlated on the basis of similarity of mineral content, insoluble residues, or mineral grain-size frequencies.

Correlation by Geophysical Techniques

Before the widespread use of modern geophysical equipment, subsurface correlations were made primarily on the basis of comparison of lithology logs representing the record of rock penetrated by the drill. This record was usually kept by the driller, who would make an educated guess about the composition of the strata encountered. Al-

Figure 4.2 Geologists at work on the mountains above Cappo Glacier in south central Alaska. In such areas where the exposures are good, correlations may be made by tracing rock units on foot. (Courtesy of Humble Oil and Refining Company.)

though his observations were often close to the actual lithology, they were not accurate enough for reliable correlations. It became necessary for geologists to "sit" on a well and provide a more accurate log of the rock cuttings or rock cores from the well. These samples are inspected in the field for lithology, structures, and fossils (Fig. 4.4) and in the laboratory for porosity, permeability, and gas content.

Techniques are now available to increase the accuracy of subsurface correlations. Wire line depth recorders aid in the measuring of the depth of the drill hole. The discovery of variations in the electrical conductivity of rocks led to the development of instruments that can measure this property in the drill hole. As an electrical current passes through different strata, variations in conductivity are recorded in the form of an *electric log*. Strata with a high fluid content generally have a high electrical conductivity. Correlations of rock units can be made by matching the peaks of conductivity (Fig. 4.5).

Figure 4.3 Sketch illustrating correlation of units along strike in an area in which the exposures are not continuous.

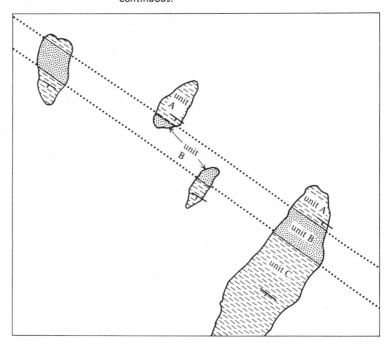

Figure 4.4 Geologists and petroleum engineers inspecting a core sample from deep in the earth during an exploratory drilling operation. (Courtesy of Humble Oil and Refining Company.)

Correlation of rock units may also be aided by measurement of the natural radiation of rocks (Fig. 4.6). The modern composite well log provides a relatively complete record of the properties of strata and helps geologists locate subsurface accumulations of oil, gas, and water.

Seismic techniques provide control in correlating rock units in areas that have not been sampled by drilling. Seismic reflection measurements have been especially useful in delineating structures such as folds, faults, salt domes, and igneous intrusions. In this type of study, seismic waves are generated by explosions of small dynamite charges (Fig. 4.7), or they may be produced by dropping large weights. The resulting seismic waves are reflected from subsurface horizons to a set of receivers on the ground surface, and they are recorded on a set of seismograms (Fig. 4.8). Individual reflecting horizons produce a series of deflections on the seismogram and these deflections are used to trace rock strata.

TIME - STRATIGRAPHIC UNITS

A time - stratigraphic unit is deposited during a specified time interval and may include a wide variety of lithologies. The Cambrian System, which was deposited during the Cambrian Period, is an example of such a unit. The time - stratigraphic units and their matching units of geologic time are given in Table 4.1.

Table 4.1 Relationship between time-stratigraphic and geologic time units

Time-stratigraphic units	Geologic time units
—	Eon
Erathem	Era
System	Period
Series	Epoch
Stage	Age

In the nineteenth century, Charles Lyell correlated some of the time - stratigraphic units of North America with those of Europe. He noted that the faunas and lithologies of Cambrian and Silurian rocks in New York State were in many respects similar to those in Wales. He did not consider that these similarities indicated a physical connection between these areas, but only that the geologic processes occurring at a given time were similar.

Time - stratigraphic correlations are mostly based on paleontological data, but recently radiometric and paleomagnetic measurements have come into widespread use. The establishment of time - stratigraphic units has permitted intercontinental correlations and has provided a means of comparing the geologic histories of all the continents.

Correlation by Fossils

Within Cambrian and younger sequences, most correlations of time - stratigraphic units are made through study of their fossil content. The unique succession of assemblages of fossils makes this possible. By identifying the fossils within a rock unit and determining their time ranges, the geologist may establish the age of the unit, which will in turn make it possible to compare that unit with

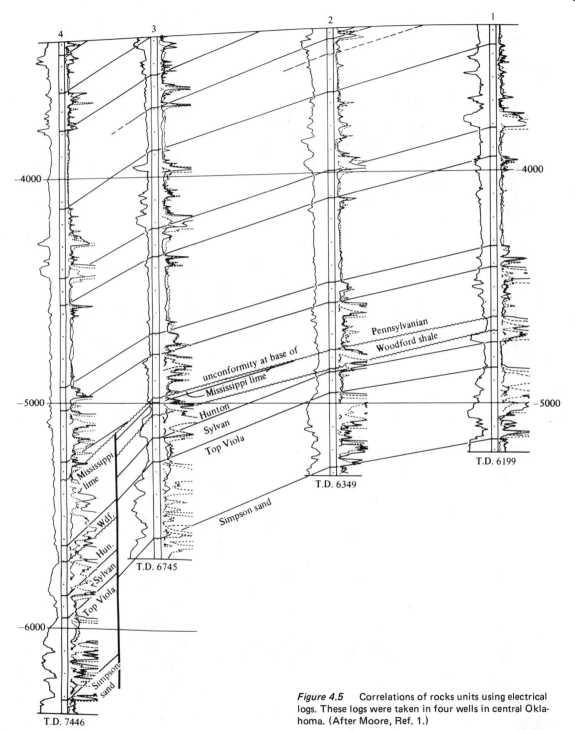

Figure 4.5 Correlations of rocks units using electrical logs. These logs were taken in four wells in central Oklahoma. (After Moore, Ref. 1.)

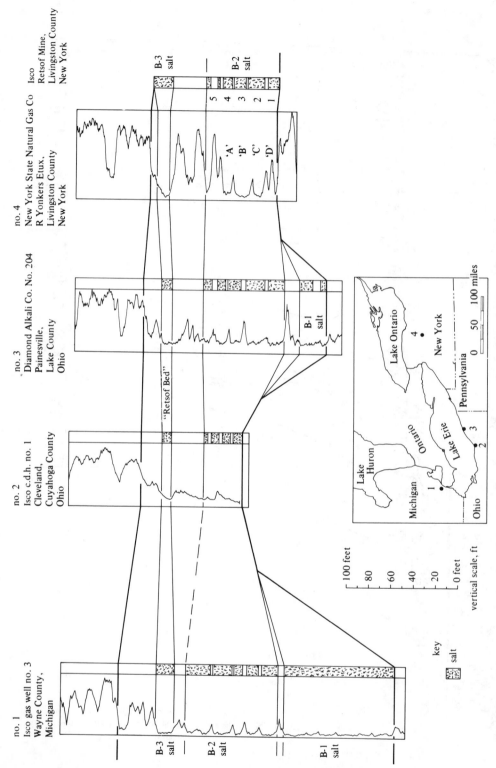

Figure 4.6 Correlation of rock units near Lake Erie using gamma-ray logs. (From Jacoby, Ref. 2.)

others of the same age. Fossils are identified by comparison with museum collections and with photographs in the geological literature. Assistance is also available from paleontologists who specialize in certain groups of fossils.

Index or *guide fossils* are particularly helpful in making correlations. A good guide fossil should:

1. have a short time range
2. have a wide geographic distribution
3. be abundant
4. be easily identifiable

The trilobite *Olenellus* is a good guide fossil, since it is restricted to Early Cambrian deposits, is found in many parts of North America and Europe, is relatively abundant, and has a distinctive morphology.

A group of several different fossil species which are consistantly found together comprise a fossil assemblage. If an assemblage is restricted in geological range, it may be used in determining the age of the rock unit. Fossil assemblages may contain microscopic as well as macroscopic forms. A comparison of the time ranges of each of the

Figure 4.7   Typical seismograph "shot" taken during seismic profiling in west Texas. (Courtesy of Humble Oil and Refining Company.)

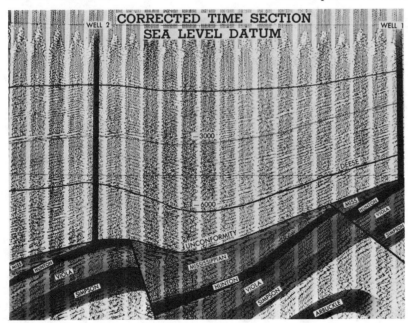

Figure 4.8   Correlation of rock units using seismograms. (From Dunlap, Ref. 3.)

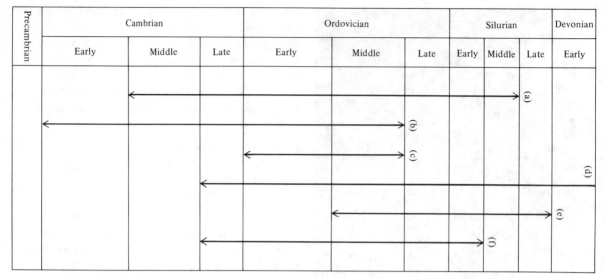

Figure 4.9   An assemblage of six fossil species was found in a rock unit. The diagram plots the time ranges during which each of the species lived. During what period of time could all these organisms have been living together?

fossils within an assemblage serves to determine when they all could have lived together (Fig. 4.9).

Correlation by Radiometric Ages

Radiometric dating is increasingly important in the correlation of time - stratigraphic units, especially in igneous and metamorphic terrains. The accuracy with which units may be correlated depends on the method(s) of dating and the materials dated. A potassium - argon date on mica from a metasedimentary rock provides a minimum age for the deposition of the original sediment. A uranium - lead or whole-rock rubidium - strontium date for the basement rock on which the sediment was deposited gives a maximum age for the deposition of the sediment. Sequences that were metamorphosed at the same time and were deposited on basements of similar age are likely to be time - stratigraphic equivalents.

Correlation by Paleomagnetism

Time - stratigraphic units may be correlated by measuring the direction of magnetization of drill cores taken in oceanic sediments (Fig. 4.10). This method commonly involves matching the pattern of normally and reversely magnetized sediments or lavas (see Chapter 7).

BIOSTRATIGRAPHIC UNITS

The zone is the basic unit of biostratigraphy. A *zone* is a bed or group of beds characterized by the presence of an index fossil or assemblage of fossils. The zone takes its name from a characteristic fossil (index species) of the zone. Zones that cross rock unit boundaries may indicate that these strata were deposited at the same time. This is especially true if zones are not repeated. If the beds were deposited contemporaneously, the faunas are said to be *synchronous*. The Jurassic ammonite zones are probably of this type. However, some of the Devonian brachiopod zones may be repeated several times in a given undisturbed section; in this case, the zones would not have been deposited contemporaneously. The term *homotaxial* designates strata that have the same fossil assemblages but are not of the same age.

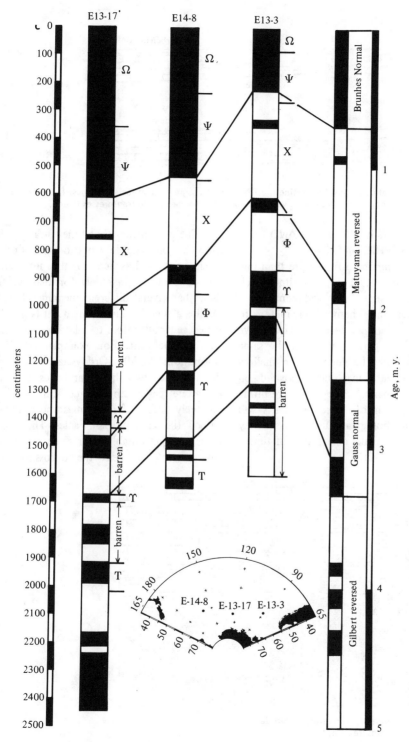

Figure 4.10 Correlation of cores taken in the deep ocean near Antarctica using paleomagnetic and fossil data. (From Hays and Opdyke, Ref. 4. Copyright © 1967 by American Association for the Advancement of Science.)

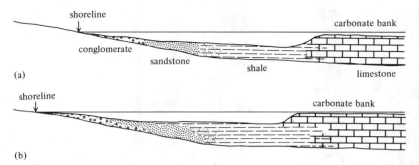

(a)

(b)

Figure 4.11 Diagrammatic sketches illustrating the effect of a transgression on facies changes: (a) sediments deposited during a period of stable sea level; (b) migration of facies as sea transgresses inland.

RELATION BETWEEN ROCK UNITS AND TIME - STRATIGRAPHIC UNITS

In any area, it is common to find more than one depositional environment at a given time. Each environment is represented by sediments of a distinctive physical and chemical character. In coastal regions, for example, beaches may receive well-sorted sands and gravels, while at the same time, clay and silt may be deposited in shallow lagoons, and carbonate may be deposited on offshore carbonate banks. The rocks resulting from the consolidation of these sediments will change laterally in lithology and in the fossils they contain. Such changes make correlation of time - stratigraphic units difficult.

The recognition that time - stratigraphic units show a considerable variation in lithology and fossil content has led to the introduction of the concept of *facies*. Unfortunately, the term facies has several different meanings, but it always implies a specific environment. In North America, facies generally refers to a lateral subdivision of a stratigraphic unit. For example, the Edinburgh Formation, Middle Ordovician, of Virginia, is divisible into two dissimilar facies—a fossiliferous limestone called the Lantz Mills Facies and a relatively unfossiliferous limestone and black shale named the Liberty Hall Facies. The fossiliferous limestone presumably represents a shallow-water environment, whereas the unfossiliferous lime-

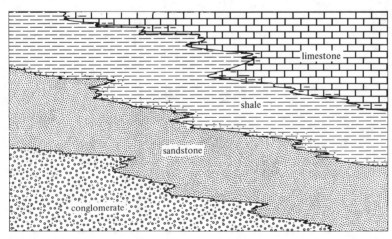

Figure 4.12 Cross section of sedimentary rocks deposited during a marine transgression.

stone and associated black shale represent a deeper water environment.

In Europe the term facies is commonly used without reference to a specific stratigraphic unit. Thus "the redbeds facies" may be used to designate any sequence of red sandstones and shales that were probably deposited in a subaerial environment. Similarly, "the black shale facies" refers to sequences containing dark shales and sandstones deposited in relatively deep water.

A *lithofacies* is the rock record of a sedimentary environment. For example, we may say that the limestone lithofacies of the Onondaga Limestone in New York State represents a shallow, clear-water environment. A lithofacies map shows the distribution of rock types formed at a specific time. Such maps are useful in the construction of paleogeographic maps.

A *biofacies* is an assemblage of fossils repre-

senting a particular sedimentary environment. For example, the Onondaga Limestone contains a coralline biofacies representing a shallow, clear-water marine environment.

The sizes of sedimentary particles that may be deposited in any one location are controlled by the depth of water, the distance from the source, and the energy of distributary currents. If the water level has not changed relative to the land surface and if the sediments have been deposited in a subsiding basin, the contacts between lithofacies would be essentially vertical. However, if the water level has risen or fallen relative to the land, changes would occur in the depositional environments, and thus in the lithofacies. With a rising or *transgressing* sea, beach sands may be deposited over lagoonal or freshwater deposits. Subsequently, finer marine sediments would overlap the beach sands (Figs. 4.11 and 4.12). The decrease in grain

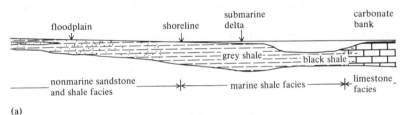

(a)

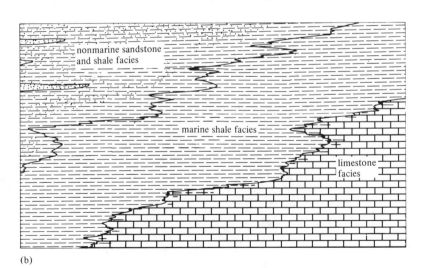

(b)

Figure 4.13 Diagrammatic sketches illustrating the effect of a regression on facies changes: (a) sediments deposited during a period of stable sea level, (b) cross section of sedimentary rocks deposited during a marine regression.

size is related to decreasing energy of the sediment-transporting currents in the deepening water.

With a lowering or *regression* of the sea, coarse coastal sediments would overlap finer marine sediments (Fig. 4.13). Times of general regression of the sea are commonly interrupted by short episodes during which the sea transgresses inland, and times of general transgression are interrupted by brief regressions of the sea. The result of such fluctuations in sea level is an irregular interfingering of lithofacies (Figs. 4.12 and 4.13b). This situation has been repeatedly encountered in ancient deltaic sequences.

During a transgression or regression of the sea, the contacts between lithofacies are at a relatively low angle to time - stratigraphic units (Figs. 4.12 and 4.13). In other words, facies changes may occur within time - stratigraphic units. The recognition of facies changes may be aided by the presence of *marker beds* or *key horizons* with broad geographic distribution. Volcanic ash beds and their weathered equivalents, *bentonites,* provide excellent time planes because they are deposited essentially instantaneously. A change in lithofacies below the ash or bentonite would indicate a difference in depositional environment at the time the ash fall occurred.

REFERENCES CITED

1. C. A. Moore, 1963, *Handbook of Subsurface Geology:* Harper & Row, New York.

2. C. H. Jacoby, 1969, Correlation, faulting and metamorphism of Michigan and the Appalachian Basin salt: *Amer. Assoc. Petrol. Geol. Bull.,* v. 53, p. 136.

3. R. C. Dunlap, 1956, Geophysical data for geologic study: *Amer. Assoc. Petrol. Geol. Bull.,* v. 40, p. 1462.

4. J. D. Hays and N. D. Opdyke, 1967, Antarctic radiolaria, magnetic reversals and climate change: *Science,* v. 158, p. 1001, Nov. 24.

SUGGESTED READINGS

C. O. Dunbar and J. Rodgers, 1957, *Principles of Stratigraphy:* Wiley, New York.

W. C. Krumbine and L. L. Sloss, 1963, *Stratigraphy and Sedimentation:* Freeman-Cooper, San Francisco.

R. C. Moore, 1949, *Meaning of facies:* Geological Society of America, Mem. *39.*

J. Rodgers, 1959, The meaning of correlation: *Amer. J. Sci.,* v. 257, p. 684.

J. M. Weller, 1960, *Stratigraphic Principles and Practice:* Harper & Row, New York.

5
Reconstruction of Ancient Environments

Sedimentary rocks reflect the chemical, physical, and biological environments in which the original sediments were deposited. With our increased understanding of modern depositional environments, such features as color, grain size, degree of sorting, lithology, and fossils may be related directly to specific environments of deposition. Such parameters as chemistry and depth of water, sediment composition, direction of depositional currents, relative distance from the source, and climatic conditions may be inferred.

Reconstruction of ancient environments over large areas provides the basis for paleogeographic maps on which environmental relationships may be represented. Changes in sedimentary facies within time - stratigraphic units may be used to locate such features as rivers, flood plains, shorelines, barrier beaches, and lagoons. These reconstructions provide the basis for interpretation of the history of an area during depositional episodes and may be economically useful in oil and mineral prospecting.

Paleontologists consider that changes in ancient environments may be related to possible causes of evolution and extinction of certain fossil groups. The comparison of ancient and modern climates may also provide evidence for changes in latitude that may have resulted from polar wandering and continental drift.

SEDIMENTARY ENVIRONMENTS

Marine environments include those below mean sea level, and continental environments are those above mean sea level. Since sea level exhibits considerable short-term variations, mainly because of tides, there is a complex transitional zone between marine and continental environments. Within each of these environments, the character of the sediments encountered depends on such factors as agents of transportation, energy of currents, source of the sediment, kinds of organisms present, and chemistry of the water. The last category includes such factors as oxygen content, salinity, pH (acidity or alkalinity), and content of dissolved solids. Organisms interact with the overall environment and produce changes in the chemistry of the seawater and changes in the character of the sediments deposited.

Transitional Environments

The transitional environments lie within the littoral zone, between average high and average low tide levels, and include beaches, tidal flats, lagoons, estuaries, and coastal marshes and swamps. The littoral zone may vary from a few feet to a few miles in width, since the extent of tidal flooding depends on the slope of the beach or tidal flat and on the height of the tides.

Beaches. In the high-energy environment of a beach, sediments may range from medium-grained sand to gravel. Sandstones deposited in such an environment may retain a variety of characteristic sedimentary structures such as ripple and swash marks (Fig. 5.1). Relatively few varieties of organisms are found in the beach zone, primarily because wave action severely restricts the inhabitable niches.

Figure 5.1 Swash marks on a sandy beach on the shore of the Gulf of Mexico. Swash marks are thin, wavy lines of sand, mica flakes, bits of seaweed and other debris left at the point where each wave began to recede from the beach. (Photo by Ralph Yalkovsky.)

Organisms that have adapted with varying success to this environment include planktonic (floating) organisms, burrowing crustaceans, heavy-shelled pelecypods and gastropods, plants rooted to holdfasts, and nektonic (free-swimming) forms such as fish. The fossil remains of many organisms are destroyed or broken and abraded by wave action, and shell and bone fragments may be found in thin beds along with coarse sand and small pebbles in this zone.

Tidal pools common to rocky coasts may contain a great variety of organisms. The rocks of the tide pool are often coated with attached invertebrates and algae. Mobile benthonic (bottom-dwelling) forms such as crabs and starfish live among the rocks. A tidal pool protected from subsequent wave action could yield abundant fossils after burial.

Table 5.1 Facies characteristics as recorded in Cambrian through Devonian carbonates of the Central Appalachians[a]

Facies characteristics	Facies suites			
	Tidal flat	Shallow, subtidal	Deep, subtidal	Organic build-ups (e.g., reefs)
Mud cracks	Typical	—	—	—
Scour and fill structures	Typical	—	—	—
Pebble conglomerates	Typical	—	—	—
Laminations	Typical	—	—	—
Early dolomite	Typical	—	—	—
Cross-bedding	Small scale	Medium scale	—	Sometimes present
Burrow-mottling	Rare	Common	Abundant	Rare
Oolites	—	Often present	—	—
Bedding	Thin-medium	Medium-thick	Thick-massive	Unbedded-massive
Algal structures	Present	Present	Absent	Present
Burrows	Vertical	Vertical and horizontal	Horizontal	Rare
Fossil abundance	Low	Very high	Variable	Very high
Fossil diversity	Low	Medium	Usually high	Medium to high
Major taxa	Trilobites and/or ostracodes	Calcareous algae, pelmatazoa, brachiopods	Brachiopods, bryozoa, trilobites, and pelmatozoa	Tabulate, rugose and stromatoporoid corals
Vertical facies variations	Sharp and frequent	Transitional and common	Very gradual and infrequent	Complex
Areal facies variations	Outcrop scale	Relatively persistent	Basinal scale	Outcrop scale to several miles
Facies strike	Variable	Parallel to basin margin	Parallel to basin axis	Variable

[a]From L. F. Laporte, personal communication.

(a)   (b)

Figure 5.2   Comparison of modern mud cracks with those in an ancient sedimentary rock: (a) mud cracks produced by the drying of a small pond (photo by David Leveson), (b) mud cracks in argillaceous siltstone of Permian age, Kansas. (Photo by Ada Swineford.)

Tidal Flats. Tidal flats generally occur in low energy littoral environments which have developed a low profile through deposition of fine-grained sediments such as silt, clay, and lime mud (Table 5.1). Clastic sediments accumulate in areas adjacent to highlands, whereas lime muds, which are the precursors of limestone and dolostone, are deposited in warm climates in the absence of nearby sources of clastics. Intermittent subaerial exposure of the tidal flats may produce mud cracks (Fig. 5.2a) that will be filled in by sediments; the resulting sedimentary rock (Fig. 5.2b) will preserve their structure.

Lagoons. Barrier islands and coral reefs may isolate quiet, shallow bodies of water from the sea. The deposits of such lagoons consist of fine-grained terrigenous (land-derived) clastic sediments or calcareous muds. Lagoons represent zones of mixing of fresh and salt water, and as semirestricted bodies of water they may retain nutrients from terrestrial sources. The nutrient level may favor high productivity in certain organisms, or it may give rise to a large variety of organisms. Restricted circulation may cause hypersaline conditions in dry climates so that evaporite deposition will occur. Alternatively, reducing conditions may develop, causing toxic substances such as hydrogen sulfide to spread throughout the lagoon. Such chemical changes may result in the mass destruction of the plants and animals living within the lagoon.

Estuaries. Mixing of marine and freshwater is also characteristic of estuaries. Estuaries are open bodies of brackish water which form at the mouths of rivers because of tidal changes and mixing of fresh and seawater. Estuarine sediments are generally terrigenous clastics and commonly contain current-produced, depositional structures such as ripple marks, cross-beds, and channels (Figs. 5.3 and 5.4).

Estuaries have distinct faunal and floral assemblages which are adapted to the brackish water environment. Recent studies of estuarine fauna have revealed distinct salinity preferences on the part of certain species. For example, although shell fishermen know from experience where to find clams and oysters in an estuary, ecological studies

(a)   (b)

Figure 5.3   Comparison of modern and ancient ripple marks: (a) ripple marks on the shore of Plum
Beach, New York (photo by David Leveson), (b) current ripple marks on sandstone, northern Scotland
(Crown Copyright Geological Survey photograph. Reproduced by permission of the Controller, Her
Britannic Majesty's Stationery Office.)

show that their location is mainly determined by
the salinity of the water. Some species are eury-
haline (i.e., able to withstand a wide range of
salinities) and can live in both fresh and seawater.
Other species are stenohaline, or restricted to fairly
narrow salinity tolerances.

Coastal Marshes and Swamps. Coastal salt marshes
lie within the tidal zone and grade into freshwater
marshes above sea level. The sediments in these
environments are essentially the same as those
found in lagoons and tidal flats. However, water
movement is so restricted that marsh plants may become established. Salt-tolerant plants that root on
the bottom act as filters trapping fine sediments.

Swamps differ from marshes in that they
contain trees and shrubs, as, for example, the
tropical and subtropical mangrove swamps. Brackish-water swamps of the transitional zone may

grade landward into freshwater swamps, but they
generally end abruptly in deeper water on the
seaward side.

Coastal marshes and swamps provide quiet
environments for smaller invertebrates and a
breeding ground for many marine vertebrates.
Some marine vertebrates have acquired tolerances
to salinities ranging from the typical marine
environment of 34 parts per thousand to nearly
freshwater of less than 10 parts per thousand.
Coastal marshes and swamps distributed throughout the world provide models of coal-forming
environments and are the probable ecologic realm
from which marine vertebrates ultimately colonized the land.

Marine Environments
Ocean bottom coring and surface sampling have

Figure 5.4   Sandstone filling channel in shale.

provided modern analogs for the interpretation of ancient marine conditions. The most reliable criteria for the recognition of subtidal marine deposits are the presence of typical marine fossils and general absence of freshwater or brackish-water forms. Modern corals, for example, live only in marine environments, and it is very probable that ancient corals were similarly restricted. The presence of extensive deposits of salt or gypsum may indicate deposition in a marine environment. However, relatively minor evaporite deposits may have been formed in enclosed terrestrial basins, such as that in which the Great Salt Lake is situated.

The sea-floor below the low tide level comprises the benthonic environment, which may be divided into the sublittoral, bathyal, abyssal, and hadal zones (Fig. 5.5). The sublittoral zone is located between low tide and a depth of about 600 ft. The shallower portions of the sublittoral environment down to about 150 ft (50 m), the limit of sunlight penetration, generally support an abundance of life. Where wave action affects the sea-floor, the shells of benthonic organisms are heavier than those found in lagoon environments. Along the northern Atlantic coast, offshore storms often sweep tons of heavy-shelled clams, snails, and oysters from their sanctuaries below normal wave base onto the beaches.

Where the influx of clastic sediment is not too great, organisms of the sublittoral zone build moundlike reefs or *bioherms*. The tops of these mounds are typically just below the low tide level. Bioherms may be comprised of a great variety of life such as algae, worms, corals, and oysters. If burial of the reef is rapid enough, the organic components of the living reef will be preserved, often in their "life" positions.

Modern algal stromatolites are also restricted to relatively shallow waters, although some ancient stromatolites may have developed at depths as great as 140 ft (45 m) below sea level (1). *Ooliths*, which are small, concentrically layered grains generally composed of calcium carbonate, form in the shallow, sublittoral waters. Their presence in a limestone would then suggest deposition in a shallow-water environment (Fig. 5.6). Oscillatory

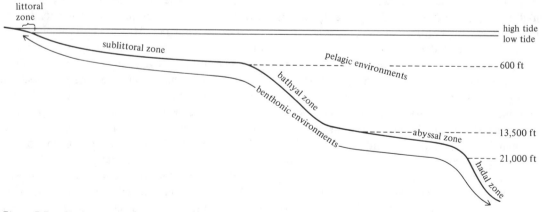

Figure 5.5   Modern marine environments.

Figure 5.6 Hand sample of an oolithic limestone. (Photo by W. T. Huang.)

ripple marks, which are produced by nearshore currents, are also indicative of shallow, sublittoral environments. Current ripple marks, however, may be formed at any depth as long as currents are available. They have even been photographed in the deep ocean.

The bathyal zone (600 - 13,500 ft, 200 - 4100 m) is approximately coincident with the continental slope. Seismic reflection profiling indicates that little sediment accumulates on the relatively steep slopes in the upper bathyal region and that thick wedges of sediment are deposited at the base of the continental slope in the lower bathyal and upper abyssal zones. Core samples from the base of the slope are comprised mostly of turbidites, which are poorly sorted deposits very similar to graywackes. Most turbidites were probably deposited by rapidly moving muddy currents, which flowed down the continental slope as a result of their relatively high density.

Abyssal sediments that have been recovered from the deep ocean at some distance from the continents include turbidites as well as clay and biogenic oozes. The oozes are comprised largely of the remains of planktonic organisms such as foraminiferans, radiolarians, and pteropods. The lithified sedimentary rocks equivalent to this environment are graywackes, claystones, limestones, and chert. Since these deposits have been laid down on a basement composed of mafic lavas,

sedimentary rocks deposited in the abyssal environment are commonly associated with basalts and their metamorphosed equivalent, greenstone.

Although sunlight does not penetrate the bathyal zone, both nektonic and benthonic organisms exist there. Even in the permanent darkness of the abyssal zone, where pressures are in excess of 2000 pounds per square inch (psi), photographs reveal the presence of brittle stars as well as the tracks and burrows of many other organisms. Such life is not abundant, however, and the principal fossils in abyssal deposits are of planktonic and nektonic forms that lived in shallower waters.

Continental Environments

The bulk of continental deposition occurs in the channels and flood plains of rivers and streams, in lake beds, swamps, and deserts, and at the margins of glaciers.

Streams and Rivers. Streams and rivers transport and deposit a variety of sediments ranging from boulders and pebbles to sand and mud. Deposition is greatest on flood plains or in deltas, where the transporting currents are sufficiently diminished for sedimentation to occur. Although generally not as extensive as marine deposits, deltaic deposits may cover thousands of square miles. Stream and river deposits are characterized by a variety of depositional structures, such as channel fillings and asymmetrical ripples (Fig. 5.4).

The color of stream and river deposits may be useful in their identification. Generally subaerial deposits are brown or red, due to oxidation of iron-bearing minerals. Marine clastic deposits, on the other hand, are normally gray, tan, or black, since ocean water is generally not strongly oxidizing. However, the presence of marine fossils in some red beds indicates that red coloring alone is not conclusive evidence of continental deposition. The red coloring in such deposits may have developed subaerially prior to redeposition of the sediments in the marine environment.

Fluvial environments are less likely to preserve fossils than marine environments, since swift currents and turbulence discourage colonization and destroy potential fossils. However, plants and animals may be abundant in the quieter backwaters. Freshwater fauna may include clams, snails, and crustaceans, as well as fish, amphibians, and mammals.

Lakes. Lacustrine deposits are generally fine-grained clastics having floral and faunal associations similar to fluvial environments. Lake muds may be inhabited by a variety of microscopic plants and animals, as well as clams, snails, and fish. The relatively short-lived nature of most lakes makes their sediments poor repositories of long geologic records, even though their slow rates of deposition are beneficial in preserving detailed evidence of events within a geologic epoch. For example, the Eocene lake deposits in the western United States have proven very useful in reconstruction of Eocene life and environments.

Swamps, Marshes, and Bogs. Poorly drained swamps, marshes, and bogs favor the accumulation of peat. These deposits are generally brown or black because the chemically reducing waters prevent oxidation of organic matter. Peat "bogs" support a surprising variety of algae, protozoans,

mollusks, and plants. They range from the sub-Arctic muskegs to tropical swamps. Swamps and marshes furnish a vast breeding ground for insects, fish, and aquatic birds. These environments retard decomposition and provide rapid burial of organic remains, thereby producing extensive freshwater coal deposits.

Glaciers. Glacial deposits vary considerably in texture and composition and range from stratified to unstratified deposits. Since stratified drift is deposited by runoff from melting glaciers, fluvioglacial deposits closely resemble other fluvial deposits. However, the glacial origin of gravels is evident when they are associated with till or with tillite, the lithified counterpart of till (Fig. 5.7). Fluvioglacial deposits generally contain few fossils owing to the low temperature, lack of nutrients, and high energy of the meltwater; tills, on the other hand, may contain fossils as a result of the rapid burial of fauna and flora during fluctuations of the ice margin. Organisms that have adapted to cold climates and have colonized outwash close to the glacier, may be buried by advancing ice and sediments, as in the case of the Two Creeks forest in Wisconsin.

Figure 5.7 Glacial till overlying deformed outwash gravels, Harbor Hill Moraine, Long Island, New York. Deformation of the gravels resulted from overriding by the glacier that deposited the till.

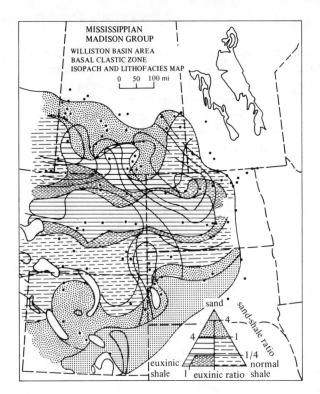

(a)

Figure 5.8 Using isopach and lithofacies maps in the construction of a paleogeographic map: (a) Isopach and lithofacies map of the Mississippian Madison Group. (From H. G. Thomas in Sloss, et al., Ref. 2. Copyright 1960 John Wiley, New York.); (b) inferred paleogeographic map at the time of deposition of the Madison Group.

(b)

PALEOGEOGRAPHIC MAPS

Paleogeographic maps depict depositional environments of contemporaneously deposited rock units. Geographic features may be located by such criteria as the transition from marine to terrestrial deposits, facies changes within marine deposits, changes in thickness of rock units, current directions, and variations in grain sizes of sediments. Knowing the thickness of time - stratigraphic units is useful in the construction of paleogeographic maps. The *isopachs,* or contours of equal thickness of a unit, generally parallel the shorelines, as do boundaries between facies (Fig. 5.8). Current directions, which are determined by study of primary sedimentary structures, may indicate sources of sediment. However, some currents are parallel to the shoreline rather than perpendicular to it.

A major problem arising in the preparation of paleogeographic maps for continental areas is the lack of subsurface data for regions in which relatively young deposits cover older sequences. This problem is especially acute near the continental margins, where sedimentary units thicken greatly. Many paleogeographers have assumed that an area was above sea level at a given time if direct evidence of marine deposits of that age have not been found. However, unless marine deposits of that age are known to be absent, this assumption can be erroneous. Even areas in which marine deposits of a given age are known to be absent may have had such deposits removed by erosion.

PALEOCLIMATOLOGY

Climate plays an important role in determining the nature and composition of sediment in a given area. Generally, glacial deposits are associated with cold climates, and coral reefs and thick limestones are associated with warm climates. Dune sands, evaporites, arkosic red beds, and dolostones are commonly deposited in arid or semiarid climates. The simultaneous occurrence of several types of deposits, each indicating a particular climate, may provide a reasonably reliable basis for determining ancient climates.

Generally, modern climatic belts are restricted to specific latitudes. When compared with paleoclimates (ancient climates), these may then indicate paleolatitudes (ancient latitudes). Thus one method of tracing changes in latitude of a given landmass depends on analysis of sediments and interpretation of changes in climates. On a global scale, studies of paleoclimates of a given age have provided evidence of large-scale displacements of the continents.

Evidence of Glacial Climate

Glacial climates are commonly indicated by the presence of such deposits as tills, tillites, varved clays, and ice-rafted detritus. Of these, tills and tillites are the most easily recognized and consequently the most commonly reported. Tills and tillites are characterized by a lack of sorting of the sediments. Commonly they exhibit a complete range from clay-sized particles to boulders. The clay-sized particles (less than 1/256 mm) may consist of mineral fragments pulverized by glaciers without having been subjected to significant chemical weathering. In other types of sediments, clay minerals formed by chemical weathering are the dominant component of the clay-sized fraction. Sand-sized mineral grains in the till may bear surface markings such as microstriations and roughly broken surfaces, due to glacial abrasion (Fig. 5.9). Pebbles, cobbles, and boulders in tills and ice-rafted deposits are commonly striated and faceted (Fig. 5.10). Glaciated bedrock surfaces may bear grooves and striae, which will be preserved when covered by till (Fig. 5.11). The presence of a poorly sorted sediment or sedimentary rock resting on a grooved and striated pavement is almost conclusive evidence of glaciation. The alignment of striations indicates the direction in which the ice moved.

Exposed bedrock may be smoothed and shaped as *roches moutonneés* (Fig. 5.12). These features are elongated parallel to the direction of ice flow and have a gently inclined *stoss* side in the direction from which the glacier flowed and a steeply inclined *lee* side in the direction of ice flow.

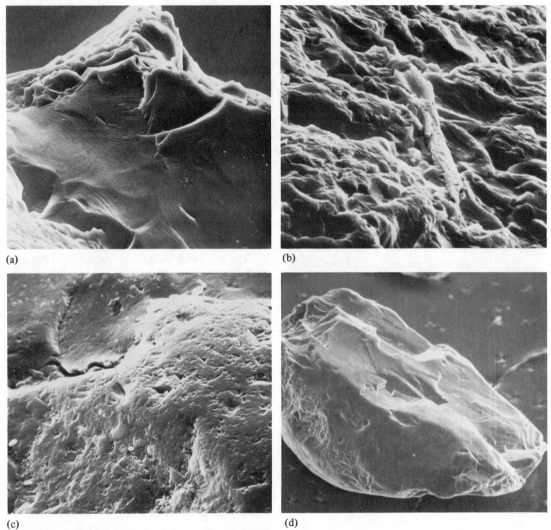

(a)

(b)

(c)

(d)

Figure 5.9 Scanning electron photomicrographs of sand grains: (a) edge of quartz sand grain showing irregular fractures probably caused by grinding in glacial ice; this grain was taken from a core of
Miocene age taken from the North Pacific. Magnification 900X; (b) quartz grain from a dune sand collected near Cairo, Egypt, upturned impact plates probably resulted from impact and crack propagation
across crystallographic planes. Magnification 1300X; (c) quartz sand grain from a high-energy beach
near La Jolla, California; this grain shows V-shaped patterns superimposed on worn breakage blocks.
Magnification 1500X; (d) quartz sand grain from Late Pleistocene deposits of Nassau County, Long
Island. The irregular fractures on the upper surface and the angular outline suggest glacial action; the
lower left-hand portion of the grain is rounded as a result of later dune action. Magnification 75X.
(Photo courtesy of David Krinsley and Stanley Margolis.)

Indications of Arid Climate

Arid climates are generally found at low to middle
latitudes, approximately 10 to 40° from the equator. However, arid climates also may be found in
the rain shadow of a mountain and in the center of

a large continent. Sand dunes, evaporites, and red
beds are characteristic of such environments.
Cross-bedding in sandstones formed from dunes is
parallel to the slip face of the dunes (Fig. 5.13b).
Although the dip may vary among beds, the

Figure 5.11 Striated bedrock in front of the Athabasca Glacier in Jasper National Park, Canada.

Figure 5.10 Striated boulders in the morainal deposits left by the Athabasca Glacier in Jasper National Park, Canada. Notice poorly sorted deposit and the end of the glacier at the top of the illustration.

average direction of dip is leeward of the prevailing wind at the time of deposition. Dune sands are well sorted and may have become frosted by repeated impacts with other grains. The microtextures of dune sands are also distinctive (Fig. 5.9b). Extensive aeolian (wind-borne) deposits generally indicate deposition in an arid or desert climate. However, in some moist regions, such as northeastern North America and southern Alaska, the combination of excessive supply of sand and high winds gives rise to dune fields in regions of moderate or high rainfall. Thus the presence of dune sandstones alone may not be conclusive evidence in determining ancient dry climates but may supplement other indicators of aridity, such as red beds and evaporites.

Red Beds. The origin of the red coloring in red

Figure 5.12 Striated and polished *roche moutonnée* in the Pyrenees. (Photo by François Arbey.)

beds has been the subject of considerable debate among geologists (3, 4). Red, lateritic soils may be produced in moist tropical and subtropical regions owing to deep weathering. Most silicates and ferromagnesian minerals are chemically weathered

(a)

Figure 5.13   Comparison of modern sand dunes with ancient dune sandstone: (a) sand dunes in the Calanscio Sand Sea, Libya (courtesy of British Petroleum Co.); (b) cross-bedding in dune sandstone of the Triassic New Red Sandstone. (Crown Copyright Geological Survey photograph. Reproduced by permission of the Controller, Her Britannic Majesty's Stationery Office.)

(b)

into red clays rich in aluminum and iron. Sediments eroded from a lateritic soil and deposited in an oxidizing environment would presumably retain the red color. However, there are no known areas in which red beds of this type are being formed at the present time. Furthermore, the absence of feldspar in lateritic soils indicates that red arkoses cannot have been formed by erosion of such soils.

Red sediments are also common in desert regions such as the Kalahari Desert of southern Africa and in northwestern Mexico (4). Near the northern end of the Gulf of California, rain is infrequent and much of what there is comes down in cloudbursts. For this reason, feldspar and iron-bearing minerals are eroded before they can be completely weathered chemically. The first step in the formation of the red coloring in the sediment is the weathering in place of the iron-bearing minerals, such as hornblende and biotite, to limonite (hydrated iron oxide) within the sediment. In a hot arid climate the limonite is slowly dehydrated to form hematite. Since feldspar is relatively stable in a dry climate it is likely that red arkoses are produced in this manner. In temperate or polar climates, red beds would not commonly form, since iron is generally carried away in solution.

Evaporites. Evaporation of large quantities of seawater may result in the precipitation of gypsum, halite, and sylvite, provided deposition occurs:

1. in an enclosed basin with limited inflow of normal marine waters

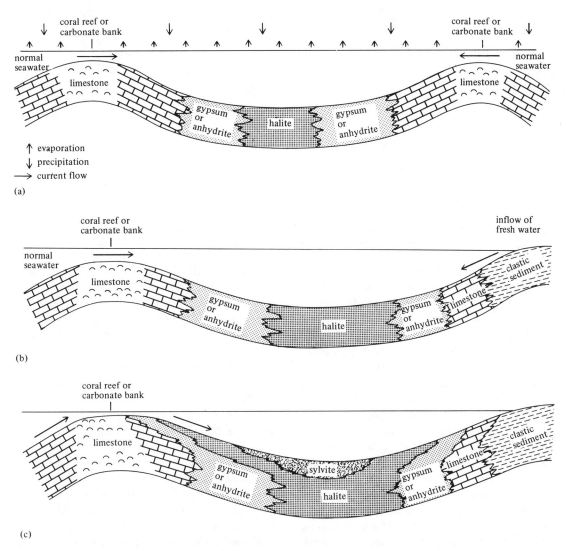

Figure 5.14 Formation of evaporites in an enclosed basin: (a) basin bordered by coral reefs or carbonate banks, (b) basin bordered partly by coral reefs and partly by land, (c) basin with very restricted inflow due to regression of the sea or up-building of carbonate bank.

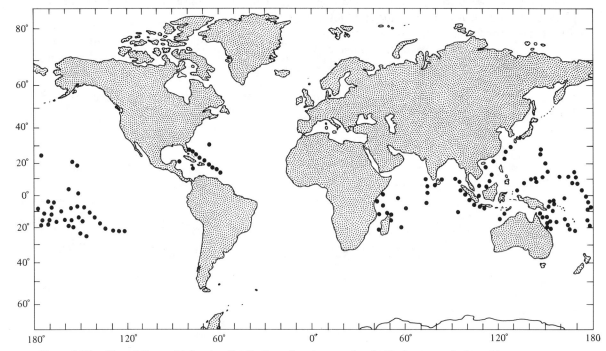

Figure 5.15 Map of the world showing distribution of modern coral reefs. Reef areas are indicated by large dots. There are no important coral reefs at latitudes beyond approximately 30°. (After Shrock and Twenhofel, Ref. 9. Copyright 1953 by McGraw-Hill, Inc. Used with permission of McGraw-Hill Book Co.)

2. in an area of relatively low rainfall
3. where the rate of evaporation of seawater is high, as in an area of low humidity and high temperature.

Most major evaporite deposits were formed in large basins that were bordered either by land or coral reefs (5) (Fig. 5.14). Such basins allowed little exchange of water with the open ocean, and thus sufficient concentrations of dissolved salts were formed. The Mediterranean Sea is an enclosed basin in an area of relatively low rainfall, but no extensive evaporites are now forming in this area because the inflow of normal marine water is balanced by an outflow of more saline water through the Strait of Gibraltar.

For precipitation of large quantities of dissolved salts to occur, the rate of evaporation must exceed the inflow of freshwater from rainfall, streams, and rivers. Such conditions are found today in arid climates of low to middle latitudes, leading climatologists to believe that large evapo-rite deposits are indicative of arid climates. Since the deposition of potassium-bearing salts (e.g. sylvite) requires almost complete evaporation of seawater, the presence of such salts may indicate deposition in a very arid climate. Thinner salt deposits, including salt crusts and halite crystal casts on bedding planes, may form in semiarid climates.

Since large, enclosed marine basins do not presently exist on the continents, there are no extensive marine evaporites forming today. However, minor amounts of marine evaporites are presently forming in the Persian Gulf, on the east and west coasts of Baja California, and on the northern border of South America (6). These areas are located between 10 and 30°N latitude and have arid climates.

Dolostones. The origin of dolostones is another widely debated topic among geologists. Experimental and observational data indicate that dolomite, which is the principal mineral of dolostone, is seldom if ever precipitated directly from sea-water. At the present time, however, dolomite is replacing calcite and aragonite in the sediments above mean high tide in areas of high aridity (7). Evaporation of the waters trapped in sediments in the supratidal zone results in an increase in the magnesium content of the brines. Under such conditions, primary calcium carbonate may be dolomitized before the sediment is lithified. Dolomitization may also occur after lithification in relatively porous rocks such as reef deposits. Concentrated brines moving through the reefs result in replacement by dolomite (8). The climatic significance of these processes is that concentrated brines are produced only where rainfall is low and evaporation is high. Consequently, dolomitization should be widespread only in arid climates. The common association of dolostone with thick evaporites and arkosic red beds supports this inference.

Indication of Warm Climates

Precipitation of large quantities of calcium carbon-ate, both organically and inorganically, requires a rather warm climate. As in the case of evaporites, large volumes of seawater are required because of the low concentration of calcium carbonate in seawater. Warm temperatures are favorable to the precipitation of calcium carbonate, since the rate of evaporation increases and the solubility of calcium carbonate decreases with increasing temperature. Limestones are being formed today in the warm waters of Bermuda, the Bahamas, and the Persian Gulf.

Modern reef-building corals have skeletons composed of calcium carbonate, and are generally found in waters warmer than 68°F (20°C). Such waters are found today in warm-moist or warm-dry climates within 30° of the equator (Fig. 5.15). Thus the presence of thick limestones and extensive coral reefs suggests deposition in a warm climate relatively near the equator. However, it should be pointed out that less extensive carbonate-rich sediments and small coral banks are forming today in relatively cool waters at latitudes as high as 71°N.

Figure 5.16 Peat bog in Ireland. The low escarpment is where peat has been removed. (Photo by Eleanor Catena.)

Indication of Moist Climates

Thick coal sequences have long been associated with the lush growth of tropical swamps and marshes. However, coal-forming swamps and marshes may form in any climatic zone in which the rainfall exceeds potential evaporation plus evapotranspiration. Peats, which are the precursors of coals, are now forming in tropical, temperate, and cold climates (Figs. 5.16 and 5.17). The muskeg swamps of the Arctic represent potential coal-forming environments. The only environment unfavorable for coal formation is a hot, dry climate. It is interesting to note that coals are not generally found in sequences containing abundant evaporites, dune sandstones, and red beds (Figs. 10.3, 11.1, and 12.1).

In some cases it is possible to differentiate coals that developed in a tropical environment from those produced in a temperate or cold climate. Many tropical plants do not have growth

(a)

(b)

Figure 5.17 Aerial view (a) and close-up (b) of mangrove swamp in southern Florida. Peat is now forming in such swamps. (From Spackman, et al. Ref. 10.)

rings, which are characteristic of trees in temperate or cold regions. The lack of growth rings has been related to continuous, rather than seasonal growth. If the fossil flora in a coal lacks growth rings, it is likely that the coal was formed in a tropical environment. For Mesozoic and Cenozoic coal deposits, climatic conditions have been inferred from a comparison of the fossil flora with modern plant assemblages.

REFERENCES CITED

1. P. E. Playford and A. E. Cockbain, 1969, Algal stromatolites: Deepwater forms in the Devonian of western Australia: *Science,* v. 165, p. 1008.

2. L. L. Sloss, E. C. Dapples, and W. C. Krumbein, 1960, *Lithofacies Maps of the United States and Southern Canada:* Wiley, New York.

3. P. D. Krynine, 1949, The origin of red beds: *N.Y. Acad. Sci. Trans.,* series II, v. 2, p. 60.

4. F. B. Van Houten, 1968, Iron oxides in red beds: *Geol. Soc. Amer. Bull.,* v. 79, p. 399.

    T. R. Walker, 1967, Formation of red beds in modern and ancient deserts: *Geol. Soc. Amer. Bull.,* v. 78, p. 353.

5. L. F. Dellwig and R. Evans, 1969, Depositional processes in Salina salt of Michigan, Ohio, and New York: *Amer. Assoc. Petrol. Geol. Bull.,* v. 53, p. 949.

    A. M. Klingspor, 1969, Middle Devonian muskeg evaporites of western Canada: *Amer. Assoc. Petrol. Geol. Bull.,* v. 53, p. 927.

    L. L. Sloss, 1969, Evaporite deposition from layered solution: *Amer. Assoc. Petrol. Geol. Bull.,* v. 53, p. 776.

6. F. B. Phleger, 1969, A modern evaporite deposit in Mexico: *Amer. Assoc. Petrol. Geol. Bull.,* v. 53, p. 824.

7. K. S. Deffeyes, F. J. Lucia, and P. K. Weyl, 1965, Dolomitization of Recent and Plio-Pleistocene sediments by marine evaporite waters on Bonaire, Netherlands Antilles, *in* L. C. Pray and R. C. Murray, eds., *Dolomitization and Limestone Diagenesis,* Society of Economic Paleontologists and Mineralogists Special Publ. 13, p. 71.

8. R. L. Jodry, 1969, Growth and dolomitization of Silurian Reefs, St. Clair County, Michigan: *Amer. Assoc. Petrol. Geol. Bull.,* v. 53, p. 957.

9. R. R. Shrock and W. H. Twenhofel, 1953, *Principles of Invertebrate Paleontology:* McGraw-Hill, New York.

10. W. Spackman, W. L. Riegel, and C. T. Dolsen, 1969, Geological and biological interactions in the swamp-marsh complex of Southern Florida, *in* E. C. Dapples ed., *Environments of Coal Deposition:* Geological Society of America Special Paper, p. 1.

SUGGESTED READINGS

W. C. Krumbein and L. L. Sloss, 1963, *Stratigraphy and Sedimentation:* Freeman-Cooper, San Francisco.

L. F. Laporte, 1968, *Ancient Environments:* Prentice-Hall, Englewood Cliffs, N. J.

L. L. Sloss, 1953, The significance of evaporites: *Jr. Sediment. Petrol.,* v. 23, p. 143.

6
Dynamics of the Earth's Interior

On November 14, 1963, fishermen observed black clouds coming from the sea south of Iceland. Upon investigation, they found a new island that apparently had been built up from the sea bottom by volcanic eruptions. The birth of the island was accompanied by a plume of steam four miles high, formed as the cold seawater came in contact with the hot lava. The volcano hurled millions of tons of rock onto the surface of the island so that after 16 days of activity, the island was 3300 ft (1000 m) long (Fig. 6.1).

The island, which was christened Surtsey, after a fire demon of Norse mythology, is located on a vast and largely submarine range of mountains known as the Mid-Atlantic Ridge. Earthquakes are common along the length of the ridge, and submarine depth profiling has located numerous volcanic seamounts covering the crest and flanks of the ridge (1). The Mid-Atlantic Ridge is part of a worldwide system of oceanic ridges more than 36,000 miles (60,000 km) long (Fig. 1.10).

Another very active area in which volcanic eruptions and earthquakes are common is the "ring of fire" around the Pacific Ocean basin. Many of the world's great mountain belts are located in this zone. The Alpine-Himalyan region comprises another unstable zone.

Observations such as these have led scientists to conclude that the interior of the earth is quite active. Geophysical investigations of the earth's interior have provided insights into processes related to this activity. These studies have indicated that there are large-scale convection currents within the earth's mantle. Moreover, geological data suggest that such convection currents have played an important role in the development of major structural features such as geosynclines, oceanic ridges, island arcs, and mountain belts, and that convection currents may be the force behind movements of large segments of the earth's crust.

HEAT FLOW

In deep gold mines in South Africa, the rocks are almost too hot to touch. The miners, who must work shortened shifts to avoid heat prostration, are well aware of the heat that escapes from within the earth. The rate of increase in temperature with depth, the geothermal gradient, as measured in mines and well holes throughout the world, has an average value of about $17°F$ per 1000 feet ($30°C$ per kilometer).

The observable heat flow represents that portion of the earth's internal heat which reaches the surface. Mathematically, heat flow is a product of the temperature gradient and the conductivity of the rock or sediment in which the measurements were made. In the ocean basins, the geothermal gradient is measured by a probe dropped into the soft sediment. Investigations in oceanic areas show that the heat flow in the vicinity of mid-oceanic ridges is significantly greater than it is in the deeper parts of the ocean basins (2). The mid-oceanic ridges are believed to be sites of upwelling convection currents within the mantle.

Figure 6.1   The violent birth of a new island, Surtsey, in November 1963. (Courtesy of Solarfilma, Reykjavik.)

THE EARTH'S VISCOSITY

Many solids behave plastically when they are near their melting temperature. The flow of glaciers is a dramatic example of this phenomenon. The deeper levels of continental and alpine glaciers are subjected to high pressures owing to the weight of the overlying ice. When the temperatures within several glaciers were measured, it was found that their deeper layers are generally close to $32°F$ ($0°C$). At these temperatures the ice is capable of plastic flow.

The mantle of the earth is thought to be similarly capable of flowing in response to relatively small directed stresses applied over long periods of time. This is well illustrated in the process of crustal rebound which follows the melting of large ice sheets. The area in the vicinity of Hudson Bay has been uplifted more than 900 ft (300 m) since the melting of the Pleistocene ice sheet. Such uplift must have been accompanied by the flow of a large volume of mantle material into the region below Hudson Bay.

The rate at which rock flows in response to pressure is a function of its apparent viscosity. Viscosity is a measure of the internal friction of a fluid in motion. The higher the viscosity of a fluid, the more sluggish its flow. Although rocks within the mantle are thought to behave as plastic solids rather than viscous fluids (3), it is possible to consider the apparent viscosity of the mantle.

The unit of measure of the viscosity of fluids is the poise (P). Water, for example, has a viscosity of approximately 0.01 P, whereas a heavy weight of oil has a viscosity of about 1 P. The apparent viscosity of the upper mantle under Hudson Bay has been calculated at 10^{22} P based on the rate of postglacial uplift. Calculations based on the rate of rebound of the area in the vicinity of the Great Salt Lake following the evaporation of Lake Bonneville provide a figure of 10^{21} P for the apparent viscosity of the upper mantle in this area.

Based on seismic studies, the viscosity of the upper mantle between the depths of 45 and 540 miles (75 and 900 km) has been estimated at between 10^{21} and 10^{23} P (4). Seismic studies indicate that the viscosity of the lower mantle is between 10^{23} and 10^{26} P (5). However, calculations based on a study of the acceleration of the earth's rotation have indicated that the apparent viscosity of the lower mantle is approximately 10^{22} P (4). It has been calculated that apparent viscosities equal to or less than 10^{25} P within the mantle will permit the development of convection currents. Thus convection is possible within the upper mantle and may be possible within the lower mantle as well.

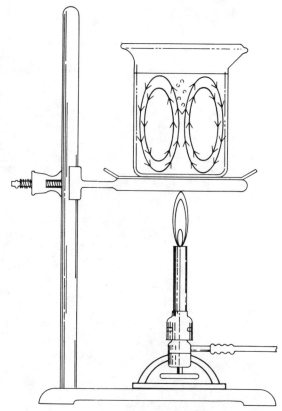

Figure 6.2 Convection currents within a beaker filled with water.

CONVECTION CURRENTS

Convection is defined as the vertical transfer of heat by the circulation or movement of a gas, liquid, or plastic solid. Such a motion results from unequal heating, which produces differences in density. Rapidly moving convection currents may be observed by heating water in a glass coffee pot or beaker (Fig. 6.2). As the water in the bottom of the container is heated, it expands and becomes less dense. The warm water rises while the cooler water at the top and sides of the container sinks. This results in a heat transfer within the container. In this case, convection takes the form of a cell in which the fluid moves in a more or less circular fashion. However, it is not necessary that all convection currents form such a pattern.

F. A. Vening Meinesz was the first to provide geophysical evidence of the existence of convection currents within the earth's mantle (6). His gravity measurements in the East Indies indicated that large negative gravity anomalies are associated with the oceanic trench bordering Indonesia (Fig. 6.3). Vening Meinesz believed that the lower-than-average force of gravity indicated that the earth's crust was being pulled down into the mantle by sinking convection currents. This he proposed, would create a mass deficiency which would cause a negative gravity anomaly.

Convective flow within the mantle is commonly represented in the form of convection cells with more or less circular cross sections (Fig. 6.4). However, rocks behave as semiplastic solids during deformation and would not necessarily flow in the same manner as fluids. Egon Orowan (3) has stated that at the temperatures and pressures obtaining within the mantle, convection within a semiplastic

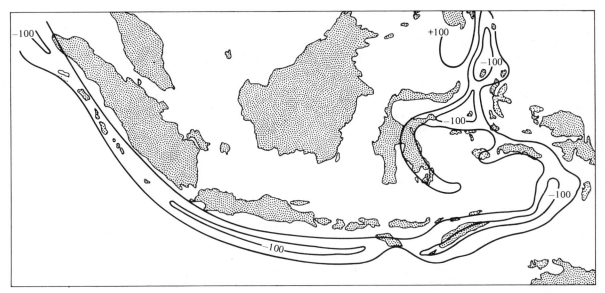

Figure 6.3   Regional isostatic gravity anomalies in the vicinity of the Indonesian Island Arc. (Contours in milligals. (Modified after Vening Meinesz, Ref. 6. Used with permission of North-Holland Publishing Co., Amsterdam.)

solid would take the form of rising hot dikes (Fig. 6.5). Such convective motion would resemble the diapiric movement of solid masses of salt during the formation of salt ridges and salt domes (8) but would be quite different from the convective motions within a liquid cell.

Diapiric intrusions of mantle rock would cause lateral separation of rigid plates comprised of crust and mantle as the current rises. These plates are referred to as *continental plates* if they include a continent and *oceanic plates* if they do not include continental crust.

During injection of dike material from the mantle into a crustal plate, the outer, more rapidly cooled, parts of the intrusion are accreted onto the sides of the plate. Any subsequent intrusions are in the center of earlier intrusions, since this zone becomes warmer and less rigid than the walls of the intrusion (Fig. 6.6). Therefore, if a dike of semiplastic mantle were to split a continent, the center of the dike would remain midway between the resulting continents. This may explain why the Mid-Atlantic Ridge and the Mid-Indian Ocean

Ridge are approximately centered between the adjacent continents.

For most continental plates, little if any material is lost as new material is added beneath mid-oceanic ridges. In fact, the Pacific margins of continental plates appear to be growing by accretion of oceanic sediments and volcanic rocks. Therefore, it appears that the continental plates are increasing in area and volume. Since the earth is not expanding significantly (9), the Pacific Ocean plates must be decreasing in area and volume at approximately the same rate as the continental plates are increasing. A transfer of mantle material from the Pacific Ocean plates to the continental plates would be required. The transfer presumably occurs through downward movement of material in the vicinity of oceanic trenches. The mantle material would then move horizontally under the continental plates, replenishing the mantle material of the rising convection current under the Mid-Atlantic Ridge (Fig. 6.5).

Measurements of the velocity of seismic waves at various depths under ocean basins indicate that

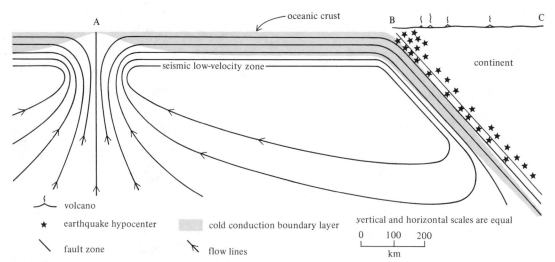

Figure 6.4 Hypothetical cross section through the crust and upper mantle showing a thin slab of lithosphere carried along by the motions of underlying convection cell within the asthenosphere. (After Turcotte, Ref. 7, courtesy of *Engineering: Cornell Quarterly.*)

the upper 45-mile (75-km) layer of the earth's interior is relatively rigid. This layer, referred to as the *lithosphere*, overlies a more plastic layer called the *asthenosphere*. The lithosphere of Pacific Ocean plates bends downward beneath the Pacific margin of continental plates and extends to a depth of 420 miles (700 km) (Fig. 6.5). This means that the transfer of material from the

Pacific Ocean plates to the continental plates must occur at a depth of more than 420 miles, since the lithosphere acts as a barrier to horizontal flow. Therefore, it is likely that the continental plates are at least 420 miles thick.

Heat flow measurements support the concept of relatively thick continental plates. The average value of heat flow for the continental plates is

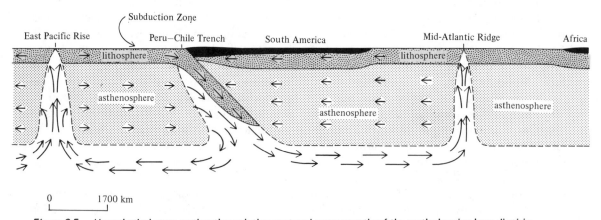

Figure 6.5 Hypothetical cross section through the crust and upper mantle of the earth showing how diapiric convective motions within the mantle may produce horizontal movements of large blocks.

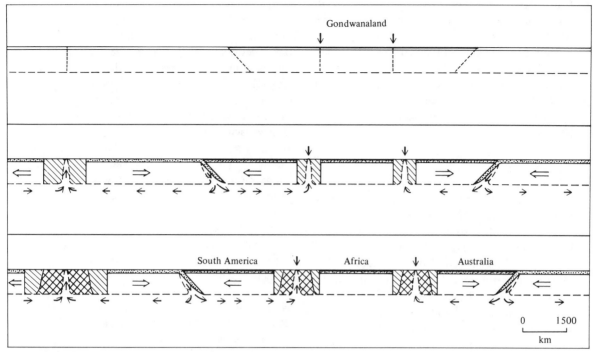

Figure 6.6   Hypothetical cross section showing accretion of mantle onto blocks as diapiric intrusions of mantle produce horizontal movements of blocks.

almost exactly the same as the average heat flow for the ocean basins (10). This is surprising, since the continental crust contains greater amounts of radioactive heat-producing isotopes, such as ^{235}U, ^{238}U, ^{232}Th, and ^{40}K, than the oceanic crust. Apparently the mantle beneath the continents has been depleted of radioactive elements relative to the mantle beneath the ocean basins. In order to maintain the similarity in heat flow during continental drift, the continents and the underlying depleted mantle must move as a single unit (11). This would be the case if the continental plates were more than 420 miles thick. The Pacific Ocean plates may be of the same thickness as the continental plates, although there is little direct evidence for this supposition. A recently discovered layer of low seismic velocity at a depth of 480 miles (800 km) (12) may be located at the bottom of both continental and Pacific Ocean plates.

The material descending at the margins of the Pacific Ocean basin must replenish both the mantle material rising under the Mid-Atlantic Ridge and that rising under the East Pacific Rise. This would require a splitting of the descending convection current, which would then result in a spiral movement of the material within the Pacific Ocean plate (Fig. 6.5). This motion contrasts with the circular movement of convection cells within viscous fluids. Furthermore, there is not a complete cycle of material within continental plates (as would be the case in a convection cell—Fig. 6.5). Thus convection with the mantle probably does not take the form of cells.

TECTONIC ELEMENTS OF THE CONTINENTS

Measurements of the thickness of time - stratigraphic units indicate that certain areas have a thicker accumulation of sediments or sedimentary rocks than others. Differential subsidence of the continents is thought to account for this. Linear belts in which subsidence has been relatively rapid are termed *geosynclines,* and broad areas that have been more stable for long periods of time are termed *cratons.* Linear belts in which uplift is relatively rapid are termed *geanticlines.* Geanticlines furnish much of the sediment deposited in geosynclines.

Geosynclines

James Hall, State Geologist of New York in the 1850s, observed that the folded sedimentary rocks in the Appalachian Mountains were much thicker than the equivalent strata that had been deposited in the midwestern part of the United States. Hall noted that the Paleozoic deposits thickened from several thousand feet (about 1500 m) in the Mississippi Valley region to 40,000 ft (12,500 m) in the Appalachians. In 1857 Hall proposed that the sediments of the Appalachian Mountains had been deposited in an elongated, slowly subsiding trough. Later in the nineteenth century, James Dana proposed the name *geosyncline* for this type of structure.

Both Hall and Dana believed that there was a connection between the geosynclines and the mountains that subsequently developed from them. Hall postulated that when the trough became filled with sediments, the rocks were folded by collapsing into the center of the trough. Dana, on the other hand, believed that the earth was cooling and contracting, and that the geosyncline was folded as a result of a horizontal compressive force produced by this contraction. Subsequent workers modified and extended the ideas of Hall and Dana. Hans Stille observed that geosynclinal deposits could be separated into two types—those in which the deposits are almost entirely sedimentary, and those in which volcanic rocks are interbedded with sedimentary rocks. He proposed that the former be called *miogeosynclines* and the latter *eugeosynclines.*

Some geologists prefer to restrict the use of the term geosyncline to an elongated depositional trough in which clastic sediments and volcanic rocks are deposited in moderately deep water. However, this definition is quite different from Hall's definition of a geosyncline. Hall used the Appalachian Miogeosyncline to illustrate the character of a geosyncline, and this geosyncline contains a great abundance of shallow-water carbonates and clastics with no associated volcanics. The term geosyncline should denote an elongate zone in which a thick sequence of sedimentary and/or volcanic deposits has accumulated, regardless of the depth of water in which they were laid down. During deposition, the basement on which the deposits accumulate is often downwarped in a broad syncline. The rate of subsidence of the basement in geosynclines is more than 40 ft (12 m) per million years. The three main types of geosynclines are discussed below.

Miogeosyncline. Miogeosynclinal deposits include limestone, shale, quartz sandstone, and conglomerate. The presence of such primary structures as mud cracks, ooliths, and algal stromatolites in these deposits indicates deposition in relatively shallow water. Igneous intrusions and volcanic deposits rarely occur within the miogeosyncline.

Eugeosyncline. The total thickness of sedimentary and volcanic rocks in a eugeosyncline is generally greater than in the adjacent miogeosyncline. Eugeosynclinal sedimentary rocks are almost entirely clastic and include shale, graywacke, and conglomerate. Well-sorted quartz sandstones are generally not very abundant. Volcanic rocks within eugeosynclines include flows and tuffs as well as shallow dikes and sills. The volcanics are most commonly andesitic in composition, but basaltic and rhyolitic volcanics are abundant in some areas. Angular unconformities are common within eugeosynclinal sequences, indicating a tectonically unstable environment of deposition.

Modern depositional environments analogous to those in ancient eugeosynclines are found in the vicinity of island arcs, such as the Indonesian Island Arc (Fig. 6.7). These regions are typified by

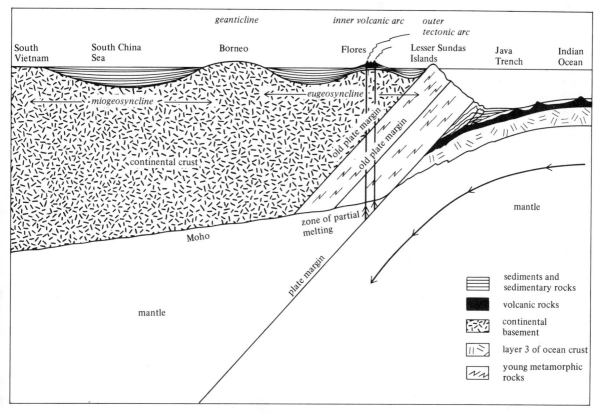

Figure 6.7 Diagrammatic cross section across the Indonesian Island Arc.

the presence of volcanoes, geanticlines (elongate upwarps), and structural troughs in which sediments are being deposited. Sedimentation is especially rapid adjacent to volcanic and tectonic islands. Since modern island arcs are restricted to continental margins, it is likely that ancient eugeosynclinal deposits formed in similar environments. This relationship may be used in reconstructing the geography of ancient continents.

Paraliageosyncline. Marshall Kay proposed the term *paraliageosyncline* for a geosyncline located on the tectonically stable margins of the continents (13). Modern paraliageosynclines border much of the area of the Atlantic and Indian oceans (Fig. 6.8). Sediments deposited in these areas are relatively undeformed and underlie

coastal plains and the waters of the continental shelf. These deposits have a great lateral extent, and their rate of sedimentation averages 100 ft (34 m) per million years. The sequences deposited in paraliageosynclines contain almost no volcanic rocks. Unconsolidated sands, muds, and gravels are abundant where there is a source of clastic sediments. Limestone and chalk are deposited on carbonate banks in areas containing little clastic sediment. Paraliageosynclinal deposits are similar to miogeosynclinal deposits except that eugeosynclinal deposits are not generally associated with them. Furthermore, compressional folding does not generally occur within this type of geosyncline.

Subsidence in Geosynclines. Geosynclines com-

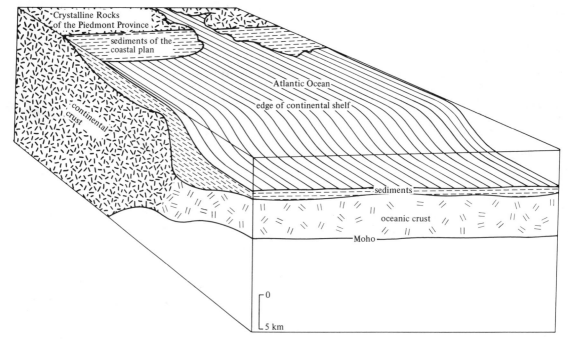

Figure 6.8   Block diagram showing paraliageosyncline on the Atlantic margin of North America. This geosyncline consists of one structural trough under the continental shelf and a second structural trough at the base of the continental slope. Sediments of the paraliageosyncline are shown as short dashes.

monly contain several tens of thousands of feet of sediments, and yet even the oldest of these sediments often shows evidence of having been deposited in shallow water. Evidently the rate of subsidence within the geosyncline was approximately equal to the rate of accumulation of its sediments. Only in this way would the surface of deposition remain at approximately the same level throughout sedimentation. Both Hall and Dana recognized this relationship, but they differed on the cause of the subsidence. Hall suggested that the weight of the sediments caused the subsidence, whereas Dana proposed that horizontal compression produced a trough that soon became filled with sediments. The cause of subsidence within geosynclines continues to be a subject for debate. The following mechanisms have been suggested:

1. When sediments and volcanics are deposited in a geosyncline, a certain amount of subsidence results from the isostatic adjustment of the crust to the weight of the deposits. This adjustment takes place by a plastic flow within the mantle. The amount of subsidence is determined by the thickness of sediment deposited, the density of the sediments, and the density of the layer in which flow takes place. If isostatic balance is maintained, the weight of sediment will be equal to the weight of seawater displaced plus the weight of mantle displaced. This relationship is expressed in the following equation:

$$D_s \times T = D_w \times d + D_m(T - d)$$

As an example, assume that the density of the sediments (D_s) is 2.4 g/cc, the density of seawater (D_w) is 1.03 g/cc, and the density of the mantle (D_m) is 3.3 g/cc. If the initial depth of seawater (d) is 100 m (60 ft), the thickness of sediments that may be deposited below sea level (T) is 242 m (152 ft).

Clearly the amount of subsidence produced by isostatic adjustment is not sufficient to explain the accumulation of tens of thousands of feet of shallow-water sediments in geosynclines. There must be another factor involved in the process. Since present-day geosynclines are in isostatic balance, isostasy must play an important role in their subsidence. The portion of subsidence due to isotatic adjustments will be equal to D_s/D_m or approximately ¾ of the total required for the rate of subsidence to be equal to the rate of sedimentation.

2. It has been suggested that convection currents produce subsidence within geosynclines by dragging down the earth's crust as the currents descend. It is also possible that convection currents erode the crust from beneath geosynclines and thus contribute to subsidence. Although convection currents may descend beneath miogeosynclines and eugeosynclines, the lack of seismic activity and folding associated with paraliageosyn-

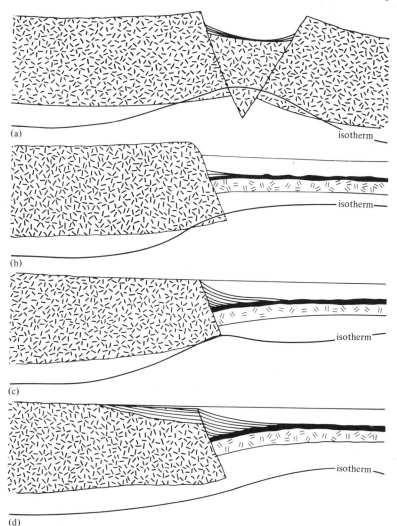

(a)

isotherm

(b)

isotherm

(c)

isotherm

(d)

Figure 6.9   Development of a geosyncline on the margin of an ocean basin formed by continental drift and sea-floor spreading: (a) uplift and faulting over an upwelling convection current, (b) beginning of erosion of uplifted margin of continent, (c) erosion of continental margin to sea level before the underlying mantle cools to the temperature of the adjacent mantle, (d) subsidence of continental margin and deposition of sediments as underlying mantle cools.

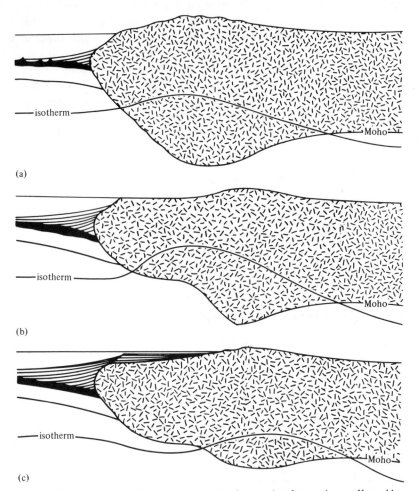

Figure 6.10 Development of a geosyncline on the margin of a continent affected by compressional folding: (a) beginning of erosion of mountain following its uplift, (b) erosion of mountain belt to sea level before the underlying mantle cools to the temperature of the mantle beneath the stable interior of the continent, (c) subsidence of margin and deposition of sediments as the underlying mantle cools.

clines suggests that this mechanism is not applicable in such areas.

3. The Mohorovicic discontinuity, or Moho, at the base of the continents may represent a transition from a mafic rock with a density of 2.9 g/cc to eclogite with a density of about 3.5 g/cc. In this case, the transition would be the result of a phase change rather than a change in chemical composition of the rock. An increase in pressure or a decrease in temperature in the vicinity of such a phase change would result in a conversion of the less dense phase to the more dense phase (eclogite). If the resulting decrease in volume occurred at the base of the crust beneath a geosyncline, it would result in subsidence of the sea-floor within the geosyncline (14). Such a phase change may be due to the weight of sediments deposited in a geosyncline, which would cause an increase in pressure within the rock below the geosyncline and thereby produce a phase change at the base of the crust.

4. Withdrawal of magma beneath a eugeosyn-

cline during volcanic eruptions or lateral migration of magma within the mantle could also account for some of the subsidence in the eugeosyncline and perhaps even some of the subsidence in the adjacent miogeosyncline (15). However, the lack of volcanism associated with deposition within paraliageosynclines indicates that this mechanism does not produce subsidence in these areas.

5. If the mantle beneath part of a continent is heated, it undergoes thermal expansion. As a consequence, the overlying continent is uplifted. The uplifted part of the continent is then subjected to increased subaerial erosion, resulting in a thinning of the earth's crust. If this erosion were followed by a cooling of the mantle, the mantle would contract and the overlying continent would subside (16). The sequence of heating - uplift - erosion - cooling - subsidence may occur over an upwelling convection current or in the vicinity of a mountain range. The paraliageosynclines bordering the Atlantic and Indian oceans may have been formed as a result of such a sequence of events (14) (Fig. 6.9).

Miogeosynclines and eugeosynclines commonly develop on the roots of ancient mountain systems. Mountains would initially have a high heat flow, and therefore the mantle under the mountains would be hotter than that under the stable interior of the continent (Fig. 6.10). If the mountainous region had been eroded before the underlying mantle cooled to the temperature obtaining under the stable interior, the crust could have been so reduced in thickness by erosion that it would be thinner than the crust of the stable interior. Further cooling of the mantle under the eroded mountain belt would cause subsidence of the mountain roots and would thus permit geosynclinal deposition.

Cratons

The interiors of continents are generally characterized by relatively thin sedimentary deposits or ancient crystalline rocks. In most areas, the sedimentary rocks are undeformed. The stable part of the continent, the *craton*, includes shields and platforms. *Shields* are extensive areas in which

Precambrian rocks are exposed, whereas the *platform* contains the sedimentary rocks. Areas of greater-than-average subsidence in the platform are called *basins*; slower subsidence occurs over *arches* or *domes*. The average rate of subsidence of platforms is generally less than 40 ft (12 m) per million years. However, higher rates have been calculated for the centers of some basins. Since these areas are not elongated, they would not be geosynclines according to Hall's definition of the term. The deposits of the platform include abundant carbonates (limestone and dolostone) as well as clastics (shale and sandstone). Volcanic rocks are rare. The craton is generally bordered by geosynclines.

REFERENCES CITED

1. L. R. Sykes, 1969, Seismicity of the mid-oceanic ridge system, *in* R. H. Hart, ed., *The Earth's Crust and Upper Mantle:* Geophysical Monograph No. 13, American Geophysical Union, Washington, D. C.

2. R. P. Von Herzen, 1965, Present status of ocean heat flow measurements, *in* L. H. Ahren et al., eds., *Physics and Chemistry of the Earth,* v. 6: Pergamon Press, London.

3. E. Orowan, 1964, Continental drift and the origin of mountains: *Science,* v. 146, p. 1003.

4. R. H. Dicke, 1969, Average acceleration of the earth's rotation and the viscosity of the deep mantle: *Jr. Geophys. Res.,* v. 74, p. 5895.

5. D. L. Anderson, 1965, Earth's viscosity: *Science,* v. 151, p. 321.

   R. K. McConnell, Jr., 1968, Viscosity of the earth's mantle, *in* R. A. Phinney, ed., *The History of the Earth's Crust—A Symposium:* Princeton University Press, Princeton, N. J., p. 45.

6. F. A. Vening Meinesz, 1964, *Developments in Solid Earth Geophysics—1—The Earth's Crust and Mantle:* Elsevier, New York.

7. D. L. Turcotte, 1970, Continental drift: *Engineering—Cornell Quarterly,* v. 5, p. 2.

8. C. K. Seyfert, 1968, Dilational convection: An explanation for the similarity of conti-

nental and oceanic heat flow during sea-floor spreading: *Trans. Amer. Geophys. Union,* v. 49, p. 202.

J. C. Maxwell, 1968, Continental drift and a dynamic earth: *Amer. Sci.,* v. 56, p. 35.

9. S. I. Van Andel and J. Hospers, 1968, Paleomagnetism and the hypothesis of an expanding earth: A new calculation method and its results: *Tectonophysics,* v. 5, p. 273.

10. K. Horai and G. Simmons, 1969, Spherical harmonic analysis of terrestrial heat flow: *Earth Planet. Sci. Lett.,* v. 6, p. 386.

11. G. F. G. MacDonald, 1964, The deep structure of continents: *Science,* v. 143, p. 921.

12. C. Wright, 1968, Evidence for a low velocity layer for P Waves at a depth close to 800 km: *Earth Planet. Sci. Lett.,* v. 5, p. 35.

13. M. Kay, 1951, *North American Geosynclines:* Geol. Soc. Amer. Mem. 48.

14. D. C. Noble, 1961, Stabilization of crustal subsidence in geosynclinal terranes by phase transition at M: *Geol. Soc. Amer. Bull.,* v. 72, p.287.

W. B. Joyner, 1967, Basalt-eclogite transition as a cause for subsidence and uplift: *Jr. Geophys. Res.,* v. 72, p. 4977

15. A. E. Scheidegger and J. A. O'Keefe, 1967, On the possibility of the origination of geosynclines by deposition: *Jr. Geophys. Res.,* v. 72, p. 6275.

16. N. H. Sleep, Thermal effects of the formation of Atlantic continental margins by continental break-up: *Geophys. Jr. Roy. Astronom. Soc.,* v. 24, p. 325.

SUGGESTED READINGS

L. Knopoff, 1969, The upper mantle of the Earth: *Science,* v. 163, p. 1277.

F. A. Vening Meinesz, 1955, Plastic buckling of the earth's crust: The origin of geosynclines, *in* A. Poldervaart, ed., *Crust of the Earth—A Symposium:* Geological Society of America Special Paper 62, p. 319.

A. H. Mitchell, and H. G. Reading, 1969, Continental margins, geosynclines, and ocean floor spreading: *Jr. Geol.,* v. 77, p. 629.

7

Polar Wandering, Continental Drift, And Sea-Floor Spreading

Until the 1960s, most geologists believed in the permanence of continents and ocean basins. This concept did not provide answers to such problems as the origin of mountain ranges, the distribution of Permo-Carboniferous glacial deposits, and the distribution of Paleozoic fauna and flora. Extensive investigation of the sea-floor in recent years has yielded evidence that large plates of the earth's crust have moved relative to one another (Fig. 7.1). Movements of the earth's crust have been attributed to polar wandering, continental drift, and sea-floor spreading. These mechanisms may help solve some of the persistent problems in earth history.

During *polar wandering,* the earth's crust and upper mantle move relative to the rotational poles of the earth. Since the spinning earth acts as a giant gyroscope, which is stabilized by the earth's equatorial bulge, large changes in the angle of inclination of the earth's axis are considered to be unlikely. Rather, it is thought that the outer shell of the earth shifts slowly over the inner layers,

Figure 7.1   Aerial view of the Thingvellir rift valley in Iceland showing a graben bounded by normal faults. Rifting of this type develops as sea-floor spreading causes the island to be pulled apart. (Courtesy of Sigurdir Thorarinsson.)

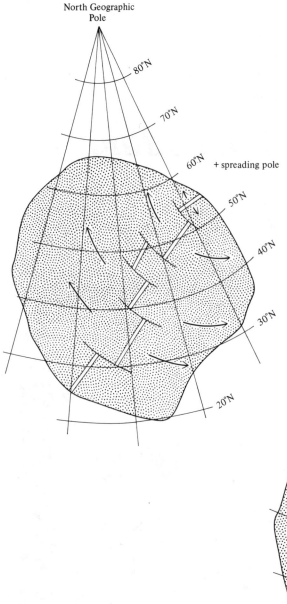

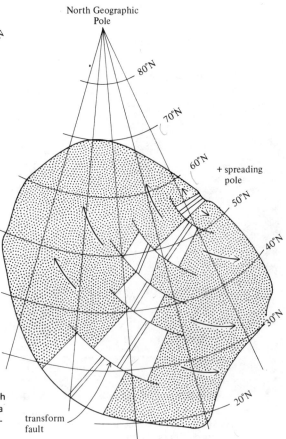

while the axis remains at a relatively constant angle to the earth's plane of revolution about the sun.

The movement of the continents relative to one another is termed *continental drift*. Unless the continents move in a precisely east - west direction, continental drift will result in some degree of polar wandering. *Sea-floor spreading*, which describes the movements of oceanic plates relative to one another, may operate independently or in association with continental drift and polar wandering. During continental drift and sea-floor spreading, the earth's crust and a significant portion of the upper mantle move together.

The term *plate tectonics* is used in describing the effects of polar wandering, continental drift, and sea-floor spreading.

Figure 7.2 The movement of two plates relative to each other on a sphere such as the Earth may be expressed as a rotation about a spreading pole. Transform faults are parallel to small circles concentric about that pole.

These movements are a result of plate motions involving the rotation of one plate relative to another (1). The relative direction of movement is parallel to faults at the junction of the plates. The axis of rotation of one plate relative to another is referred to as the *pole of spreading* (Fig. 7.2). It is now apparent that the processes associated with mountain building, such as folding, faulting, volcanism, and plutonism, occur principally at plate margins. Evidence of polar wandering, continental drift, and sea-floor spreading is found even in the most ancient rocks of the earth and, therefore, plate tectonics has played an integral role in earth history.

MATCHING OF CONTINENTAL OUTLINES

The reconstructions of the continents made by Wegener and Carey are based largely on visual matching of continental outlines (Fig. 1.9). Sir Edward Bullard, J. E. Everett, A. Gilbert Smith,

and A. Hallam matched the continents bordering the Atlantic and Indian oceans with the help of a computer (2). This technique minimizes the overlaps and gaps between the continents. They found that the best fit between the continents could be obtained if the 1000-m (3300-ft) depth contour is considered to be the edge of the continent (Fig. 7.3).

PALEOCLIMATOLOGY

Much of Wegener's evidence in support of continental drift came from his study of ancient climates. He noted that certain kinds of sedimentary rocks are found in areas where present climates are not conducive to their deposition. Coral reefs and coals derived from tropical plants in the Arctic and Antarctic are notable examples. Applying a uniformitarian approach, Wegener assumed that these rocks had been deposited under the same climatic conditions in the past as

Figure 7.3 Late Paleozoic reconstruction of the continents bordering the Atlantic and Indian oceans. The margins of the continents are shown at the 1000 fathom depth contour. (From Briden, Smith, and Sallomy, Ref. 3.)

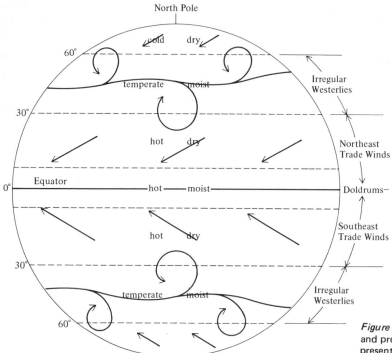

Figure 7.4 Idealized climatic belts and prevailing wind directions for the present time. (Adapted from Neuberger and Cahir, Ref. 4. Copyright © 1969, Holt, Rinehart & Winston, Inc.)

they are today and that the climatic zones of the earth have always been about as they are today (Fig. 7.4). Although climatic belts may shift and change in size, it is likely that the earth has always had a cool or cold climate at the poles and a warm, moist climate near the equator.

Wegener determined the approximate paleo-latitudes for each continent by placing a rotational pole near the center of glaciation and an equator along a belt of coral reefs and evaporites of that age. Comparison of the paleolatitudes with present latitudes led Wegener to conclude that polar wandering had occurred. Moreover, comparison of paleolatitudes of various continents indicated that continental drift had taken place.

Further Paleoclimatic Observations

Late Ordovician glacial deposits have recently been discovered in the Sahara Desert in northern Africa and in several localities in southern Africa (5).

Figure 7.5 Boulder tillite overlying a *roche moutonnée*. Glacial striae are parallel to the hammer handle. (From Hamilton and Krinsley, Ref. 5.)

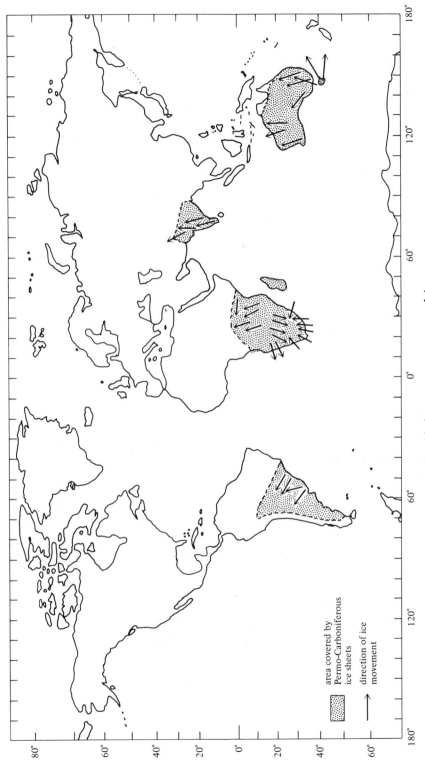

Figure 7.6 Distribution of Permo-Carboniferous glacial deposits with the present arrangement of the continents. The movements of the ice sheets away from the present ocean basins is a strong argument for continental drift.

area covered by Permo-Carboniferous ice sheets

direction of ice movement

These deposits are so extensive that they must have been laid down by continental rather than alpine glaciers. The orientation of *roches moutonnées,* glacial striations, chatter marks, and glacial valleys indicates that the ice moved *away* from the present equator (Fig. 10.3b).

Widespread glacial deposits of Permo-Carboniferous age occur in South America, Africa, India, Australia, and Antarctica (Fig. 7.5). Some of these deposits are on or near the present equator (Fig. 7.6). During this glacial episode, the ice moved away from what are now the Atlantic and Indian ocean basins. This relationship is puzzling, since the Pleistocene ice sheets almost always moved toward the ocean basins.

Many ancient evaporites and red beds are found in the middle and high latitudes where present climates are wet and/or cold. For example, Paleozoic salt deposits are found within 20° of the present North Pole in northern Canada and Siberia, and Paleozoic red beds occur at 75°N latitude in Asia. Today, however, extensive red beds and evaporites are forming in hot, dry climates, at latitudes between 15 and 45°.

Ancient latitudes may also be approximated from the study of the direction of prevailing winds. At the present time, the prevailing winds blow from the west in the mid-latitudes and from the east near the equator (Fig. 7.4). Changes in the direction of prevailing wind may indicate a change in latitude. Since cross-beds in aeolian sandstones dip to the leeward, the average dip of such beds indicates the direction of prevailing winds at the time of deposition. Such measurements show that, in the southwestern United States and in the British Isles, the prevailing winds were easterly during the late Paleozoic (Fig. 7.7). Both these areas are now in the zone of prevailing westerlies.

Although modern coral reefs are restricted to warm water environments within about 30° of the equator, Paleozoic reefs are found at much higher latitudes in Greenland, northern Canada, and Spitsbergen. Similarly, thick carbonates are generally deposited in warm waters, but extensive Paleozoic carbonates occur in areas bordering the Arctic Ocean. Carboniferous coal deposits contain fossil plants that lack growth rings. Such plants are typical of tropical or subtropical environments. However, these deposits are now located in the middle and high latitudes.

Interpretation of Paleoclimatic Data

It is evident that, in many cases, rock units were deposited under climate conditions quite different from those which exist presently in the regions where those rocks occur. The incidence of Paleozoic continental glaciers near the present equator, and evaporites, coral reefs, thick carbonates, coals, and red beds near the present poles, indicates that large-scale shifts in climatic belts have occurred since the Paleozoic. Some geologists have proposed worldwide climatic changes in order to account for this. However, such climatic changes do not provide a complete explanation for paleoclimatic data. For example, the Late Ordovician and Permo-Carboniferous glacial deposits are restricted to the southern hemisphere and India. Contemporaneously with the glaciation of the southern continents, evaporites, coral reefs, limestones, red beds, and coals were being deposited near the present north pole. These observations cannot be explained by worldwide climatic changes.

It has also been proposed that paleoclimatic data may be accounted for by polar wandering alone, rather than by a combination of polar wandering and continental drift. However, regardless of where the poles are placed, late Paleozoic glaciation would have occurred within 10° of the paleoequator. Furthermore, unless continental drift is postulated, there can be no continental source for the Permo-Carboniferous ice sheets. Geophysical studies indicate that the basins of the Atlantic and Indian oceans are underlain by oceanic crust.

Paleoclimatic data can best be explained by a combination of polar wandering and continental drift. If the southern continents had been together in one large landmass, the Permo-Carboniferous ice sheets would have radiated away from a center in

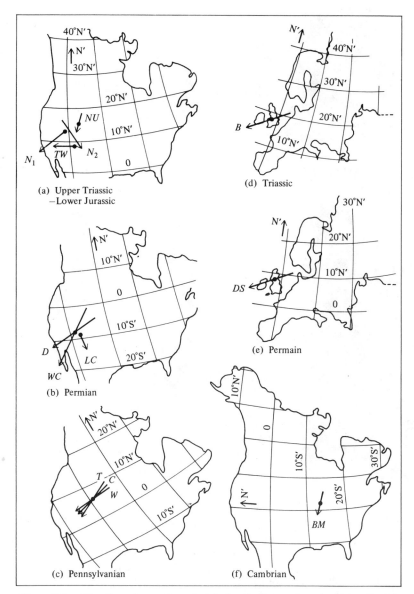

Figure 7.7 Direction of prevailing wind during the Paleozoic and as determined from the average direction of dip of cross-beds in dune sandstones. (After Irving, Ref. 6.)

southern Africa and eastern Antarctica (Fig. 11.26). The center of glacial accumulation was very close to the Permo-Carboniferous South Pole as determined from paleomagnetic data (discussed in a later section), and tropical coals, red beds, evaporites, and reefs of Permian and Carboniferous age would lie at low paleolatitudes. Furthermore, both Britain and the southwestern part of the United States would then fall within the belt of northeast trade winds.

(a)

(b)

Figure 7.8   Late Paleozoic fossil plants from Antarctica:
(a) *Glossopteris communis*, (b) *Gangamopteris obovata*
(scale at right is in 0.1 in. graduations). (Print negatives
3099-1x and 2877A-2x, U.S. Geological Survey, Colum-
bus, Ohio, courtesy of James M. Schopf.)

PALEONTOLOGICAL EVIDENCE

Wegener believed that continental drift provided the best explanation for the striking similarities in fossil faunas and floras of continents now separated by deep oceans. Some examples that impressed Wegener and later investigators are:

1. Sixty-four percent of Carboniferous and 34 percent of Triassic reptile faunas are the same for the southern continents.

2. The freshwater reptile *Mesosaurus* has been found only in South America and Africa. The bone structure of this salamanderlike organism is such that it would not have been able to swim a large ocean.

3. Numerous bones of terrestrial reptiles have been discovered in Triassic rocks in Antarctica (7), which is now completely isolated from other continents.

4. In many cases early Paleozoic marine invertebrate assemblages are quite similar in geosynclinal deposits on both sides of the Atlantic Ocean.

5. Fossil leaves of cycads and seed ferns belonging to the *Glossopteris* flora of Permian age have been found on all the southern continents and India, but not elsewhere (Fig. 7.8).

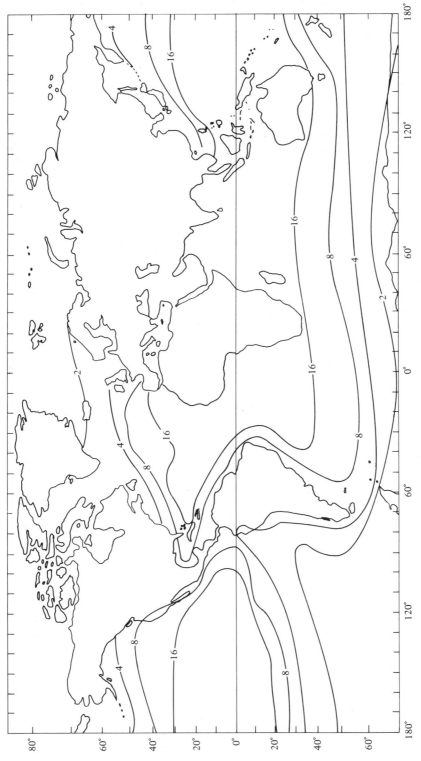

Figure 7.9 Representation of the diversity of recent planktonic Foraminifera by the contouring of the number of species. (From Stehli and Helsley, Ref. 8. Copyright © 1963 by the American Association for the Advancement of Science.)

Many paleontologists believed that such faunal and floral similarities may have resulted from normal dispersal of organisms by wind and ocean currents. Spores, seeds, and certain larvae may be carried by currents, but it is unlikely that entire assemblages of plants or animals could be transported across a wide ocean. Furthermore, although small animals may be rafted long distances on logs or mats of vegetation, the oceans are a barrier to the migration of most terrestrial reptiles and mammals.

Opponents of continental drift also cite paleontological evidence in their arguments:

1. Comparison of living species of mammals in eastern Europe and Asia indicates that a much greater similarity existed there than between the Paleozoic terrestrial faunas of South America and Africa.

2. Certain groups of modern marine organisms show greater diversity toward the equator and less diversity near the poles (Fig. 7.9). A similar distribution has been demonstrated for a group of Permian brachiopods (Fig. 7.10a). This has been interpreted as an indication that little polar wandering or continental drift has occurred since Permian time.

3. The Carboniferous flora of central Europe is thought to have developed in tropical or subtropical conditions, whereas that of Siberia developed in a cool or cold environment (9). Since

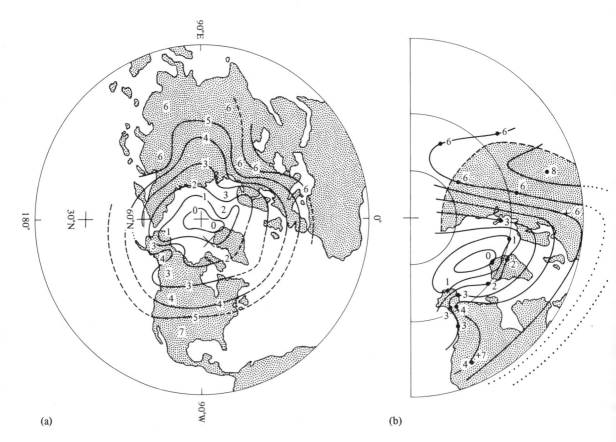

(a)   (b)

Figure 7.10 Diversity of two families [(a) and (b), respectively] of Permian brachiopods on a paleomagnetically determined model of the earth. (From Stehli and Helsley, Ref. 8. Copyright © 1963 by the American Association for the Advancement of Science.)

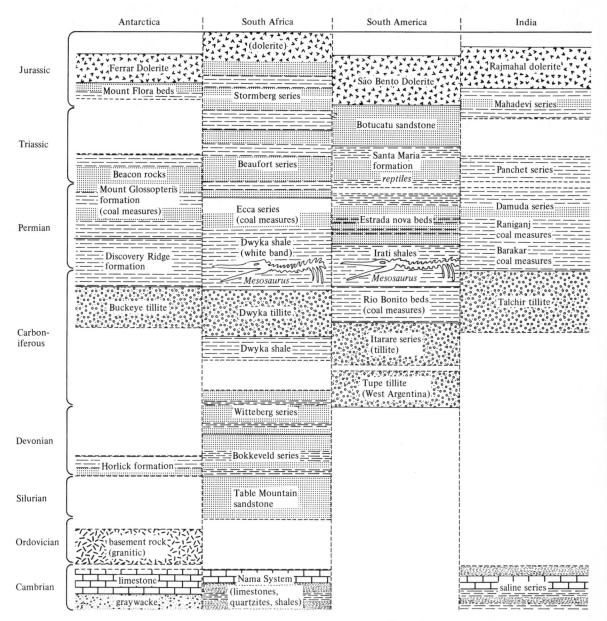

Figure 7.11 Correlation of stratigraphic columns from the southern continents. (From Doumani and Long, Ref. 10. Copyright © 1962 by Scientific American, Inc., all rights reserved.)

these climates somewhat resemble those of today, there is no need to postulate a shift in the earth's poles of rotation.

Proponents of continental drift and polar wandering propose alternative explanations for these three arguments:

1. Paleozoic terrestrial faunas in South America and Africa appear to be dissimilar because fossils of Paleozoic terrestrial animals have been found in only a few localities.

2. Diversity gradients of Permian brachiopods also decrease toward the Permian pole as deter-

mined by paleomagnetism. Only two localities have significantly greater faunal diversities than would be expected (Fig. 7.10b). Thus the faunal diversity of Permian brachiopods does not constitute proof against either polar wandering or continental drift.

3. The distribution of Carboniferous floras is in accord with the position of Carboniferous latitudes determined from paleomagnetic studies. Siberia was located between 30 and 60°N latitude, but central Europe was located near the Carboniferous equator.

STRATIGRAPHIC EVIDENCE

Wegener was also impressed by the strong similarity of the stratigraphy of South America, Africa, India, and Australia. Recent studies have revealed that the stratigraphy of Antarctica is remarkably similar to that of the other southern continents. The correlation of Permo-Carboniferous tillites, Permian shales, Lower Triassic red beds, and Jurassic lavas is quite impressive (Fig. 7.11). Futhermore, in the northern hemisphere, the Paleozoic stratigraphic sequences of the Appalachian Geosyncline are similar to those of the Caledonian Geosyncline of Europe.

STRUCTURAL EVIDENCE

Modern geosynclines are located on continental margins and tend to encircle the continents. Ancient geosynclines were probably distributed in a similar pattern. However, several geosynclinal belts appear to end abruptly. The Appalachian Geosyncline ends just northeast of Newfoundland, and the Caledonian Geosyncline is terminated just west of Ireland. The mountains formed from these geosynclines also end at the continental margins and cannot be traced into the Atlantic Ocean basin. This relationship led Wegener and others to suggest that the Appalachian Mountain belt was once continuous with the Caledonian Mountain belt of Europe (Fig. 7.12).

In the southern hemisphere, the Tasman Geosyncline ends south of Tasmania and the Transantarctic Geosyncline ends near Victoria Land. Geosynclines on the northwestern and southern borders of Africa and on the western border of South America are also abruptly terminated. In the reconstruction of the continents suggested in this text, ancient geosynclines formed continuous belts that encircled the shield areas (Fig. 7.13).

PALEOMAGNETISM

All rocks have magnetic fields owing to the presence of magnetic minerals, such as magnetite. By measuring the direction and inclination of this magnetic field, it is possible to calculate the approximate position of the magnetic poles of the earth at the time of crystallization or deposition of the rock. Thus rock magnetism furnishes a fossil "compass."

The magnetic memory of rocks, that is, their *natural remanent magnetism,* may be due to one or more factors:

1. *Thermoremanent magnetism* develops when an igneous rock such as basalt cools past the Curie temperature of its magnetic minerals. The Curie temperature is that temperature above which a substance is no longer magnetic. Magnetite has a Curie temperature of 578°C.

2. *Chemical remanent magnetism* is produced during the formation of iron-bearing minerals at low temperatures. As mentioned in Chapter 5, hematite in arkosic red beds probably forms by weathering in place of such minerals as biotite and hornblende. Such red beds would have a magnetic field oriented parallel to the earth's magnetic field at the time of formation of the hematite.

3. *Depositional remanent magnetism* results from the alignment during sedimentation of magnetic particles. This alignment is the result of a physical rotation of the magnetic particles as they sink to the bottom.

4. *Viscous remanent magnetism* develops after the initial cooling or lithification of the rock. Such magnetism may be produced during the weathering of iron-bearing minerals or as a result of

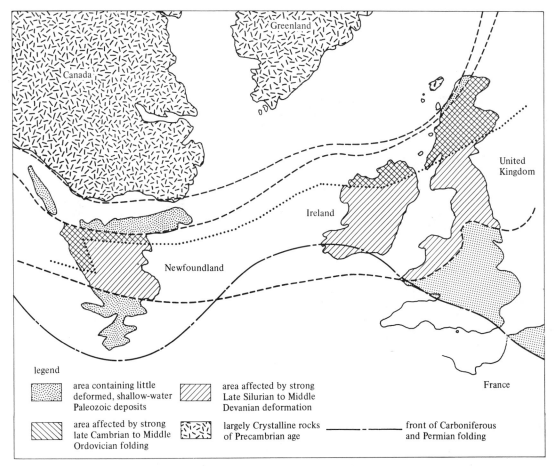

legend

	area containing little deformed, shallow-water Paleozoic deposits		area affected by strong Late Silurian to Middle Devanian deformation
	area affected by strong late Cambrian to Middle Ordovician folding		largely Crystalline rocks of Precambrian age

————— — ————— front of Carboniferous and Permian folding

Figure 7.12 Comparison of the Appalachian and Caledonian fold belts. (Modified after Dewey and Kay, Ref. 11. Copyright © 1968 Princeton University Press, used with permission.)

lightning strikes. Also, the magnetic field of a rock may be reoriented parallel to the present field. Fortunately, viscous remanent magnetism is a relatively weak component of magnetization and it may be removed by partial demagnetization. In this process, samples are placed in an alternating magnetic field or heated to a temperature of several hundred degrees centigrade.

Procedure

In order to determine the location of the paleomagnetic poles, oriented samples are collected from several horizons within a formation and at several different localities. This procedure serves to average out minor variations in the earth's magnetic field. The direction and inclination of the axis of the magnetic field of each sample is measured with a very sensitive magnetometer. The inclination is used to determine the magnetic latitude at which the rock was formed. For example, a high magnetic inclination in a basalt would indicate that the rock crystallized near either the magnetic North Pole or magnetic South Pole; low inclination would indicate crystallization near the magnetic equator (Fig. 7.14). The magnetic latitude, λ, may be determined by measuring the inclination of the magnetic axis, I, and substituting in the equation:

$$\tan I = 2 \tan \lambda$$

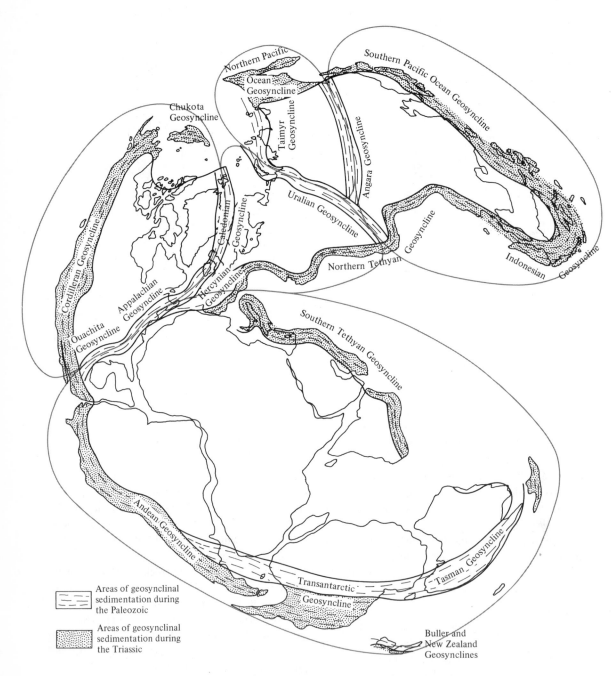

Figure 7.13 Location of Paleozoic and Triassic geosynclines plotted on a predrift reconstruction of the continents.

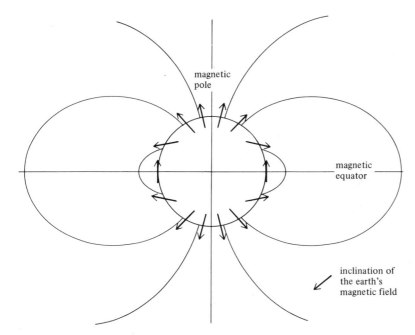

magnetic
pole

magnetic
equator

inclination of
the earth's
magnetic field

Figure 7.14 Magnetic field of a uniformly magnetized sphere. The magnetic field of the earth is very similar, but is considerably distorted because of solar wind.

Assumptions

The interpretation of paleomagnetic data is based on the following assumptions:

1. The magnetic axis of the rock was oriented parallel to the earth's magnetic field when the rock formed. Field and laboratory investigations indicate that this assumption is probably valid. For example, the pole position calculated from the orientation of the magnetic field of the 1669 lava flow on Mt. Etna agrees to within one degree of the location of the magnetic poles as determined from magnetic observations in 1669.

2. The average position of the magnetic pole is assumed to approximate that of the rotational pole. Although the present geomagnetic pole is inclined 11.5° to the rotational pole, paleomagnetic studies of rocks less than 25 million years old indicate that the average position of the magnetic pole was close to the rotational pole (Fig. 7.15). Moreover, there is a close correlation between paleolatitudes determined from paleomagnetic and pale-

oclimatic data. Paleomagnetic studies of glacial deposits indicate deposition at high or middle latitudes, whereas paleomagnetic studies of thick limestones, red beds, and evaporites generally indicate deposition at low paleolatitudes. This correlation is also shown by the agreement of the paleomagnetic pole with the center of the Late Ordovician and Permo-Carboniferous ice sheets.

3. The earth's magnetic field is assumed to have always been dipolar, that is, with one north and one south magnetic pole. This assumption is consistent with the generally accepted *dynamo theory* of the origin of the earth's magnetic field. Briefly, the dynamo theory states that a conductor (in this case the liquid, nickel-iron core of the earth) moving in a preexisting magnetic field produces an electric current, and an electric current flowing in a conductor generates a magnetic field. In this way, the earth's magnetic field is continuously being generated. One problem, however, is the source of the original magnetic field, which is necessary to start the whole process.

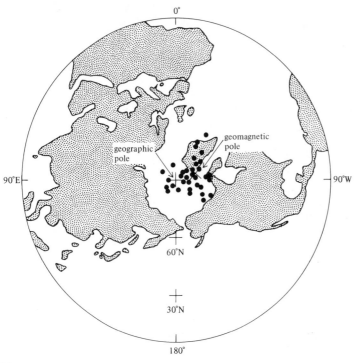

(a)

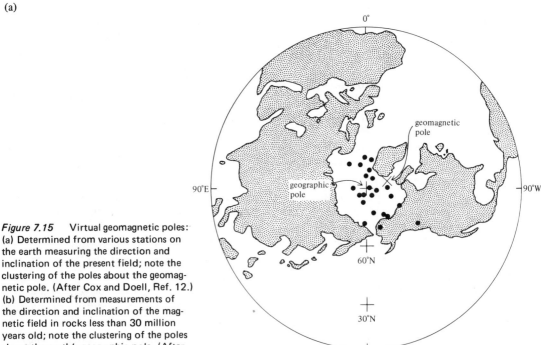

Figure 7.15   Virtual geomagnetic poles:
(a) Determined from various stations on
the earth measuring the direction and
inclination of the present field; note the
clustering of the poles about the geomagnetic pole. (After Cox and Doell, Ref. 12.)
(b) Determined from measurements of
the direction and inclination of the magnetic field in rocks less than 30 million
years old; note the clustering of the poles
about the earth's geographic pole. (After
Irving, Ref. 6.)   (b)

Table 7.1   Locations of paleomagnetic North Poles for various continents

Age	North America	South America	Europe	Africa	Australia	Antarctica	Siberia	India		Eurasia	China
Tertiary	86N 158W		75N 179E	83N 115E	70N 56W	86N 178E	81N 131W	35N	82W	76N 178W	
Cretaceous	69N 168W		85N 176E	71N 109W	53N 31W	81N 157W	70N 168E	15N	59W	77N 175E	70N 155W
Jurassic	73N 156E	87N 110W	77N 139E	67N 109W	45N 37W	55N 36E	70N 148E			74N 144E	
Triassic	65N 96E	84N 102E	52N 141E	69N 104W	53N 28W		62N 153E	11N	54W	58N 147E	
Permian	49N 120E	75N 152E	49N 169E	43N 97W	41N 41W		63N 177W	15N	60W	52N 167E	
Carboniferous	37N 125E	58N 157E	37N 166E	40N 145W	62N 29W		45N 170E	32S	46W	41N 168E	
Devonian	30N 121E[a]	5N 141E	32N 162E		47N 136W[a]		28N 162E			31N 162E	
Silurian			21N 159E				19N 156E			19N 157E	
Ordovician		8S 144E	13N 165E	50S 169E	15N 155W	28N 170W	25S 135E				
Cambrian	0S 147E	27S 130E	20 168E	51S 173E	9N 160E		37S 133E	28N	148W		11S 137E
570-800 M.Y.[b]	2S 160E		11N 147E	28S 165E			5S 154W	82N	106E		45S 90W
800-940 M.Y.			30S 135W	15N 111E		62N 157E	40N 62W				
940-1199 M.Y.	34N 167W			38S 162W							
1200-1399 M.Y.	10N 173W		10N 65W	8S 137W							
1400-1599 M.Y.	5N 140W		17N 170E	66S 133W							
1600-1799 M.Y.	53N 115W	43S 32E		18S 171E							
1800-1999 M.Y.	36N 123W		48N 127W	20S 152W	29S 134W						
2000-2199 M.Y.	9S 101W			55S 5W	21S 124W						
2200-2399 M.Y.	22N 97W			34S 77W							
2400-2500 M.Y.	50N 68E			16S 114W		8S 157E					
2600-2899 M.Y.	62N 112E			33N 149W							

[a]Silurian and Devonian poles averaged together.
[b]M.Y. = million years.

Results

Thousands of paleomagnetic pole positions have been determined for formations of various ages and from various continents. However, only several hundred of these positions have been determined from samples that were tested for magnetic stability by partial demagnetization. Only the latter results can be considered reliable, since partial demagnetization removes the unstable viscous remanent magnetism. The reliable pole positions have been averaged for each continent and for each geological period (Table 7.1).

It it is assumed that the average position of the magnetic pole is coincident with the rotational pole. It is apparent that the positions of the rotational poles have changed relative to the continents. The early Paleozoic poles were generally located near the equator, but late Paleozoic poles were generally in the mid-latitudes. Mesozoic and Cenozoic poles were located near the present geographic poles (Fig. 7.16). A line drawn through the paleomagnetic poles of any one continent for successive periods is the *polar wandering curve* of that continent.

When polar wandering curves are compared, it is evident that they do not coincide (Fig. 7.17). The North American polar wandering curve lies to the west of the European curve, suggesting that North America has moved westward relative to Europe. The polar wandering curves for Africa, India, and Australia lie to the east of the curve for Europe. Apparently these continents have moved eastward relative to Europe.

In order to further test the possibility that the continents have drifted, the outlines of the continents may be traced from a globe onto a plastic overlay. The continents may then be moved to their relative predrift positions and the positions of the paleomagnetic poles may be recalculated. Using the reconstruction of the continents of Fig. 7.13, the locations of the recalculated pole positions for the Carboniferous, Permian, and Triassic periods are shown to coincide rather closely (Fig. 7.18). Although alternative interpretations of

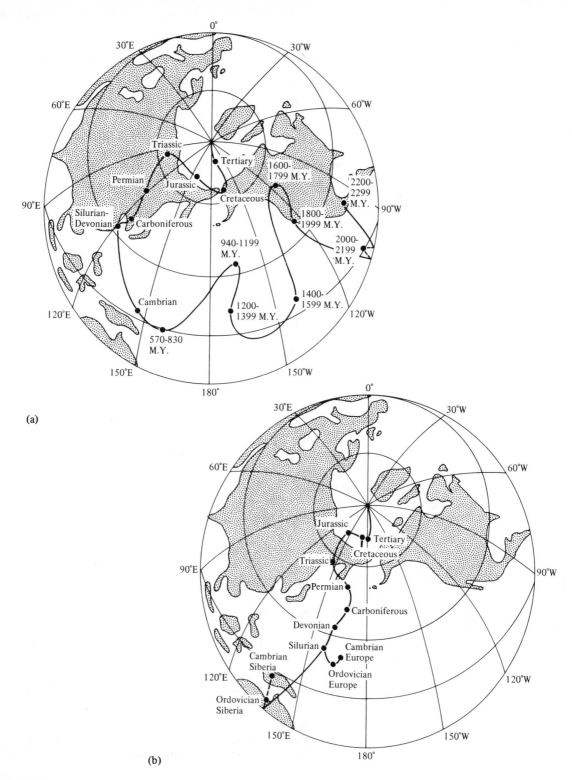

(a)

Triassic
Tertiary
1600-1799 M.Y.
Permian
Jurassic
2200-2299 M.Y.
Cretaceous
Silurian-Devonian
Carboniferous
1800-1999 M.Y.
940-1199 M.Y.
2000-2199 M.Y.
Cambrian
1200-1399 M.Y.
1400-1599 M.Y.
570-830 M.Y.

(b)

Jurassic
Tertiary
Cretaceous
Triassic
Permian
Carboniferous
Devonian
Silurian
Cambrian Europe
Cambrian Siberia
Ordovician Europe
Ordovician Siberia

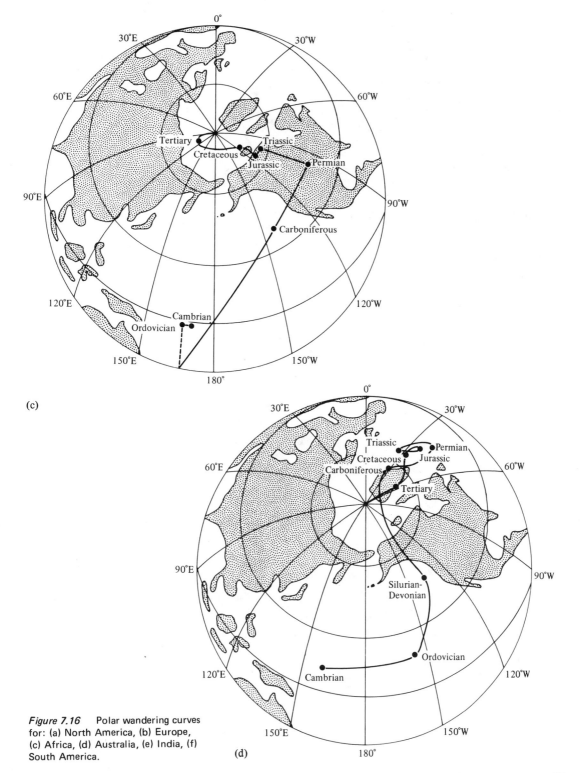

(c)

Figure 7.16 Polar wandering curves for: (a) North America, (b) Europe, (c) Africa, (d) Australia, (e) India, (f) South America.

(d)

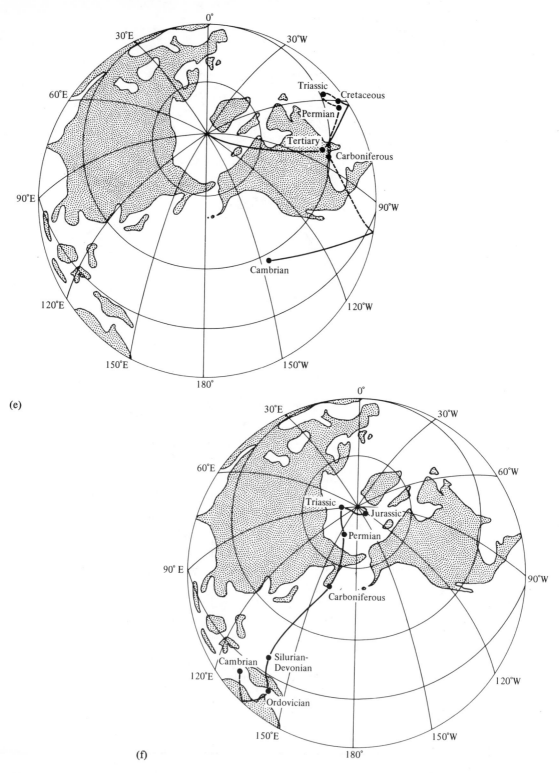

(e)

(f)

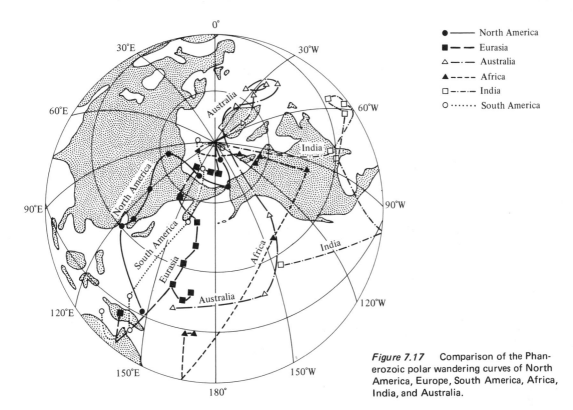

Figure 7.17 Comparison of the Phanerozoic polar wandering curves of North America, Europe, South America, Africa, India, and Australia.

paleomagnetic data have been suggested (13), the agreement of recalculated paleomagnetic poles provides a strong indication that all the continents formed a single landmass during the Carboniferous, Permian, and Triassic.

RADIOMETRIC EVIDENCE

Areas in which the radiometric ages of the basement rocks are similar are termed *geologic provinces.* In western Africa, a 2.0 billion year old province and a 600 million year old province are separated by a boundary that trends southwesterly into the Atlantic Ocean near Accra, Ghana. Detailed radiometric studies in northern Brazil have defined two adjoining provinces of similar ages. When Africa and South America are moved back to their supposed predrift positions, the boundary between the provinces in Africa is perfectly aligned with the boundary between the provinces of the same age in South America (Fig. 7.19).

SEISMOLOGY

Studies of the earth's seismicity have provided new insights into the movements of continents and the sea-floor. A plot of recent earthquake epicenters indicates that earthquakes are most common in linear belts along mid-oceanic ridges, beneath island arcs, and under mountains (Fig. 7.20). Most of these epicenters are located at the margins of large plates. From his studies of the patterns of seismic waves from earthquakes in many parts of the world, Lynn Sykes found that normal faulting is the dominant type of faulting over mid-oceanic ridges (16). Normal faults are thought to be caused by extension of the earth's crust. Presumably, the normal faults in the vicinity of mid-oceanic ridges are the result of separation of plates or blocks over rising convection currents.

Earthquakes also occur along faults that offset mid-oceanic ridges. Sykes determined that the sense of motion along these faults is strike-slip, but in the opposite direction to the apparent

Figure 7.18 Carboniferous, Permian, and Triassic paleomagnetic pole positions recalculated for a Pangaea reconstruction of the continents. NA = North America, EA = Eurasia, Af = Africa, SA = South America, Au = Australia, In = India.

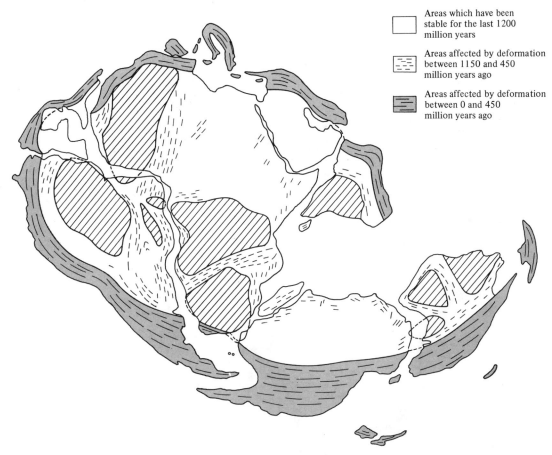

	Areas which have been stable for the last 1200 million years
	Areas affected by deformation between 1150 and 450 million years ago
	Areas affected by deformation between 0 and 450 million years ago

Figure 7.19　Tentative matching of geological provinces with South America and Africa in their predrift positions. (After Hurley, Ref. 14. Copyright © 1968 by Scientific American, Inc., all rights reserved.)

offset of the ridge (Fig. 7.21). If a mid-oceanic ridge appears to be offset in a left-lateral sense, the movement along the fault would be right-lateral. The movement along faults that offset mid-oceanic ridges is exactly what one would expect if the ridges were underlain by upwelling convection currents (Fig. 7.21). The existence of such faults, termed *transform faults,* was predicted by J. Tuzo Wilson before they were observed (17).

The movement of plates in the vicinity of oceanic trenches is rather complex. Here earthquakes occur along more or less planar zones which dip under a continent or island arc and extend to depths as great as 420 miles (700 km)

(18). These are named *Benioff zones* after the seismologist Hugo Benioff (Fig. 7.22).

Studies of the attenuation of earthquake waves indicate that the earthquakes occur within relatively cool, brittle slabs of lithosphere that have been thrust under continents and island arcs. The earthquakes are the result of both compressional and extensional stresses on the slabs (19). Linear belts along which the sea-floor is being consumed are called *subduction zones.*

SEA-FLOOR SPREADING

Harry Hess was the first to propose that blocks of sea-floor are moving relative to one another in

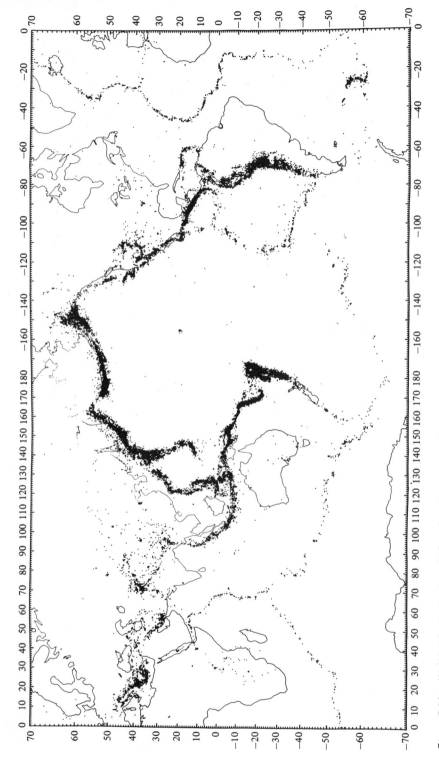

Figure 7.20 Worldwide distribution of all earthquake epicenters for the period 1961-1967, as reported by the U.S. Coast and Geodetic Survey. (From Barazangi and Dorman, Ref. 15.)

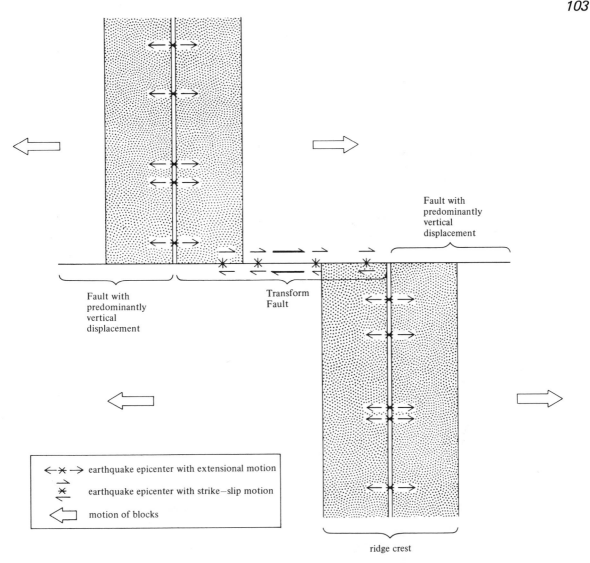

Figure 7.21 Idealized map view of a transform fault. Notice that the ridge *appears* to be offset in a left-lateral sense, but the motion on the transform fault between the ridge crests is right-lateral.

response to the motion of convection currents within the mantle (20). The evidence in support of the concept of sea-floor spreading has come from many sources, including the magnetism of the ocean floor, the age of oceanic volcanic rocks, and the age of deep-sea sediments.

Marine Magnetics

Since 1950 there have been extensive surveys of the magnetism of the ocean floor. Studies of the earth's magnetic field in the Pacific Ocean west of California have revealed the presence of a series of north - south trending magnetic anomalies (21) (Fig. 7.23). A *magnetic anomaly* is a deviation from the average intensity of the earth's magnetic field. In an area with a positive anomaly, the earth's magnetic field has a greater than average intensity, whereas in an area with a negative

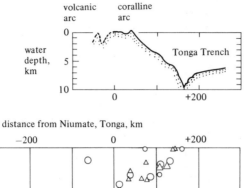

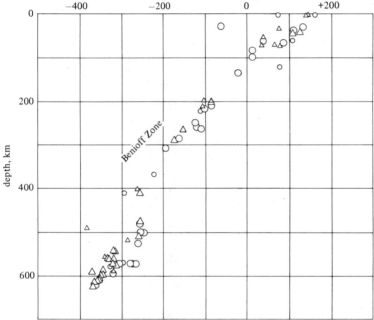

Figure 7.22 Vertical cross section perpendicular to the Tonga Island Arc northeast of New Zealand showing the spatial distribution of earthquakes of small magnitude. Circles and triangles represent the foci of earthquakes monitered in 1965 projected onto the plane of section. Vertical exaggeration of the upper inset showing the topography is approximately 13 : 1. Note that the seismic zone is less than 20 km wide for most of the section. (After Isacks et al., Ref. 16.)

magnetic anomaly, the intensity is below average.

A survey of the Reykjanes Ridge south of Iceland revealed a symmetrical pattern of magnetic anomalies (Fig. 7.24). Since these pioneering studies were made, numbers have been assigned to distinctive anomalies. The broad positive anomaly over the crest of the Mid-Atlantic Ridge is Anomaly Number 1, and two broad positive anomalies equidistant on either side of the ridge are each numbered 5 (Fig. 7.25).

The origin of the linear magnetic anomalies in the oceans remained a mystery for several years after their discovery. Then in 1963, Fred Vine and D. H. Matthews proposed that the anomalies had resulted from periodic reversals in the earth's magnetic field (25).

Studies of the direction of magnetization within sequences of rocks have shown that in some strata, the direction of magnetization is exactly the opposite of that in adjacent strata (Fig. 7.26). Almost all samples of volcanic rocks that have been dated radiometrically as less than 690,000 years have the north-seeking pole of their magnetic axis pointing approximately in the

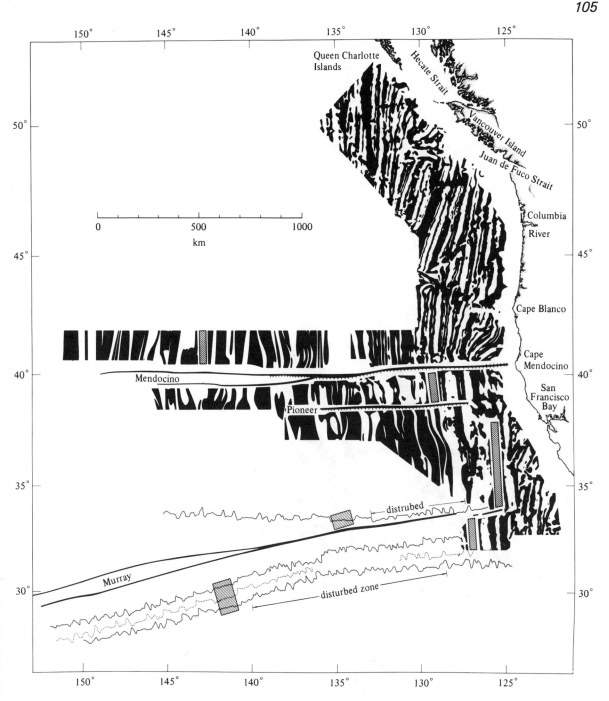

Figure 7.23 Magnetic anomalies and major faults in the northeast Pacific Ocean Basin. Positive magnetic anomalies are in black, negative anomalies within ocean basin are white. (From Menard, Ref. 22. Copyright © 1964 by McGraw-Hill, Inc. Used with permission of McGraw-Hill Book Co.)

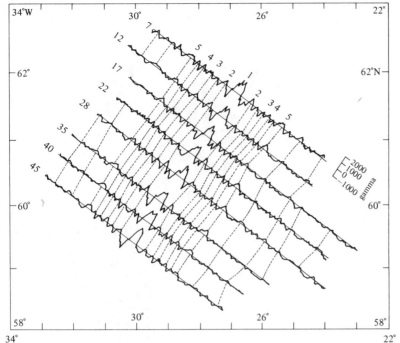

Figure 7.24 Magnetic anomalies on the Reykjanes Ridge south of Iceland. Notice the symmetry of the profiles about Anomaly No. 1 at the crest of the ridge. Eight magnetic profiles were taken across the ridge. (After Heirtzler et al., Ref. 23. Used with permission of Maxwell International Microforms Corp.)

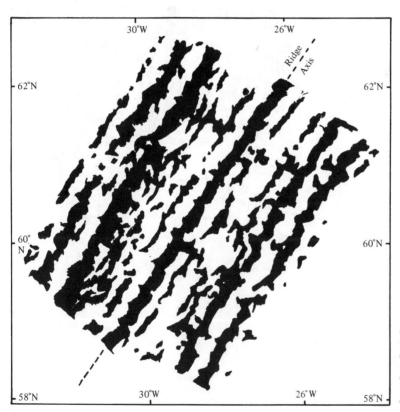

Figure 7.25 Map showing the inferred pattern of magnetic anomalies on the Reykjanes Ridge. (After Vine, Ref. 24. Copyright © 1968 by Princeton University Press. Reproduced with permission.)

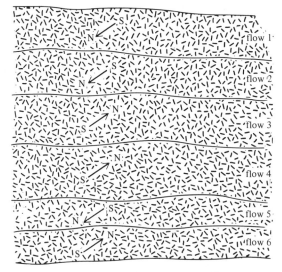

Figure 7.26 Directions of magnetization of a sequence of volcanic flows. Flows 1, 2, and 5 are reversely magnetized and flows 3, 4, and 6 are normally magnetized.

direction of the magnetic North Pole. Such rocks, which have a *normal polarity*, were formed when the earth's magnetic field was approximately the same as it is today. Rock samples between 690,000 and 890,000 years old have an opposite direction of magnetization (i.e., a *reversed polarity*) and were formed during a time when the earth's magnetic field was the reverse of what it is today (Fig. 7.27). During such an interval, the needle of a compass would have pointed toward the south.

Vine and Matthews proposed that the magnetic anomalies of the ocean basin were the result of alternating strips of normally and reversely magnetized lavas that had been extruded at the crests of mid-oceanic ridges. The normally magnetized lavas would produce a positive anomaly, since their magnetism is added to the magnetism produced in the core of the earth. The reversely magnetized lavas would produce a negative anomaly, since their magnetism would cancel some of the earth's magnetism.

The Vine - Matthews hypothesis provides an explanation for the striking magnetic symmetry that appears on profiles taken across the mid-oceanic ridges. As new sea-floor is added at the

ridge crests, older magnetized lavas are carried in both directions away from the crests. A symmetrical pattern of magnetic anomalies would be produced only if volcanism occurred principally at the ridge crest and the sea-floor was spreading during alternating periods of reversed and normal polarity.

In 1965 Vine and J. Tuzo Wilson showed that the pattern of magnetic anomalies over the Juan de Fuca Ridge off the coast of Washington State conformed exactly to the pattern predicted by the Vine - Matthews hypothesis. According to Vine and Wilson, the large positive anomaly over the crest of the ridge (Anomaly Number 1) was produced by lavas erupted during the Brunhes normal-polarity epoch, whereas the two adjacent negative anomalies were caused by reversely magnetized lavas erupted during the Matuyma reversed-polarity epoch (Fig. 7.28).

Comparison of the widths of the magnetic anomalies with the geomagnetic time scale indicates that the width of positive anomalies is proportional to the length of time that the earth's magnetic field was normal. Similarly, the width of the negative anomalies is proportional to the duration of reversed polarity. The pattern of anomalies is essentially the same as the pattern of the geomagnetic time scale (Fig. 7.29).

Using a computer, Vine constructed a model that would predict the pattern of magnetic anomalies for a series of thin blocks of oceanic crust that had been alternately magnetized in normal and reversed directions. He made the widths of the blocks proportional to the length of time between reversals of the earth's magnetic field. Vine found an excellent correlation between the observed and computed magnetic profiles (Fig. 7.30).

Magnetic measurements of several samples of basaltic pillow lava dredged from near the crest of the Mid-Atlantic Ridge have provided additional support for the Vine - Matthews hypothesis (28). Two samples collected from an area within a positive magnetic anomaly were normally magnetized, and one sample from an area within a negative anomaly had a reverse magnetization.

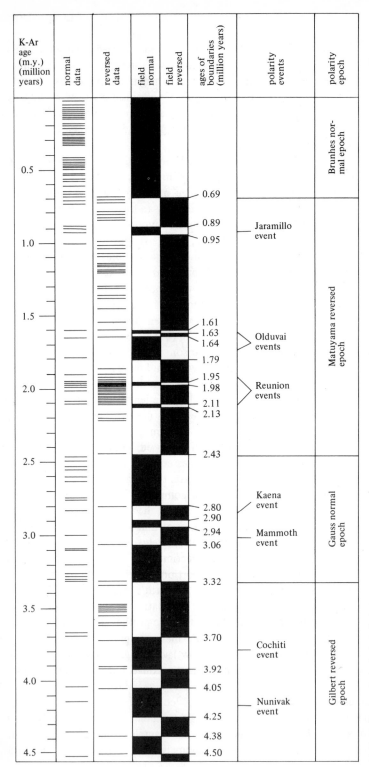

K-Ar age (m.y.) (million years)	normal data	reversed data	field normal	field reversed	ages of boundaries (million years)	polarity events	polarity epoch
							Brunhes normal epoch
					0.69		
					0.89	Jaramillo event	
					0.95		
					1.61		Matuyama reversed epoch
					1.63	Olduvai events	
					1.64		
					1.79		
					1.95	Reunion events	
					1.98		
					2.11		
					2.13		
					2.43		
					2.80	Kaena event	Gauss normal epoch
					2.90		
					2.94	Mammoth event	
					3.06		
					3.32		
					3.70	Cochiti event	Gilbert reversed epoch
					3.92		
					4.05	Nunivak event	
					4.25		
					4.38		
					4.50		

Figure 7.27 Time scale for geomagnetic reversals. Each short horizontal line shows the age of a lava as determined by potassium-argon dating. Those flows with normal polarity are put in the first column, those which are reversely magnetized are recorded in the second column. The duration of polarity events within each epoch is based in part on the study of the magnetization of sediment cores from the deep ocean. (From Cox, Ref. 26. Copyright © 1968 by American Association for the Advancement of Science.)

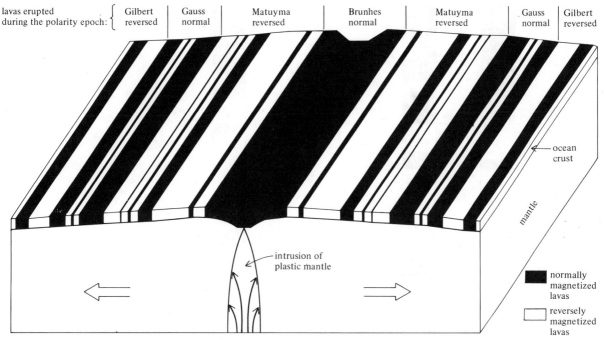

| lavas erupted during the polarity epoch: | Gilbert reversed | Gauss normal | Matuyma reversed | Brunhes normal | Matuyma reversed | Gauss normal | Gilbert reversed |

Figure 7.28   Block diagrams of a mid-oceanic ridge showing the inferred distribution of normally and reversely magnetized lavas according to the Vine-Matthews hypothesis.

Comparison of magnetic profiles from the Atlantic, Pacific, and Indian oceans shows that the anomalies may be correlated from one ocean to another (Fig. 7.31). Evidently, very similar processes produced the anomalies in each of these oceans. Changes in the earth's polarity would, of course, be worldwide.

As discussed earlier, it is possible to date magnetic anomalies by comparing the pattern of anomalies with the geomagnetic time scale. If the age of an anomaly is known, the average rate of sea-floor spreading may be determined by measuring the distance of that anomaly from the crest of the adjacent mid-oceanic ridge, since distance equals rate times time.

Correlation of magnetic anomalies within a single ocean basin reveals that the distance between anomalies varies with latitude (Fig. 7.32). Presumably this is an indication of a change in the rate of sea-floor spreading along a mid-oceanic ridge. We expect the rate of sea-floor spreading to decrease toward the poles of spreading (Fig. 7.2).

Oceanic Volcanoes

One of the earliest indications of sea-floor spreading was the general increase in age of volcanic islands and seamounts away from the Mid-Atlantic Ridge (Fig. 7.33). Several islands with active volcanoes, such as Tristan da Cunha and Iceland, are located on or very close to the crest of the ridge. The older volcanic islands, such as Bermuda, are well off the crest. This relationship is consistent with the theory that most volcanic islands originated on the crests of mid-oceanic ridges and have been carried away from the crests by sea-floor spreading.

Age of Deep Sea Sediments

For a number of years, oceanographers have been sampling the sediments of the ocean floor by

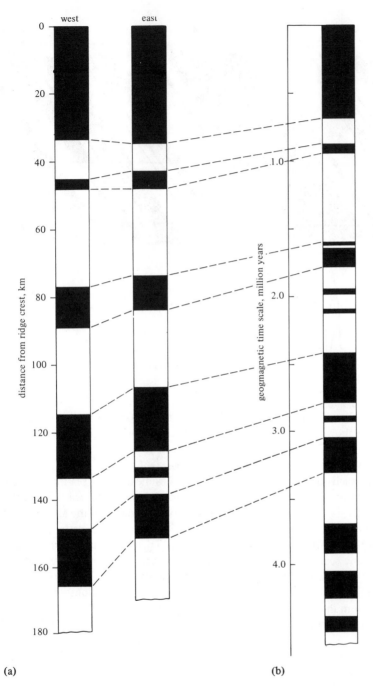

Figure 7.29 Correlation of magnetic anomalies with the geomagnetic time scale: (a) The width of the anomalies was measured from Fig. 7.30 to Fig. 7.36 and plotted as a function of distance from the crest of the East Pacific Rise; (b) comparison with the geomagnetic time scale of Cox (40) indicates that the width of the magnetic anomalies is proportional to the length of time of periods of normal or reversed polarity of the earth. Positive anomalies correlate with episodes of normal polarity and negative anomalies correlate with episodes of reversed polarity.

(a)

(b)

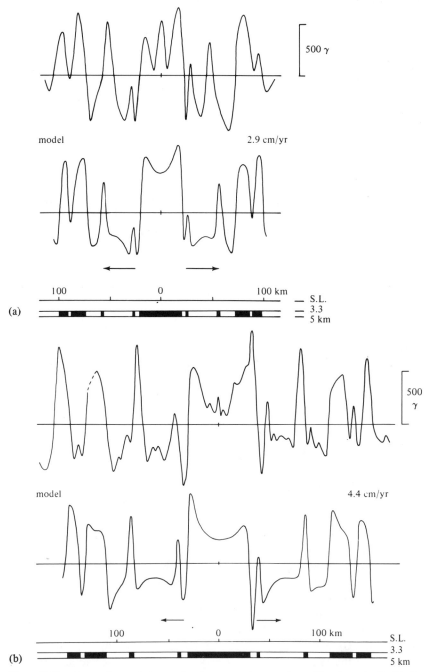

model

2.9 cm/yr

100   0   100 km   — S.L.
                                        = 3.3
(a)   = 5 km

model

4.4 cm/yr

100   0   100 km   S.L.
                                        3.3
(b)   5 km

Figure 7.30 Comparison of observed and computed magnetic profiles from the Pacific Ocean: (a) Juan de Fuca Ridge, 46° N, (b) East Pacific Rise, 51° S. The computed magnetic profiles are determined using the magnetization of slabs of normally and reversely magnetized lavas whose width is proportional to the lengths of time of normal and reversed polarity of the earth. (From Vine, Ref. 27. Copyright © 1966 by American Association for the Advancement of Science.)

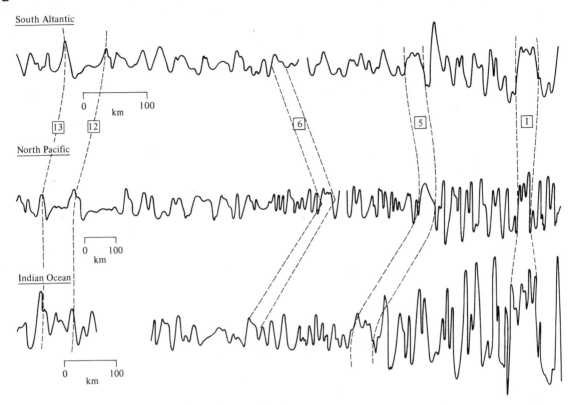

Figure 7.31 Comparison of magnetic profiles from various oceans. The correlation of the profiles indicates that the anomalies were probably formed at the same time by similar processes. (After Heirtzler et al., Ref. 29.)

means of piston coring devices (Fig. 7.34a). At first they were puzzled by the fact that none of the cores was older than Cretaceous. Even if the floor of the Atlantic Ocean was relatively young, as the "drifters" proposed, it was not clear why the Pacific Ocean did not contain very ancient sediments. Perhaps the ancient sediments were covered by thick sequences of younger sediments and therefore were not penetrated by the relatively short piston corers. However, deep cores taken recently by the drilling ship, *Glomar Challenger* (Fig. 7.34b), have shown that this is not the case. In holes drilled to the basement, the oldest sediments recovered in both the Atlantic and Pacific basins are Late Jurassic in age (Fig. 7.35).

The fact that the age of the oldest sediments recovered at each station by the *Glomar Chal-* *lenger* generally increases with distance from the crests of the oceanic ridges has certainly strengthened the case for sea-floor spreading (Fig. 7.34). Sediments older than Tertiary have not been found within 700 miles (1200 km) of the crest of the Mid-Atlantic Ridge and 2400 miles (4000 km) of the crest of the East Pacific Rise.

Undeformed Sediments

Not all of the oceanographic studies have provided evidence supporting the concept of sea-floor spreading. In fact, some observations have been used in support of the idea of the permanency of continents and ocean basins. For example, there is almost no deformation of the sediments of the ocean floor. This would be difficult to understand if sea-floor spreading involved movements of large but relatively thin plates in response to underlying convection currents. However, sediment deformation would not be expected if the blocks were more than several hundred kilometers thick.

Proponents of sea-floor spreading believe that the sea-floor descends under continents and island arcs in the vicinity of oceanic trenches. Until recently, however, there was no evidence of deformation associated with oceanic trenches. Some authors concluded that the sea-floor was not descending in these areas (37). They felt that large volumes of deformed sediments would have been scraped from the sea-floor as it descended. Although recent seismic reflection studies have indicated the presence of deformed sedimentary units

landward of some trenches, seismic refraction measurements in these areas show that this sediment layer is relatively thin (Fig. 7.36).

The absence of large accumulations of deformed sediments associated with oceanic trenches may be due to the conversion of sediments into sedimentary and metamorphic rocks as the lithosphere descends beneath a continent or island arc. During this process, the sediments would be folded, faulted, and thickened, which in turn would result in their being dewatered, lithified, and perhaps even metamorphosed (Fig. 7.37). Muds and turbidites would be converted into shale, graywacke, slate, and metagraywacke. Such rocks would presumably underlie the lower part of

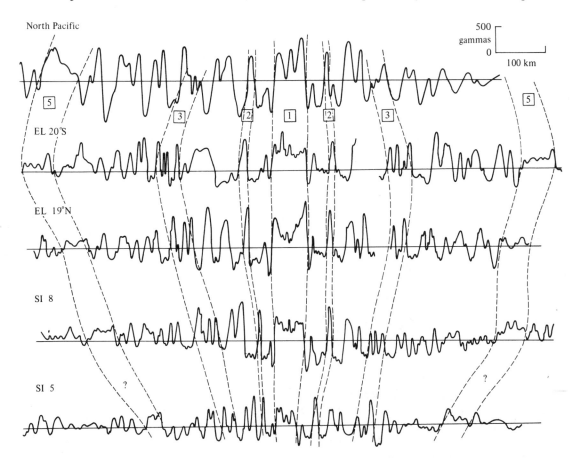

Figure 7.32 Magnetic profiles for the Pacific Ocean. The anomalies are symmetrical about Anomaly No. 1 on the crest of the East Pacific Rise. (After Pitman et al., Ref. 30.)

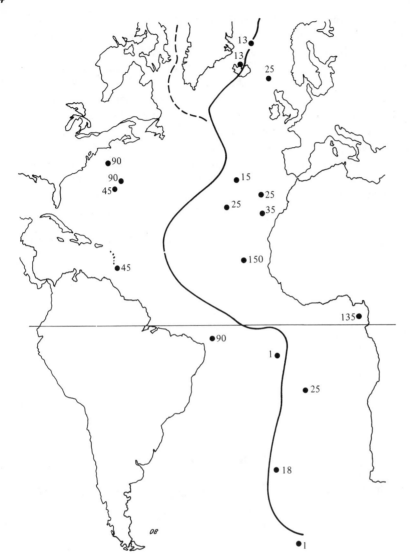

Figure 7.33 Variation in the age of volcanoes in the Atlantic Ocean with distance from the crest of the Mid-Atlantic Ridge. (From Wilson, Ref. 31.)

the continental slope and would be equivalent to and continuous with the undeformed sediments in the trench. Seismic refraction studies of the lower part of the continental slope landward of oceanic trenches indicate the presence of rock with the same seismic velocity as that of shale, slate, and graywacke.

Seismic reflection profiling of the Peru - Chile Trench has revealed a very sharp contact between undeformed sediments in the trench and deformed sediments at the base of the continental slope (Fig. 7.38). The contact may be a tectonic front analogous to the Appalachian structural front in eastern Pennsylvania. Here, the transition from

(a)

relatively undeformed Paleozoic sedimentary rocks to their strongly deformed equivalents is quite abrupt.

RECONSTRUCTING THE CONTINENTS

We have proposed a Carboniferous through Triassic reconstruction of the continents (Fig. 7.13) which generally agrees with those proposed by other authors. The reconstruction is based on the following controls:

1. Wherever possible, the continents should fit together at the 6000-ft (2000-m) depth contour with no large gaps and no large overlaps.

2. Geosynclines should border all continents.

3. A continental source for the Permo-Carboniferous ice sheets must be provided.

4. The recalculated Carboniferous through Triassic polar wandering curves should match reasonably well.

(b)

Figure 7.34　(a) Piston corer used for the sampling of sediments from the deep ocean. (Lamont-Doherty Geological Observatory, Columbia University, photos.) (b) The *Glomar Challenger*. (Scripps Institution of Oceanography Photo.)

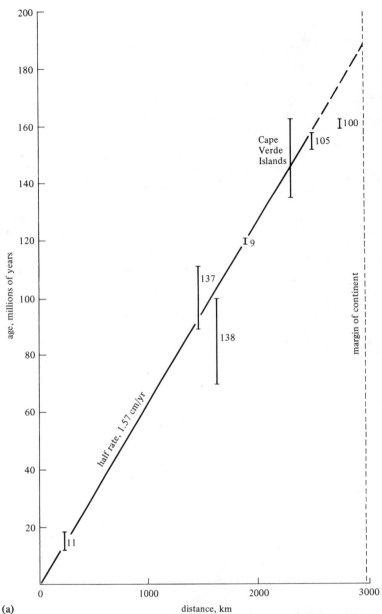

Figure 7.35   The results of drilling by the *Glomar Challenger* in the Atlantic Ocean and Gulf of Mexico: (a) Age of oldest sediment recovered by the drilling ship *Glomar Challenger* in the North Atlantic Ocean plotted against distance of the drilling site from the crest of the Mid-Atlantic Ridge. The age of the oldest sedimentary rocks on the Cape Verde Islands is also plotted (32). Other data from Ref. 33. (b) Age of the oldest sediment recovered by the *Glomar Challenger* in the South Atlantic Ocean plotted against distance from the crest of the Mid-Atlantic Ridge. (c) Age of the oldest sediment recovered by the *Glomar Challenger* in the Pacific Ocean plotted against distance from the crest of the East Pacific Rise. Data are from Ref. 35. The ages of basalt dredged from the floor of the Pacific are in agreement with the age of the oldest sediment (36).

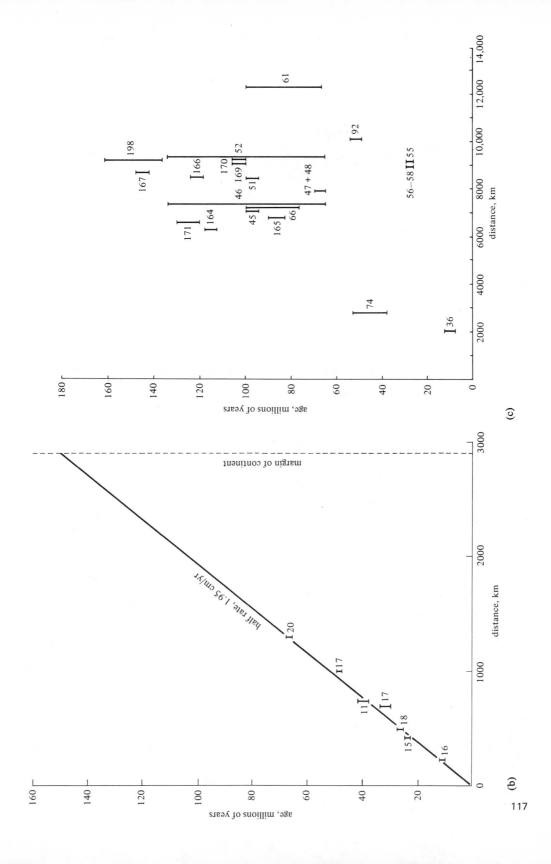

(c)

(b)

117

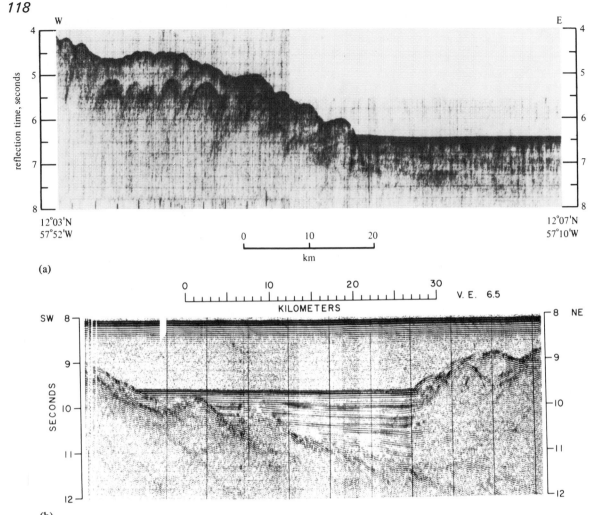

Figure 7.36 Seismic reflection profile illustrating deformed sediments in the vicinity of plate margins. (a) Profile across the eastern margin of the Barbados Ridge. Vertical exaggeration approximately 5 : 1. (From Chase and Bunce, Ref. 38.) (b) Interpolation of profile across the Aleutian Trench. (From Holmes et al., Ref. 39.)

5. Geosynclines and fold belts should be aligned so that they do not suddenly end in an ocean basin or shield area.

Paleomagnetic and paleoclimatic data provide an indication of the ancient latitude, but not longitude, of an area. Therefore, using only paleomagnetic poles for a single geologic period, there are an infinite number of possible positions of one continent relative to another (Fig. 7.39). However, if two or more points on a polar wandering curve are known for two continents, and if no relative movements took place between them during the time of polar wandering, the relative positions of the two continents may be established (41) (Fig. 7.40).

PRE-CARBONIFEROUS CONTINENTAL DRIFT

Although some authors have suggested that no continental drift occurred prior to the Mesozoic

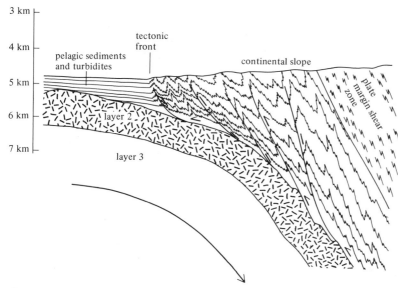

Figure 7.37 Hypothetical cross section through the Peru-Chile Trench at approximately 36°S latitude. Vertical exaggeration is approximately 2 : 1. Shear zone occurs at the margin between the Pacific Plate and the America Plate. (From Seyfert, Ref. 40.)

Era (42), the evidence presented here and in subsequent chapters indicates that there were displacements between continents during the Precambrian and the Paleozoic (43). From paleomagnetic, structural, and geochronological data it can be shown that there were five continents separated by deep ocean during the Proterozoic Era. In addition, there is evidence to show that these continents were joined into a single landmass during the Paleozoic Era.

Paleomagnetism

In order to test for the possibility of continental drift prior to the Carboniferous, the Pre-Carboniferous paleomagnetic poles were recalculated for a reconstruction of the continents in which all the continents were joined together (Fig. 7.41). It is apparent from the lack of agreement of these recalculated paleomagnetic poles that the continents were not part of a contiguous landmass prior to the Carboniferous (44).

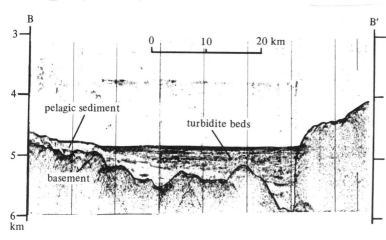

Figure 7.38 Seismic reflection profile across the Peru-Chile Trench. The turbidite sediments in this diagram were carried down the continental slope by density currents. The underlying pelagic sediments may have been deposited when the sea-floor was at greater distance from the margin of the South American continent. (From Scholl et al., Ref. 37. Copyright © 1968 by American Association for the Advancement of Science.)

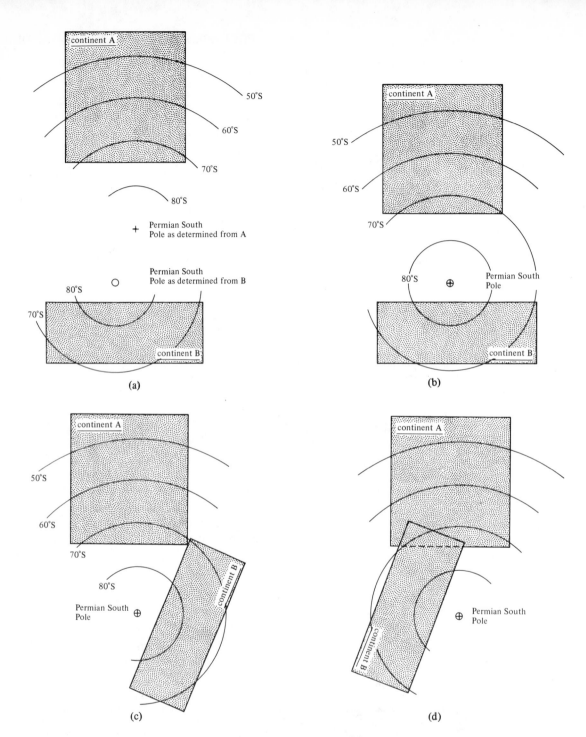

Figure 7.39 Using paleomagnetic data to reconstruct the continents: (a) before reconstruction, two continents with separated paleopoles; (b) one of many possible reconstructions in which paleopoles coincide; (c) another reconstruction in which paleopoles coincide; (d) reconstruction not permitted because of overlap of the continents, even though paleopoles agree.

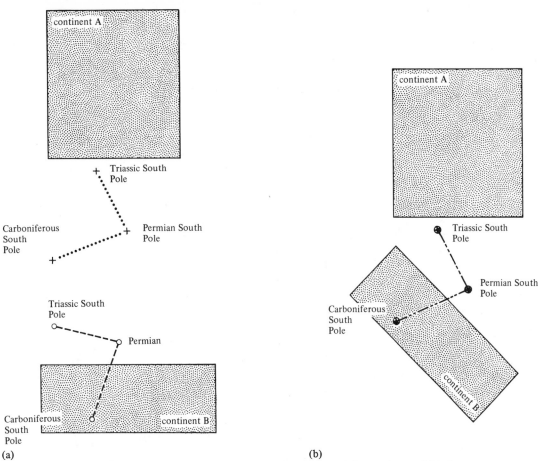

Figure 7.40 Using more than one paleopole for making reconstructions of the continents: (a) before reconstruction, two continents with separated polar wandering curves; (b) the only reconstruction in which the two polar wandering curves agree.

Geosynclines

The presence of thick sequences of deformed sedimentary and andesitic volcanic rocks suggests deposition in a geosyncline associated with an island arc. Modern geosynclines and island arcs are always located on the margins of continents. However, many Paleozoic geosynclines and island arcs are located near the center of continents (Fig. 7.13). This relationship suggests that these geosynclines and island arcs mark the borders of former continents which were subsequently joined into one "supercontinent."

Radiometric Dates

The pattern of radiometric dates suggests that there was more than one continental landmass during the Precambrian and early Paleozoic. In North America, the dates become progressively younger in an eastward direction toward the Appalachian Geosyncline, whereas in Europe the dates decrease westward toward the Caledonian Geosyncline (Figs. 9.5 and 9.21). If, as is generally believed, the continents have grown by a process of accretion, North America was growing eastward when Europe was growing westward. Therefore, these two continents were probably separate during this episode of growth.

Shape of Reconstructed Continent

When the continents are joined into a single landmass, there are two large reentrant angles on the margins of the continent. One separates the

Figure 7.41 Recalculated polar wandering curves for the continents comprising Pangaea. The Precambrian paleomagnetic poles for North America and Europe are widely separated, which indicates that these continents were apart during the Precambrian. The polar wandering curves don't agree until the Ordovician, suggesting that the continents joined at that time. Poles for the Silurian through the Triassic agree rather well, indicating that the continents were not significantly separated during this interval. The Early Paleozoic curves for China and Siberia do not agree with either the North American or the European curves, which suggests that China and Siberia were separated from Europe and North America at that time. Comparison of the polar wandering curve of Gondwanaland with the polar wandering curves of the other continents indicates that Gondwana-land was separated from the other continents prior to the Carboniferous and joined to the other continents from the Carboniferous through the Triassic.

southern continents from Eurasia and the other separates Eurasia from North America (Fig. 7.13). The process of growth of a continent by accretion at the margins of the continent would be expected to produce a continent with a more or less circular outline. The shape of Wegener's continent of Pangaea suggests that it was formed by the joining together of at least three approximately circular continents.

SUMMARY

We have presented evidence from a large variety of sources both for and against the concepts of polar wandering, continental drift, and sea-floor spreading. At the present time the weight of the evidence clearly supports these concepts and the encompassing concept of plate tectonics. The evidence for movement of the poles relative to the continents includes:

1. Thick limestones, red beds, evaporites, and coral reefs are often found at high latitudes, whereas ancient glacial deposits commonly occur near the present equator.

2. The direction of movement of Late Ordovician glaciers was away from the present equator in Africa.

3. Both the southwestern part of the United States and the British Isles were in the belt of prevailing easterlies during the Late Paleozoic; now, however, these areas are in the zone of prevailing westerly winds.

4. Paleomagnetic data indicate a large shift in the position of the magnetic poles since the Precambrian. Presumably this is associated with a shift in the geographic poles.

The evidence for movement of the continents relative to each other includes:

1. The outlines of the continents bordering the Atlantic and Indian oceans can be matched with only small areas of overlap or gap.

2. The Paleozoic and early Mesozoic fauna and flora of Europe are similar to those of North America. Moreover, the fauna and flora of the same age from each of the southern continents are also quite similar to those of the other southern continents.

3. The Permo-Carboniferous ice sheets appear to have moved from the site of the present Atlantic and Indian oceans. A continental source for the ice sheets is provided if the continents are joined.

4. Geosynclines and mountain belts often end abruptly at the continental margins. They form continuous belts encircling the continents if the continents are joined into a single landmass.

5. Polar wandering curves of the continents do not agree with one another. However, recalculated Carboniferous, Permian, and Triassic paleomagnetic poles agree reasonably well for a Pangaea reconstruction of the continents.

6. Radiometric age provinces may be traced from one continent to another if the continents are joined.

The evidence indicating that the sea-floor is moving away from mid-oceanic ridges includes:

1. The age of volcanic islands generally increases away from mid-oceanic ridges.

2. The age of the oldest sediment recovered at a given site generally increases away from mid-oceanic ridges. Generally, when the age of the oldest sediment is plotted against distance from the ridge, the points fall along a straight line, which indicates a relatively constant rate of sea-floor spreading.

3. Magnetic anomalies are symmetrical about the crest of mid-oceanic ridges, and the widths of the anomalies are proportional to the time between reversals of the earth's magnetic field.

Paleomagnetic, structural, and geochronological data indicate that there were five continents separated by deep ocean during the Precambrian and that these continents were joined into a single continent during the Paleozoic.

REFERENCES CITED

1. W. J. Morgan, 1968, Rises, trenches, great faults, and crustal blocks: *Jr. Geophys. Res.,* v. 73, p. 1959.

2. E. C. Bullard, J. E. Everett, and A. G. Smith, 1965, The fit of the continents around the Atlantic, *in* P. M. S. Blacket, E. C. Bullard, and S. K. Runcorn, eds., "A symposium on continental drift," *Phil. Trans. Roy. Soc. (London),* v. 1088, p. 41.

   A. G. Smith and A. Hallam, 1970, The fit of the southern continents: *Nature,* v. 225, p. 139.

3. J. C. Briden, A. G. Smith, and J. T. Sallomy 1970, The geomagnetic field in Permo-Triassic time: *Geophys. Jr. Roy. Astron. Soc.,* v. 23, p. 101.

4. H. Neuberger and J. Cahir, 1969, *Principles of Climatology:* Holt, Rinehart & Winston, New York.

5. R. W. Fairbridge, 1970, The ice age in the Sahara: *Geotimes,* v. 15, no. 6, p. 18.

   W. Hamilton and D. Krinsley, 1967, Upper Paleozoic glacial deposits of South Africa and southern Australia: *Geol. Soc. Amer. Bull.,* v. 78, p. 783.

6. E. Irving, 1964, *Paleomagnetism and its Application to Geological and Geophysical Problems:* Wiley, New York.

7. J. Lear, 1970, The bones on Coalsack Bluff: A story of drifting continents: *Sat. Rev.,* v. 53, no. 6, p. 46.

8. F. G. Stehli and C. E. Helsley, 1963, Paleontologic technique for defining ancient pole positions: *Science,* v. 142, p. 1057, Nov. 22.

9. D. I. Axelrod, 1963, Fossil floras suggest stable, not drifting continents: *Jr. Geophys. Res.,* v. 68, p. 3257.

10. G. A. Doumani and W. E. Long, 1962, The ancient life in the Antarctic: *Sci. Amer.,* v. 207, no. 3, p. 168.

11. J. F. Dewey and G. M. Kay, 1968, Appalachian and Caledonian evidence for drift in the North Atlantic, *in* R. A. Phinney, ed., *The History of the Earth's Crust, A Symposium:* Princeton University Press, Princeton, N.J., p. 161.

12. A. Cox and R. R. Doell, 1960, Review of paleomagnetism: *Geol. Soc. Amer. Bull.,* v. 71, p. 645.

13. A. A. Meyerhoff, 1970, Continental drift: Implications of paleomagnetic studies meteorology, physical oceanography and climatology: *Jr. Geol.,* v. 78, p. 1.

    J. Hospers and S. I. Van Andel, 1968, Paleomagnetic data from Europe and North America and their bearing on the origin of the North Atlantic Ocean: *Tectonophysics,* v. 6, p. 475.

14. P. M. Hurley, 1968, The confirmation of continental drift: *Sci. Amer.,* v. 218, no. 4, p. 53.

    P. M. Hurley et al., 1967, Test of continental drift by comparison of radiometric ages: *Science,* v. 157, p. 495.

15. M. Barazangi and J. Dorman, 1969, World seismicity map of ESSA Coast and Geodetic Survey epicenter data for 1961 - 1967: *Bull. Seismol. Soc. Amer.,* v. 59, p. 369.

16. B. Isacks, J. Oliver, and L. R. Sykes, 1968, Seismology and the new global tectonics: *Jr. Geophys. Res.,* v. 73, p. 5855.

17. J. T. Wilson, 1965, A new class of faults and their bearing on continental drift: *Nature,* v. 207, p. 343.

18. B. Gutenberg and C. F. Richter, 1954, *Seismicity of the Earth,* 2nd ed.: Princeton University Press, Princeton, N.J.

19. B. Isacks and P. Molnar, 1969, Mantle earthquake mechanisms and the sinking of the lithosphere: *Nature,* v. 223, p. 1121.

20. H. H. Hess, 1962, History of the ocean basins, *in* S. E. J. Engle et al., eds., *Petrological Studies: A Volume in Honor of A. F. Buddington:* Geological Society of America, New York, p. 559.

21. V. Vaquier, A. D. Raff, and R. E. Warren, 1961, Magnetic survey off the west coast of North America, 40° N. latitude to 50° N. latitude: *Geol. Soc. Amer. Bull.,* v. 72, p. 1267.

22. H. W. Menard, 1964, *Marine Geology of the Pacific:* McGraw-Hill, New York.

23. J. R. Heirtzler, X. Le Pichon, and J. G. Baron, 1965, Magnetic anomalies over the Reykjanes Ridge: *Deep-Sea Res.,* v. 13, p. 427.

24. F. J. Vine, 1968, Magnetic anomalies associated with mid-ocean ridges, *in* R. A. Phinney, ed., *The History of the Earth's Crust—A Symposium:* Princeton University Press, Princeton, N. J., p. 73, Fig. 6 facing p. 82.

25. F. J. Vine and P. M. Matthews, 1963, Magnetic anomalies over ocean ridges: *Nature,* v. 199, p. 947.

26. A. Cox, 1969, Geomagnetic reversals: *Science,* v. 163, p. 239, Jan. 17.

27. F. J. Vine, 1966, Spreading of the ocean floor: New evidence: *Science,* v. 154, p. 1405, Dec. 16.

28. J. de Boer, J. G. Schilling, and D. C. Krause, 1969, Magnetic polarity of pillow basalts from Reykjanes Ridge: *Science,* v. 166, p. 996.

29. J. R. Heirtzler, G. O. Dickson, E. M. Herron, W. C. Pitman, and X. Le Pichon, 1968, Marine magnetic anomalies, geomagnetic field reversals, and motions of the ocean floor and continents: *Jr. Geophys. Res.,* v. 73, p. 2119.

30. W. C. Pitman, E. M. Herron, and J. R. Heirtzler, 1968, Magnetic anomalies in the Pacific and sea-floor spreading: *Jr. Geophys. Res.,* v. 73, p. 2069.

31. J. T. Wilson, 1965, Evidence from ocean islands suggesting movement in the earth, *in* P. M. S. Blacket, E. C. Bullard, and S. K. Runcorn, eds., "A symposium on continental drift": *Phil. Trans. Roy. Soc. (London),* v. 1088, p. 145.

32. G. Colom, 1955, Jurassic-Cretaceous sediments of the western Mediterranean zone and the Atlantic area: *Micropaleontology,* v. 1, p. 109.

33. M. Ewing et al., 1969, *Initial Reports of the Deep Sea Drilling Project, Vol. I:* Government Printing Office, Washington, D. C.

M. N. A. Peterson et al., 1970, *Initial Reports of the Deep Sea Drilling Project, Vol. II:* Government Printing Office, Washington, D. C.

Scientific staff for Leg 11, 1970, Deep Sea drilling project leg 11: *Geotimes,* v. 17, no. 7, p. 14.

Scientific staff for Leg 14, 1971, Deep Sea drilling project leg 14: *Geotimes,* v. 16, no. 2, p. 14.

34. A. E. Maxwell et al., 1970, Deep sea drilling in the South Atlantic: *Science,* v. 168, p. 1047.

35. Scientific staff for Leg 5, 1969, Deep Sea drilling project leg 5: *Geotimes,* v. 14, no. 7, p. 19.

Scientific staff for Leg 6, 1969, Deep Sea drilling project leg 6: *Geotimes,* v. 14, no. 8, p. 13.

Scientific staff for Leg 7, 1969, Deep Sea drilling project leg 7: *Geotimes,* v. 14, no. 10, p. 12.

Scientific staff for Leg 8, 1970, Deep Sea drilling project leg 8: *Geotimes,* v. 15, no. 2, p. 14.

Scientific staff for Leg 9, 1970, Deep Sea drilling project leg 9: *Geotimes,* v. 15, no. 4, p. 11.

Scientific staff for Leg 16, 1971, Deep Sea drilling project leg 16: *Geotimes,* v. 16, no. 6, p. 12.

Scientific staff for Leg 17, 1971, Deep Sea drilling project leg 17: *Geotimes,* v. 16, no. 9, p. 12.

Scientific staff for Leg 20, 1972, Deep Sea drilling project leg 20: *Geotimes,* v. 17, no. 4, p. 10.

36. J. Dymond and H. L. Windom, 1968, Cretaceous K - Ar ages from the Pacific Ocean seamounts: *Earth Planet. Sci. Lett.,* v. 4, p. 47.

C. Papaliolios et al., 1968, Ages of Pacific deep-sea basalts, and spreading of the sea floor: *Science,* v. 160, p. 1106.

37. D. W. Scholl, R. von Huene, and J. B. Ridlon, 1968, Spreading of the ocean floor: Undeformed sediments in the Peru - Chile Trench: *Science,* v. 159, p. 869, Feb. 23.

38. R. L. Chase and E. T. Bunce, 1969, Underthrusting of the eastern margin of the Antilles by the floor of the western north Atlantic Ocean and origin of the Barbados Ridge: *Jr. Geophys. Res.,* v. 74, p. 1413.

39. M. L. Holmes, R. von Huene, and D. A. McManus, 1972, Seismic reflection evidence supporting underthrusting beneath the Aleutian Arc near Amchitka Island: *Jr. Geophys. Res.,* v. 77, p. 959.

40. C. K. Seyfert, 1969, Undeformed sediments in oceanic trenches with sea-floor spreading: *Nature,* v. 222, p. 70.

41. K. W. T. Graham, C. E. Helsley, and A. L. Hales, 1964, Determination of the relative positions of the continents from paleomagnetic data: *Jr. Geophys. Res.,* v. 69, p. 3895.

42. P. M. Hurley and J. R. Rand, 1969, Pre-drift

continental nuclei: *Science*, v. 164, p. 1238.

43. J. T. Wilson, 1966, Did the Atlantic close and then re-open?: *Nature*, v. 211, p. 676.

J. F. Dewey and J. Bird, 1970, Lithosphere plate—continental margin tectonics and evolution of the Appalachian Orogen: *Geol. Soc. Amer. Bull.*, v. 81, p. 655.

J. F. Dewey, 1969, Evolution of the Appalachian/Caledonian orogen: *Nature*, v. 222, p. 124.

H. Williams, 1964, The Appalachians in northeastern Newfoundland—a two sided symmetrical system: *Amer. Jr. Sci.*, v. 262, p. 1137.

P. N. Krapotkin, 1969, Eurasia as a composite continent, *in* J. T. Wilson, ed., *Symposium on Continental Drift at Montevideo:* UNESCO, Paris (in press).

J. W. Skehan, 1969, Tectonic framework of Southern New England and Eastern New York, *in* M. Kay, ed., *North Atlantic—Geology and Continental Drift:* American Association of Petroleum Geologists Memoir 12, p. 931.

J. F. Dewey and B. Horsfield, 1970, Plate tectonics, orogeny and continental growth: *Nature*, v. 225, p. 521.

44. H. Spall, 1972, Did Southern Africa and North America drift independently during the Precambrian?: *Nature*, v. 236, p. 219.

SUGGESTED READINGS

P. M. S. Blacket, E. C. Bullard, and S. K. Runcorn, eds., 1965, "A symposium on continental drift," *Phil. Trans. Royal Soc. (London)*, v. 1088.

S. W. Carey, ed., 1958, *Continental Drift—A Symposium:* Hobart, Geology Department, University of Tasmania.

J. R. Heirtzler, 1968, Sea-floor spreading: *Sci. Amer.*, v. 219, no. 6, p. 60.

M. Kay, ed. 1969, *North Atlantic-Geology and Continental Drift*, American Association of Petroleum Geologists, Mem. 12.

R. A. Phinney, ed., *The History of the Earth's Crust—A Symposium:* Princeton University Press, Princeton, N.J.

S. K. Runcorn, 1962, *Continental Drift:* Academic Press, New York.

H. Takeuchi, S. Uyeda, and H. Kanamori, 1967, *Debate about the Earth:* Freemon-Cooper, San Francisco and London.

J. T. Wilson, 1963, Continental drift: *Sci. Amer.*, v. 208, no. 4, p. 86.

8

The Origin of the Earth and its Early History

*T*he age of the earth is estimated to be about 4.6 billion years. Various methods of dating have placed the age of the oldest known terrestrial rocks at about 4.0 billion years. This leaves 600 million years of earth history unaccounted for by the rock record. The origin of the earth and development of its crust, atmosphere, and oceans can thus be reconstructed only by inference. The origin of the earth can be inferred from observation of celestial phenomena. Large optical and radio telescopes have provided the means of observing the birth and death of stars light years distant and, by applying the principle of uniformitarianism, we can suggest processes by which our solar system may have been formed. The origin of the earth's crust, atmosphere, and oceans may be deduced from a comparison of the conditions that may have existed 4.0 billion years ago, as evident from the rock record, and the conditions that must have prevailed at the time the earth was formed.

Early man was undoubtedly aware of a variety of cosmic events which inspired his imagination or fear and were the origin of many myths and superstitions. However, since these events did not directly influence his day-to-day existence, he paid little attention to them. It was not until the sixth century B.C., when the Greek philosophers began to speculate about the earth's shape and place in the universe, that astronomy became a science. Although many models of the universe have since been proposed, early ideas relied heavily on mythology and there was no scientific basis for

astronomy until Sir Isaac Newton postulated the laws of gravitation and thus laid the basis for modern astronomy (Table 8.1).

Table 8.1   Early theorization on the universe

Early man: The earth consisted of a saucer balanced on the back of a monstrous reptile.
Thales (550 B.C.), Greek: The earth was a flat disk floating on water.
Anaximander (550 B.C.), Greek: The universe was a cylinder in a great void.
Aristarchus (third century B.C.), Greek: the heliocentric universe.
Hipparchus (second century, A.D.), Greek: the geocentric universe.
Ptolemy (second century, A.D.), Egyptian: The geocentric model with stars and planets on fixed spheres around the earth.

The sun-centered or heliocentric universe had been considered more than 200 years before the birth of Christ, but the concept of an earth-centered or geocentric universe persisted in scientific thought until the beginning of the sixteenth century. In 1543 Copernicus revived the heliocentric model of the solar system and proposed that the earth was one of several planets orbiting the sun. He demonstrated that the movement of the sun and stars was more simply explained as a consequence of the rotation of the earth on its axis. In 1609 Galileo constructed the first astronomical telescope. With it he gathered evidence in support of the Copernican model of the solar system and opened the way for the modernization of the science of astronomy.

Figure 8.1   Spiral galaxy M 74 in the constellation *Pisces.*
(Mount Wilson and Palomar Observatories.)

THE EXPANDING UNIVERSE AND
THE ORIGIN OF THE ELEMENTS

With the subsequent development of large telescopes, it became evident that some starlike objects were actually comprised of billions of stars. Such assemblages of stars, called galaxies, are spiral, elliptical, or irregular in shape (Fig. 8.1). Our sun is located in one of the outer arms of a spiral galaxy, the Milky Way Galaxy. Examination of the spectra of distant galaxies revealed a systematic shift of the spectral lines toward the red end of the spectrum. Such a red shift has been shown to exist in the spectra of objects that are receding at a great speed from the point of observation. Astronomers have thus reached the conclusion that the galaxies they observed were receding from our own galaxy. Edwin Hubble estimated the distance between the Milky Way and a number of galaxies and found that the apparent rate of their recession is proportional to their distance (Fig. 8.2). In other words, the more distant galaxies appear to be receding at a faster rate than nearby galaxies. This relationship has been interpreted as evidence that the universe is expanding.

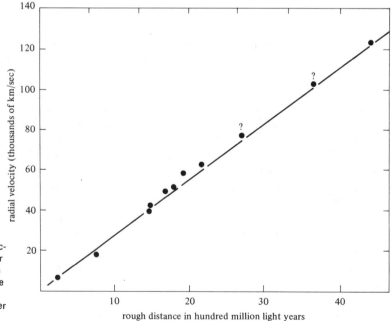

Figure 8.2   The relation of velocity of recession of galaxies to their distance, based on the assumption that the absolute magnitude of the brightest galaxy in a cluster is the same as in any other cluster. (After Abell, Ref. 1.)

rough distance in hundred million light years

The concept of an expanding universe allows us to estimate the minimum age of the universe. The slope of the curve in Fig. 8.2 indicates that the expansion began about 18 billion years ago. However, this figure is only a minimum age for the formation of the universe, since the expansion may have been preceded by an earlier phase of contraction.

George Gamow, among others, postulated that the apparent expansion of the universe was a result of a gigantic explosion. According to this theory, often referred to as the "Big Bang Theory," all of the matter of the universe was at one time contained in a relatively small, incredibly dense mass in which protons and electrons were combined into neutrons. Expansion of this mass may have produced temperatures in excess of 1,000,000,000,000°C. Strong support for the Big Bang Theory comes from the discovery of microwave radiation which is received on earth with equal intensity from all sides. It has been suggested that this radiation may be the remnant of the primeval fireball which now has cooled down almost to absolute zero (2).

Most astrophysicists now believe that the universe began as a gigantic fireball. As the fireball expanded, neutrons decayed to form protons, which are the nuclei of hydrogen atoms, and electrons. Most of the helium, which comprises approximately one quarter of the mass of the universe, probably originated at this time (Fig. 8.3, equations 1-3). In an early version of this theory, Gamow proposed that most of the elements heavier than helium were produced during the

initial explosion. Subsequently, however, Gamow and others determined that only a small percentage of these elements could have been produced in the primeval fireball.

STELLAR EVOLUTION

In seeking answers to the origin of the elements, astronomers have embarked on long-term studies of the intrinsic properties of the stars; they have also begun to investigate the element-forming processes that seem to be going on in certain areas of space. These studies have led to the observation that the color of stars varies from bluish-white on one extreme to reddish on the other extreme, with our yellow-colored sun in the middle. The color of a star is a function of the temperature of its visible surface. The blue-white stars have a very high surface temperature, but the reddish stars are comparatively cool. It has been found that the intrinsic luminosity of a star (i.e., its actual brightness) is related to the star's color, or spectral class. This would be expected, since surfaces at higher temperatures radiate more energy. When the intrinsic luminosities of stars are plotted against their surface temperatures, it is found that most intrinsically bright stars have high surface temperatures; conversely, stars that are intrinsically dim generally have low surface temperatures. Most stars in such a plot fall along a line known as the *main sequence* (Fig. 8.4). The Hertzsprung-Russell (H-R) diagram may be used to illustrate the evolution of a star (Fig. 8.5). Since the intrinsic luminosity and spectral class of every star changes as the star moves

$$^{1}_{1}H + ^{1}_{1}H \rightarrow ^{2}_{1}H + \text{neutrino} \quad (1)$$

$$^{2}_{1}H + ^{1}_{1}H \rightarrow ^{3}_{2}He + \text{gamma ray} \quad (2)$$

$$^{3}_{2}He + ^{3}_{2}He \rightarrow ^{4}_{2}He + 2^{1}_{1}H \quad (3)$$

$$^{4}_{2}He + ^{4}_{2}He \rightarrow ^{8}_{4}Be \quad (4)$$

$$^{8}_{4}Be + ^{4}_{4}He \rightarrow ^{12}_{6}C \quad (5)$$

$$^{12}_{6}C + ^{4}_{2}He \rightarrow ^{16}_{8}O \quad (6)$$

$$^{16}_{8}O + ^{4}_{2}He \rightarrow ^{20}_{10}Ne \quad (7)$$

$$^{20}_{10}Ne + ^{4}_{2}He \rightarrow ^{24}_{12}Mg \quad (8)$$

$$^{24}_{12}Mg + ^{4}_{2}He \rightarrow ^{28}_{14}Si \quad (9)$$

$$^{28}_{14}Si + ^{4}_{2}He \rightarrow ^{32}_{16}S \quad (10)$$

$$^{28}_{14}Si + ^{28}_{14}Si \rightarrow ^{56}_{28}Ni \quad (11)$$

$$^{56}_{28}Ni \rightarrow ^{56}_{27}Co + \text{positron} \quad (12)$$

$$^{56}_{27}Co \rightarrow ^{56}_{26}Fe + \text{positron} \quad (13)$$

Figure 8.3 Equations representing nuclear reactions occurring in the interior of stars.

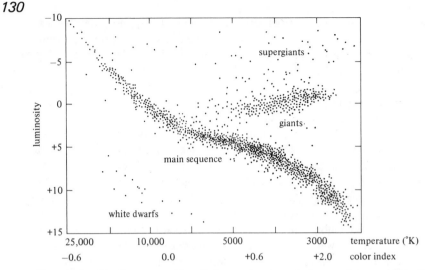

Figure 8.4   The Hertzsprung-Russell diagram for stars of a known distance. (After Abell, Ref. 1.)

through its life cycle, its position on the H-R diagram also changes with time.

It is now believed that stars originate as slowly contracting masses of gas and dust. Some astronomers have suggested that globules—dark, spherical bodies associated with certain irregular clouds of gas and dust—represent an early stage in the formation of a star (Fig. 8.6). Globules have

approximately the same mass as our sun, but their radii are thousands of times that of the entire solar system.

As the gas and dust contract, the globule heats up owing to the release of gravitational energy. Helium begins to form as the result of the fusion of hydrogen nuclei when the interior of the star reaches a temperature of 10,000,000°C and the

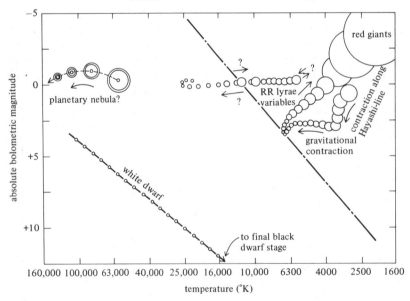

Figure 8.5   Evoluntionary track of a star of 1.2 solar masses on the Hertzsprung-Russell diagram. (After Abell, Ref. 1.)

From the red giant stage, stars may follow several possible evolutionary paths. A star that has approximately the same mass as the sun may become a *variable star* or *nova*. Eventually, the nuclear reactions in novae cease, and they cool and contract into a variety of stars known as *white dwarfs*; these bodies are believed to consist of tightly packed masses of electrons and ionized gases.

If the mass of a red giant is significantly greater than that of the sun, it may explode violently and become a *supernova*. Whereas a nova may increase in luminosity by a factor of several tens of thousands, a supernova may flare up to hundreds of millions of times the brightness of the star from which the supernova formed (Fig. 8.7). It has been suggested that the elements heavier than iron were produced during or immediately preceding a supernova explosion (3). During a

Figure 8.6   A nebula showing irregular and spherical dust clouds. The spherical dust clouds, called globules, are thought to be stars in the beginning stages of contraction. (Mount Wilson and Palomar Observatories.)

density approaches 100 g/cc (Fig. 8.3, equations 1 - 3). The conversion of hydrogen to helium provides the source of energy of main sequence stars. When much of the hydrogen in the core of a star has been converted to helium, the core of the star contracts, transforming gravitational energy into heat, which again causes the core to heat up. This in turn makes the outer layers of the star expand and causes the surface temperature of the star to decrease. At this stage, the star moves off the main sequence and becomes a *red giant* (Fig. 8.5). When the temperature of the core of a red giant reaches 100,000,000°C, helium in the core fuses to produce carbon, oxygen, magnesium, and silicon (Fig. 8.3, equations 4 - 9). Further contraction and heating of the core would produce nickel and iron (3) (Fig. 8.3, equations 11 - 13). All the elements produced during the red giant stage are important components of the earth's crust, mantle, and core.

Figure 8.7   The spiral galaxy NGC 7331 before and during the eruption of a supernova. (Lick Observatory photograph.)

Figure 8.8   The Crab Nebula photographed in red light. This nebula is the remains of a supernova that occurred in 1054. (Mount Wilson and Palomar Observatories.)

supernova explosion, most of the star's mass may be dispersed into space (Fig. 8.8). Other stars may form from the matter ejected during or after the expansion of a supernova.

Following a supernova, the remainder of the star may collapse into an extremely dense mass of neutrons, called a *neutron star,* only a few tens of miles in diameter. *Pulsars,* which emit short bursts of energy every second or so, may be rapidly rotating neutron stars. One such pulsating energy source, found near the center of the Crab Nebula by radio telescope, may be the result of a supernova explosion (4) (Fig. 8.8).

QUASARS AND THE HEAVY ELEMENTS

In 1963 Maarten Schmidt discovered a new class of starlike objects which emit large amounts of energy in the form of radio waves. These have been called quasi-stellar radio sources, or *quasars* (5). The spectra of quasars indicate that they are receding from our galaxy at very large velocities. Using Hubble's velocity - distance relation (Fig. 8.2) it has been calculated that many quasars are probably more distant than any known galaxy. If these calculations are correct, then the energy output of quasars must be greater than that of the most luminous known galaxy (Fig. 8.9). For this reason, some astrophysicists have expressed doubt that quasars are very distant objects.

On the other hand, if the Hubble relation is applicable to quasars, a large percentage of them would be located between 7 and 8 billion light years from our galaxy (6). It is quite possible that quasars were very abundant 7 or 8 billion years ago and that their numbers have decreased with time. The lack of quasars beyond a distance of 9 billion light years may indicate that there were no quasars prior to 9 billion years ago.

It has been suggested that quasars are the extremely luminous nuclei of galaxies whose outer portions we are not able to observe because they are so far away. Many of the distinctive characteristics of quasars were also observed by Carl Seyfert (Senior) in an interesting class of spiral galaxies named *Seyfert galaxies* after their discoverer (6, 7). These galaxies are distinctive in that they possess highly luminous nuclei (Fig. 8.10). Seyfert galaxies may bridge the gap between quasars and normal galaxies, since their luminosities overlap those of quasars and normal galaxies (Fig. 8.9).

It has been suggested that the enormous amounts of energy released by both quasars and Seyfert galaxies originate in the simultaneous occurrence of large numbers of novae and supernovae in the nuclei of galaxies. In our galaxy, supernovae occur once every 30 years or so. In an early stage in the history of a galaxy, the rate of occurrence of supernovae may have been much greater than it is today. The rapid variations in energy output of quasars and Seyfert galaxies may be due to differences in the number of supernovae occurring at any one time.

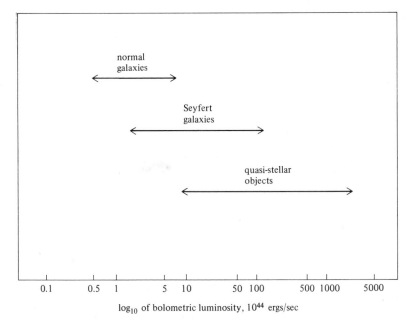

log₁₀ of bolometric luminosity, 10^{44} ergs/sec

Figure 8.9 Comparison of the luminosities of normal galaxies, Seyfert galaxies, and quasars. (Modified after Colgate, Ref. 8.)

In the Milky Way galaxy, there is a wide variation in the age of the stars. However, most of the main sequence stars have essentially the same composition, regardless of their age (9). Evidently, most of the elements heavier than helium were created at an early stage in the history of our

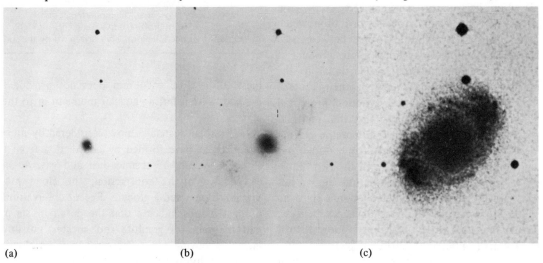

(a)   (b)   (c)

Figure 8.10 Photographs of negative plates of a Seyfert galaxy with increasing exposure time: (a) Only the nucleus is visible (at this point, the nucleus resembles a quasar); (b) and (c) with a longer time of exposure, the stars in the spiral arms of the galaxy become visible. (Copyrighted by the National Geographic Society-Palomar Observatory Sky Survey: original photographs from the Hale Observatories; montage prepared by W. W. Morgan, Yerkes Observatory.)

galaxy. C. M. Hohenberg placed the age of the Milky Way at between 8 and 8.8 billion years and suggested that a large percentage of the elements formed by neutron capture (possibly in supernovae) were produced at a very early stage in its history (10).

It would appear that large quantities of heavy elements were generated in massive stars soon after the formation of our galaxy. Since massive stars evolve rapidly, large numbers of supernovae may have also developed at this time. Perhaps during this heavy element-forming stage, our galaxy would have appeared as a quasar to a distant observer. Thus the Milky Way and perhaps many other galaxies may have originated as quasars.

ORIGIN OF THE SOLAR SYSTEM

During the latter part of the sixteenth century, the Danish mathematician Tycho Brahe made very detailed measurements of the positions of several planets over a period of many years. On the basis of these observations, Johannes Kepler, Brahe's assistant, derived the laws of planetary motion. Based on the work of Galileo and Kepler, Newton formulated the law of universal gravitation, which states that the gravitational force between two bodies is directly proportional to the product of their masses and inversely proportional to the square of the distance between their centers.

The law of universal gravitation prompted theorists to consider the possibility that the solar system had originated as a result of the gravitational attraction between the sun and another star during a close encounter between the two (Table 8.2). However, astronomers have objected to this idea because close encounters between stars are unlikely, owing to the vast distances between them. The sun's closest neighbor is more than 4 light years (24,000,000,000,000 miles) away. Furthermore, the angular momentum (the product of the mass of a body times its velocity times its radius of orbit) of the sun is only a small fraction of the angular momentum of the solar system. The remainder is contained in the planets. The hypo-

Table 8.2 Hypotheses on the origin of the solar system

Hypotheses based on close encounter

Protoplanet theory. Buffon, 1749: a passing star or comet caused the sun to shed rings of matter, which ultimately condensed to form planets orbiting the sun.

Planetismal hypothesis. T. C. Chamberlin and F. R. Moulton, 1895: a passing star and eruptive activity in the sun caused solar material to go into orbit and form the planets.

Tidal hypothesis. James A. Jeans and Harold Jeffreys, 1917: a near collision between the sun and another star drew out of the sun a tide of gases which broke up to form the planets.

Double star hypothesis. Lyttleton, 1936: a companion star of the sun collided with a third star to form the solar system.

Hypotheses based on contraction of a nebula

Nebular hypothesis. Kant, 1755; Laplace, 1796: the nebula contracted gravitationally and by conservation of angular momentum began to rotate, separated into rings with centers of gravitation which became the planets and their satellites.

Turbulent eddies hypothesis. Carl von Weizsäcker, 1944: the cosmic dust and gas contracted into a rotating disk of turbulent eddies, of which the smaller, denser eddies produced the planets.

Dust cloud hypothesis. Fred Whipple, 1946: the solar system formed through contraction of a globule.

Protoplanet hypothesis. Gerard Kuiper, 1950: a condensing and rotating globule became flattened and its dense central mass formed the sun. The protoplanets, which were larger than the modern planets, developed from orbiting eddies, and their lighter gases (hydrogen, helium) were driven off by the sun.

thesis of a close encounter does not provide a method of transferring angular momentum to the planets.

The solar system is now considered by most scientists to have formed by contraction from a cloud or nebula of cosmic dust and gas. At an early stage of its contraction, the cloud presumably resembled a globule. Recent observations of dust clouds indicate that the dust consists in part of grains of graphite and silicates, possibly including olivine (11). The gases in these clouds are largely hydrogen and helium, but water molecules and ammonia have also been detected (12). All these materials are important components of the earth and sun. It has been suggested that the planets formed as a result of eddies, which concen-

1. The moon was once part of the earth and separated from it as a result of rapid rotation of the earth (Fig. 8.11).

2. The earth and moon formed nearby in space at approximately the same time and from the same raw materials (Fig. 8.12).

3. The moon formed elsewhere in the solar system and was subsequently captured by the earth (Fig. 8.13).

George Darwin, the son of Charles Darwin, suggested that the moon came out of what is now the Pacific Ocean basin at a very early stage in the earth's history. It was later proposed that the separation of the earth and the moon was associated with the migration of nickel - iron compounds toward the earth's center during the formation of its core (13). Since angular momentum must always be conserved, the migration of heavy compounds toward the center of the earth would have caused an increase in the earth's rate of rotation. Subsequently, the earth might have been rotating fast enough to cause a large part of its mantle to be thrown outward and into orbit around the planet. The density of the moon is, in fact, almost identical to that of the earth's mantle.

Alternatively, the earth and moon may have formed simultaneously from the same cloud of gas and dust. In this case, the earth's greater density would have been accounted for by an initial segregation of nickel - iron toward the center of the condensing cloud. In a cloud of gas and dust, denser particles would tend to fall toward a center of mass at a faster rate than less dense particles.

Last, the moon may have been formed at some distance from the earth, subsequently being captured. Thus the difference in density between the earth and moon would be a result of their originating in different parts of the solar system. However, the chances are slight that an object the size of the moon would have an orbit that would allow its capture by a planet.

The samples of rock and soil brought back by the Apollo lunar expeditions have shed some light on the history of the earth - moon system. In the areas studied, a layer of soil mantles the bed-

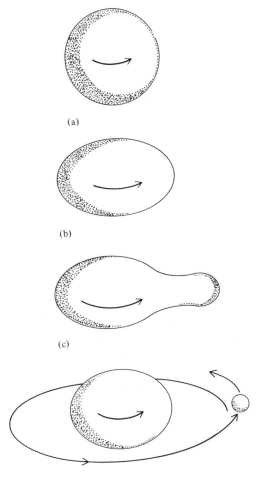

(a)

(b)

(c)

Figure 8.11 Diagrammatic sketch illustrating the formation of the moon by separation from the earth.

trated the dust into bodies that were subsequently joined to form the planets. Alternatively, the planets may have formed by the disruption of the original dust cloud due to its rapid rotation.

Earth-Moon Relationships

Possibly some of the questions of planetary origins may be resolved in exploration of the moon. Recent investigations of the moon conducted in the Apollo series have rekindled interest in the origin of the earth - moon system. The theories previously proposed fall into three categories:

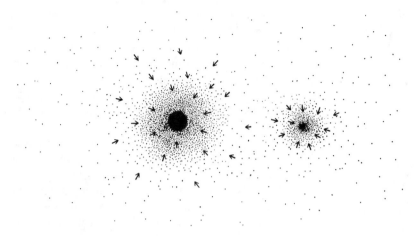

Figure 8.12   Diagrammatic sketch illustrating the formation of the moon from the condensation of a cloud of gas and dust during the formation of the earth.

rock (14). Small rocks and boulders are common in this soil. Many of the rocks are texturally and mineralogically similar to terrestrial basalts (15) (Fig. 8.14). These similarities and the presence of phenocrysts and vesicles suggest a volcanic origin. Photographs taken by lunar orbiters have also revealed the existence of volcanic activity (Fig. 8.15).

Other lunar rocks are breccias, which are composed of fragments of older rocks. Chemically, the lunar samples show significant differences from terrestrial rocks. For example, some samples contain unusually high concentrations of refractory elements, such as titanium, zirconium, and several rare earth elements, and unusually low concentra-

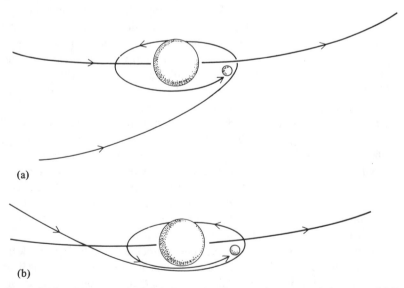

(a)

(b)

Figure 8.13   Diagrammatic sketch illustrating the possible capture of the moon: (a) From outside the earth's orbit, (b) from inside the earth's orbit.

(a) (b)

Figure 8.14 (a) Vesicular lava collected during the *Apollo eleven* flight to the moon. (Courtesy of *NASA*.) (b) Photomicrograph of thin sections of a lunar rock taken in polarized light with uncrossed nicol prisms. The colorless mineral is plagioclase, the light gray is pyroxene, and the black is ilmenite. Magnification approximately 250X. (Courtesy of NASA.)

tions of the more volatile elements, such as sodium and potassium. These differences make it unlikely that the moon was once part of the earth.

Most lunar rocks have been dated at 3.3 to 3.8 billion years by rubidium - strontium whole-rock and uranium - lead determinations, whereas samples of the soil yield ages of 4.4 to 4.6 billion years by the same methods (15, 16). It has been suggested that much of the soil has been transported from the lunar highlands and that the loose rocks were derived from nearby bedrock now covered by soil. The ages indicate that the moon is more than 4.6 billion years old and that extensive volcanic activity occurred in at least two of its maria about 3.6 billion years ago.

Asteroids, Meteorites, and Earth History

Between the orbits of Mars and Jupiter there is a belt of small bodies called asteroids, the largest of them, Ceres, being 480 miles in diameter. Many scientists believe that most of the large meteorites that strike the earth, the moon, and Mars originated in the asteroid belt. The asteroids probably represent the remains of a disrupted planet or planets.

Meteorites are of two principal varieties. Stony meteorites are composed largely of silicates, whereas metallic meteorites consist mostly of nickel - iron. The difference in composition may be due to the separation of a metallic core from a silicate mantle within the planet(s) prior to disruption. Radiometric dates on meteorites have provided a minimum age for the solar system. Uranium - lead, potassium - argon, and rubidium - strontium dates fall between 4.5 and 4.7 billion years for most meteorites (17). This is in good agreement with the age of the oldest lunar samples (4.66 billion years) (18, 19) and the age of the earth based on lead-isotopic ratios in oceanic basalts (4.53 billion years) (20).

Separation of the Earth's Core, Mantle, and Crust

According to the prevalent theory, as the earth

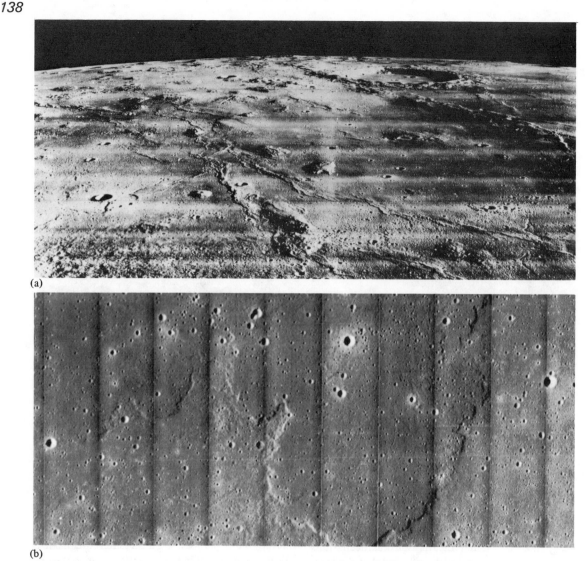

(a)

(b)

Figure 8.15 Evidence of volcanic activity on the moon: (a) Oblique photograph made by *Lunar Orbiter II* showing domes on the lunar surface. The domes are from 2 to 10 miles in diameter and from 1000 to 1500 ft high. They are shaped much like terrestrial volcanic domes. For this reason, it has been suggested that the lunar domes have a volcanic origin. (b) Lobate margin of a volcanic flow. (Courtesy of *NASA*.)

contracted in its formative stages, its interior began to heat up through release of gravitational energy and decay of radioactive isotopes. Ultimately, the nickel - iron melted, and the resulting liquid sank toward the center of the earth. The Stillwater Igneous Complex in Montana is 2.75 billion years old (21), and since it exhibits thermoremanent magnetism, the earth probably had a magnetic field at that time. This indicates that the formation of a liquid core occurred more than 2.75 billion years ago. It has been calculated that the earth's core probably formed less than 100 million years after the accretion of the earth, and that the core may have formed simultaneously with accretion (22).

The infall of nickel - iron into the core of the earth would have resulted in a temperature increase of as much as 3000°C within the earth, due to the release of gravitational energy. The resulting temperature increase may have been sufficient to cause partial melting within the mantle. A significant portion of the earth's crust may have been produced by volcanic activity as this melt rose to the surface of the earth. Alternatively, most if not all of the earth's crust may have been added at a relatively uniform rate throughout geologic time. Direct analysis of mantle rock from beneath the oldest continental areas, and perhaps analogies drawn from lunar studies, may eventually resolve the question of the origin of the earth's outer shells.

PLANETARY ATMOSPHERES

The composition of the sun is 86 percent hydrogen, 13.7 percent helium, and less than 1 percent elements heavier than helium. However, the earth and probably the other terrestrial planets as well (Mercury, Venus, and Mars) contain very little hydrogen and helium. This observation led Gerard Kuiper to propose that the earth may originally have had a mass hundreds of times its present mass and that most of the lighter gases were lost at an early stage in the earth's history. However, it is more likely that the gases were lost during the accretion of solid particles to form the earth, as a result of solar radiation and solar wind. The earth's gravitational field is too strong for large amounts of volatile gases to escape into space. The Jovian planets (Saturn, Jupiter, Uranus, and Neptune), being further from the sun and therefore colder, have lost less of their original mass. This explains why the density of the terrestrial planets is much greater than the Jovian planets (Table 8.3). Space probes have revealed that the atmosphere of Venus, which is largely carbon dioxide, may be 75 to 90 times as dense as that of the earth. The atmosphere of Mars, on the other hand, is very rare and very low in free oxygen, and its polar "ice" caps are thought to be only thin, seasonal deposits of carbon dioxide or frozen water.

The Jovian planets have relatively dense atmospheres composed of hydrogen and helium (23). Thus the earth, with its oxygen-rich atmosphere, surface water, and moderate surface temperatures, is indeed a unique planet in the solar system.

Evolution of the Earth's Atmosphere and Hydrosphere

The earth's atmosphere is greatly depleted in hydrogen, helium, and the noble gases (e.g., argon, neon, and xenon) relative to the sun. In the earth today, the ratio of neon to silicon is only 1/10,000,000,000 of that existing in the sun (24). Evidently the earth has lost virtually all its original

Table 8.3 Solar system data

Planets and others	Average distance from sun, millions of kilometers	Equatorial diameter, km	Density, g/cc	Composition of atmosphere	Approximate temperature range, °C
Mercury	58	4,840	5.42	Little if any	−162°-347°
Venus	108	12,200	5.25	CO_2 (90%), H_2O, N_2	467°
Earth	150	12,756	5.51	N_2 (78%), O_2 (21%), CO_2, H_2O	−75-50°
Mars	228	6,760	3.96	CO_2 (80%), H_2O, O_2	
Jupiter	778	142,700	1.33	H_2, He, CH_4, NH_3	−140°
Saturn	1426	121,000	0.68	⎫ probably	
Uranus	2868	47,000	1.60	⎬ the same as	
Neptune	4494	45,000	1.65	⎭ Jupiter	
Pluto	5896	5,000	3??		

atmosphere. Therefore, the present atmosphere must have developed from materials that were subsequently released from the earth's interior through volcanic activity. If the volcanic activity was much greater early in the earth's history, as has been suggested, the release of gases into the atmosphere would have been greater than it is today.

The volcanic gases released very early in the history of the earth were probably rich in hydrogen, methane, ammonia, and water vapor, which would have created a chemically reducing environment. As iron moved from the mantle to the core, the volcanic gases probably became more like modern volcanic gases. In Hawaii these gases are composed principally of water (79%), carbon dioxide (12%), sulfur dioxide (6%), and nitrogen (1%).

The earth's present atmosphere is largely composed of nitrogen and oxygen. The water and sulfur dioxide of the early volcanic gases went into the developing ocean basins. The sulfur dioxide eventually formed sulfuric acid, which reacted with the earth's crust and ocean bottom to form sulfates. Ultimately, free oxygen was released from carbon dioxide by photosynthesis. Since the oldest red beds, which are comprised of hematite (ferric oxide), are about 2.5 billion years old, the atmosphere has probably contained abundant free oxygen for the past 2.5 billion years.

Earth, the Inhabitable Planet

The unique environment of the earth has enabled it alone in the solar system to support life as we know it. Probes of our neighboring planets have indicated that they are probably incapable of sustaining life, and the gaseous Jovian planets are deemed physically unsuitable for this purpose.

Interest in the origin of life on earth gained momentum in the first half of the twentieth century. In 1936 the Russian biochemist A. I. Oparin proposed his theory that the early atmosphere of the earth was reducing and was rich in methane and ammonia (25). Oparin suggested that the first step in development of living organisms

was the formation of organic molecules. According to his theory, under reducing conditions, the organic molecules were not destroyed by oxidation, as they would be in the modern atmosphere. These organic molecules were concentrated in the primitive ocean in a dilute "organic soup." Some of the molecules then combined chemically to form more and more complex molecules. Those molecules which had a favorable composition or favorable internal arrangement acquired new molecules more readily than others. Ultimately, organic molecules capable of reproduction were formed.

In 1953 Stanley L. Miller produced organic molecules experimentally from the components that had been suggested by Oparin as comprising the earth's primitive atmosphere (26). Miller passed an electric spark through a closed chamber filled with a mixture of ammonia, methane, water vapor, and hydrogen. The spark produced the necessary energy to bring about the reactions. In the earth's primitive atmosphere, this energy may have been supplied by lightning or ultraviolet rays. The resulting organic molecules were amino acids, the basic components of proteins.

Philip Abelson, in 1957, added carbon dioxide, carbon monoxide, and nitrogen to Miller's original mixture and was able to produce all the amino acids commonly found in living matter, plus some proteins. It should be remembered, however, that even the simplest forms of life are far more complex than the proteins that have been synthesized thus far, and man is still a long way from creating living molecules in the laboratory.

Although scientists disagree about where and how molecules capable of reproduction were formed, most theorists consider an aqueous environment to be most probable. Since fluids of plants and animals are chemically close to the composition of seawater, a chemically similar body of water would appear to be a likely environment. In a warm lagoonal environment, clay minerals and quartz particles could have concentrated organic molecules through the process of adsorption. However, P. K. Weyl has theorized that organic molecules might have concentrated at a depth of several hundred feet in the open ocean at the top of a thermocline (27). This would have allowed life to develop and evolve within a zone protected from harmful ultraviolet radiation. When this all might have happened has not yet been revealed in the geologic record. However, the oldest rocks containing probable fossils are approximately 3.4 billion years old (see Chapter 9). Therefore, the oceans, atmosphere, and life all had to develop in the first billion years of the earth's existence.

the mantle. As the magma rose to the earth's surface, widespread volcanic activity occurred. Gases released during the eruption of the lavas formed the earth's primitive atmosphere and hydrosphere. Ultimately, the action of lightning, ultraviolet radiation, or both in the presence of the carbon, oxygen, hydrogen, and nitrogen compounds produced organic molecules of increasing complexity, and these were eventually joined into molecules capable of reproducing themselves. Life as we know it evolved in the intervening eras from such self-reproducing molecules.

SUMMARY

In the prevailing theory, the universe originated approximately 18 billion years ago in a gigantic explosion. As the matter in the universe moved away from the center of expansion, some of it coalesced into galaxies. Within the galaxies there were secondary centers of contraction, and these became stars. The first stars were probably composed almost entirely of hydrogen and helium. During this early stage in the history of our galaxy (8 - 9 billion years ago), there were probably large numbers of very massive, rapidly evolving stars in which the heavier elements were generated. At this time, numerous supernovae would have dispersed heavy elements into the interstellar medium as gas and dust.

About 5 billion years ago, a cloud of gas and dust which was to become our solar system began to contract. Most of the material of the cloud went into its center to form the sun. Other centers of contraction became the planets and their moons. The sun was initially heated as a result of the release of gravitational energy, but when the core heated sufficiently, hydrogen began to fuse to form helium.

Almost all the original hydrogen and helium were lost from the earth during or shortly after its accretion. The material that remained was for the most part a mixture of nickel - iron, which sank to form the earth's core, and silicates. The earth's crust was probably produced by partial melting of

REFERENCES CITED

1. G. Abell, 1969, *Exploration of the Universe*: Holt, Rinehart, Winston, New York

2. R. B. Partridge, 1969, The primeval fireball today: *Amer. Sci.*, v. 57, p. 37.

3. W. A. Fowler, 1964, The origin of the elements: *Chem. Eng. News*, v. 42, no. 1, p. 90.

4. G. Fritz et al., 1969, X-Ray pulsar in the Crab Nebula: *Science*, v. 164, p. 709.

5. G. Burbidge, 1967, The quasi-stellar objects: *Amer. Sci.*, v. 55, p. 282.

6. J. H. Oort, 1970, Galaxies and the universe: *Science*, v. 170, p. 1363.

7. R. J. Weymann, 1969, Seyfert galaxies: *Sci. Amer.*, v. 220, no. 18, p. 28.

8. S. A. Colgate, 1969, Quasi-stellar objects and Seyfert galaxies: *Phys. Today*, v. 22, p. 27.

9. A. O. J. Unsold, 1969, Stellar abundances and the origin of the elements: *Science*, v. 163, p. 1015.

10. C. M. Hohenberg, 1969, Radioisotopes and the history of nucleosynthesis in the galaxy: *Science*, v. 166, p. 212.

11. F. Hoyle and N. C. Wickramasinghe, 1970, Dust in supernova explosions: *Nature*, v. 226, p. 62.

12. M. M. Litvak, 1969, Hydroxyl and water masers in protostars: *Science*, v. 165, p. 855.

13. D. U. Wise, 1969, Origin of the moon from the earth: Some new mechanisms and comparisons: *Jr. Geophys. Res.*, v. 74, p. 6034.

14. E. M. Shoemaker et al., 1970, Lunar regolith at Tranquility Base: *Science,* v. 167, p. 452.

15. J. Arnold et al., 1970, Summary of Apollo 11 Lunar Science Conference: *Science,* v. 167, p. 449.

16. D. H. Anderson et al., 1970, Preliminary examination of the lunar samples from Apollo 12: *Science,* v. 167, p. 1325.

    D. A. Papanstassiou and G. J. Wasserburg, 1970, Rb - Sr ages from the Ocean of Storms: *Earth Planet. Sci. Lett.,* v. 8, p. 269.

    D. H. Anderson et al., 1971, Preliminary examination of lunar samples from Apollo 14: *Science,* v. 173, p. 681.

    L. Husain, J. F. Sutter, and O. A. Schaeffer, 1971, Ages of crystalline rocks from Fra Mauro: *Science,* v. 173, p. 1235.

17. S. K. Kanushal and G. W. Wetherill, 1970, Rubidium-87 - strontium-87 age of carbonaceous chondrites: *Jr. Geophys. Res.,* v. 75, p. 463.

18. L. T. Silver, 1970, Uranium - thorium - lead isotope relations in lunar materials: *Science,* v. 167, p. 468.

19. W. Compston et al., 1970, Rubidium - strontium chronology and chemistry of lunar material: *Science,* v. 167, p. 474.

20. V. M. Oversby and P. W. Gast, 1968, Oceanic basalt leads and the age of the earth: *Science,* v. 162, p. 925.

21. P. D. Nunes and G. R. Tilton, 1971, Uranium-lead ages of minerals from the Stillwater Igneous Complex and associated rocks, Montana: *Geol. Soc. Amer. Bull.,* v. 82, p. 2231.

22. V. M. Oversby and A. E. Ringwood, 1971, Time of formation of the earth's core: *Nature,* v. 234, p. 463.

23. T. Owen, 1970, The atmosphere of Jupiter: *Science,* v. 167, p. 1675.

24. H. Brown, 1952, Rare gases and the formation of the earth's atmosphere, *in* G. P. Kuiper, ed., *The Atmospheres of the Earth and Planets:* University of Chicago Press, p. 258.

25. A. I. Oparin, 1962, *Life: Its Nature, Origin, and Development*: Academic Press, New York.

26. S. L. Miller, 1953, A production of amino acids under possible primitive earth conditions: *Science,* v. 117, p. 528.

27. P. K. Weyl, 1968, *in* R. Siever, ed., Report of Environment of the Primitive Earth Congress: *Science,* v. 161, p. 712.

9

The Precambrian

The Precambrian includes all geologic time prior to the beginning of the Cambrian Period. Since the earth is about 4.6 billion years old and the Cambrian began 570 million years ago, the Precambrian represents 87 percent of geologic time. Rocks of Precambrian age, principally gneisses, schists, and granites, are found on all continents and underlie stable areas of low relief called *shields*. The shields are the nuclei of the continents around which the younger rock systems have developed (Fig. 9.1). In the older portions of the shields, the roots of the most ancient mountain systems are exposed. These areas experienced several episodes of sedimentation, mountain building, and erosion during Precambrian time.

Except for algal stromatolites, very few fossils have been found in deposits of Precambrian age. Thus the reconstruction of Precambrian geologic history and geography, including such major structures as geosynclines and mountain belts, has relied heavily on radiometric ages. Before the widespread use of radiometric dating, the degree of metamorphism served as the basis for subdividing the Precambrian. Intensely metamorphosed rocks were thought to be the oldest units and were assigned to the Archean or Archeozoic Era, whereas rocks of lower metamorphic grade were placed in the Proterozoic Era. In some instances, however, radiometric dating has demonstrated that slightly metamorphosed rocks are older than more intensely metamorphosed rocks. Thus intensity of metamorphism has proven to be an unsatisfactory method of subdividing the Precambrian.

Uranium - lead, potassium - argon, and rubidium - strontium dating of basement rocks has indicated that several major orogenic episodes occurred at approximately the same time on all continents (1). The frequency distribution of age determinations shows peaks of metamorphism and intrusion at 1.0, 1.7, and 2.5 billion years ago (Fig. 9.2). Using the ages of these orogenies as boundaries, the Precambrian may be subdivided into four eras which are, from oldest to youngest, the Archean, the Early Proterozoic, the Middle Proterozoic, and the Late Proterozoic.

PRECAMBRIAN GEOGRAPHY

Paleomagnetic, structural, and geochronologic data suggest that five separate continents existed during the Proterozoic Era. These are designated here as Ancestral North America, Ancestral Europe, Gondwanaland, Ancestral Siberia, and Ancestral China (Fig. 9.3). Ancestral North America includes most of the present continent of North America except for parts of the southeastern United States and Mexico, as well as the northwestern British Isles, the western coast of Norway, Spitsbergen, and part of Siberia. Most of central and northern Europe are included in Ancestral Europe, and Ancestral Siberia is comprised of most of Siberia, northernmost China, and Outer Mongolia. Ancestral China includes most of China, Japan, North Korea, South Korea, Southeast Asia, and the bordering islands, as well as portions of Siberia, India, Pakistan, and the Middle East. Finally,

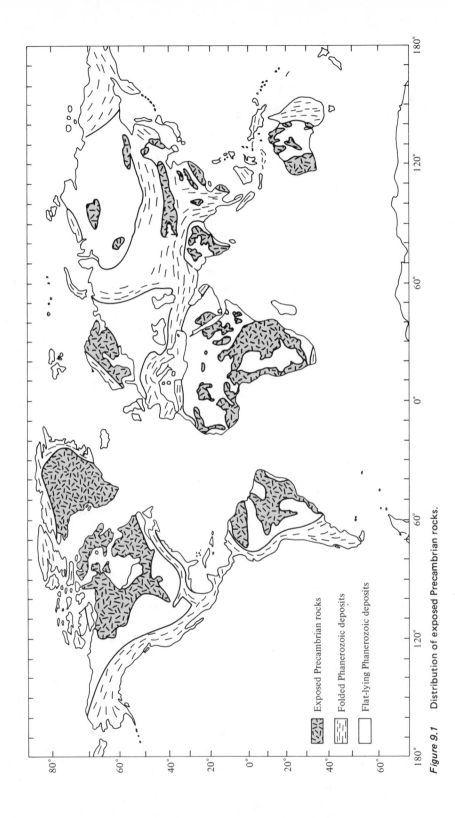

Figure 9.1   Distribution of exposed Precambrian rocks.

144

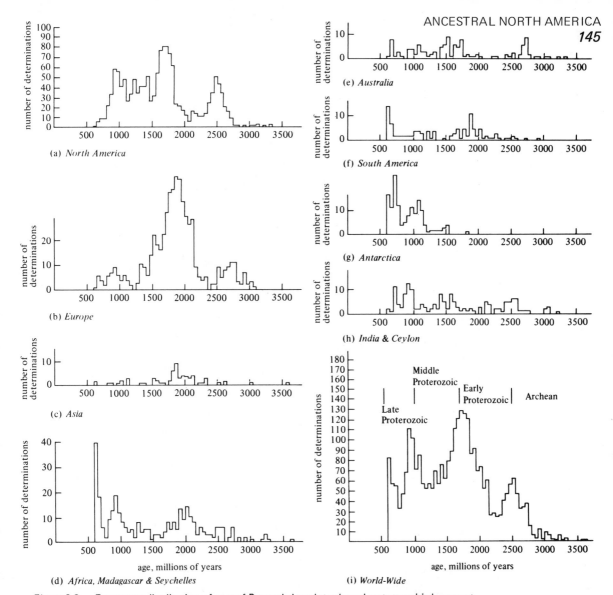

Figure 9.2 Frequency distribution of ages of Precambrian plutonic and metamorphic basement rocks: (a) North America; (b) Europe; (c) Asia; (d) Africa, Madagascar, and Seychelles; (e) Australia; (f) South America; (g) Antarctica; (h) India and Ceylon; (i) worldwide. Notice the smaller vertical scale for the North American and worldwide data.

Gondwanaland consists of Africa, South America, Antarctica, Australia, most of the Middle East and India, and segments of southern Europe, the southeastern United States, Mexico, and Central America. Each of the five continents increased in area and volume during the Precambrian through the addition of mantle - derived volcanic rocks.

ANCESTRAL NORTH AMERICA

Precambrian rocks are exposed in the Canadian Shield, northern, central, and western areas of the United States, the Appalachian fold belt, western Canada, Greenland, northern Scotland, and Spits-bergen. Rocks of this age have also been penetrated in oil wells in central United States and western

(a)

Figure 9.3 Ancestral continents: (a) Ancestral North America, (b) Ancestral Europe, (c) Ancestral Siberia, (d) Ancestral China, (e) Gondwanaland.

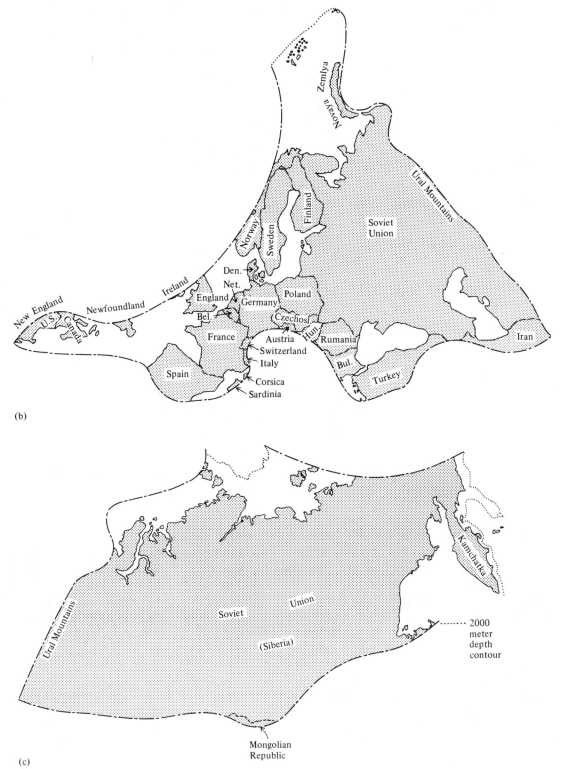

(b)

(c)

2000
meter
depth
contour

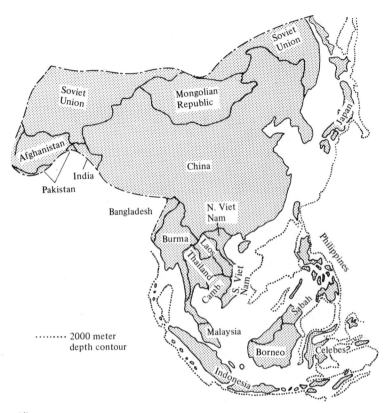

(d)

Canada, where they are covered by a relatively thin sequence of Phanerozoic sedimentary rocks.

The Canadian Shield, which covers most of eastern and central Canada, is one of the largest and most extensively studied areas of Precambrian rocks in the world (Fig. 9.2). Where Pleistocene glaciation has removed the overlying soil, the rocks of the shield are exceptionally well exposed (Fig. 9.4). Numerous radiometric dates on surface and subsurface Precambrian rocks in Ancestral North America indicate a general decrease in the age of these samples with increasing distance from the center of the continent (Fig. 9.5). This pattern shows that Ancestral North America grew by a process of accretion during the Precambrian. However, the possibilities that some old basement has been incorporated into younger mountain belts or that it has been covered by younger deposits make it difficult to estimate the rate of continental growth.

Geologic Provinces

Potassium - argon dating (2) and structural studies indicate that the Canadian Shield may be divided into a number of distinct geologic provinces in which the rocks were last deformed, metamorphosed, and intruded at approximately the same time (Fig. 9.5). Potassium - argon dates determined on basement rocks in the United States and western Canada indicate that the rest of Ancestral North America, as well, may be divided into geologic provinces (Fig. 9.6).

The oldest rocks in Ancestral North America occur in the Superior, Slave, eastern Nain, and Wyoming provinces. These rocks were last strongly deformed, metamorphosed, and intruded approxi-

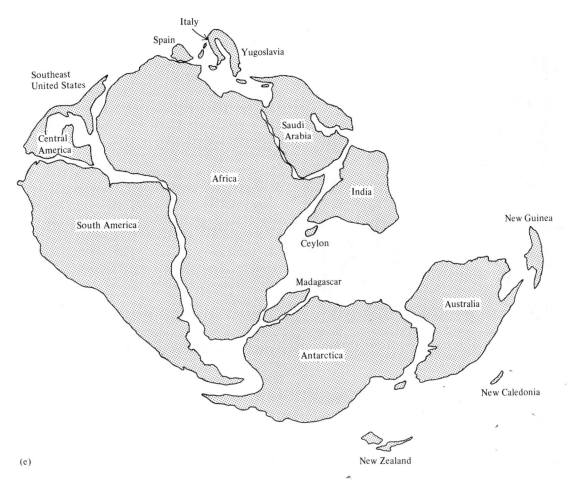

(e)

mately 2.5 billion years ago, and the provinces have been relatively stable since that time. The Slave, eastern Nain, and Wyoming provinces are separated from the Superior Province by belts of younger deformation. All these provinces may have been part of a single, large landmass at the end of the Archean. Alternatively, they may have been part of four relatively small continents that were joined together at a later date during an episode of relative movement of crustal plates.

The Churchill Province nearly encircles the Superior, Slave, and Wyoming provinces. Rocks of the Churchill provinces were last deformed, metamorphosed, and intruded approximately 1.8 billion years ago. The last strong tectonic activity in the western Nain Province occurred between 1.3

and 1.5 billion years ago. Rocks with similar potassium - argon ages are also found in the Central Province and in northern Scotland (3) (Figs. 9.5 and 9.6).

The Grenville Province is the youngest Precambrian province in Ancestral North America. It cuts across the eastern borders of the Central, Churchill, Superior, and Nain provinces. Rocks of the Grenville Province were last deformed, metamorphosed, and intruded approximately 1.0 billion years ago. Whole-rock rubidium - strontium and uranium - lead dates of samples from the Grenville Province in southeastern Ontario indicate that the first intrusive and metamorphic episode in this province occurred approximately 2.4 billion years ago, and a second orogenic event occurred

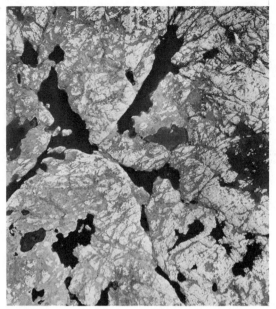

Figure 9.4   Precambrian rocks of the Canadian Shield exposed by the action of Pleistocene glaciers. (Courtesy of the National Air Photo Library, Dept. of Energy, Mines and Resources (Canada)).

1.8 billion years ago (4). Furthermore, folded iron formation of the Churchill Province in Labrador has been traced into the Grenville Province, where it has been subjected to at least one additional deformation. Iron formation is a sedimentary rock containing a significant percentage of iron-rich minerals, such as hematite, associated with chert. Thus parts of the Superior and Churchill Provinces have been incorporated into the Grenville Province during a subsequent deformation.

The Archean Era
(more than 2.5 billion years ago)

Archean rocks occur in the Rocky Mountains, in the Superior, Slave, and eastern Nain provinces, and in northern Scotland. These rocks include an older, highly metamorphosed sequence and a younger, less metamorphosed sequence. The chemical compositions of the rocks and their primary structures indicate that they were sandstones, shales, iron formation, and volcanic rocks before metamorphism.

The oldest known rocks in Ancestral North America are a sequence of amphibolites and felsic gneisses in West Greenland. These rocks have been dated at 3.98 billion years by the whole-rock rubidium - strontium method (5). Metamorphosed sedimentary and volcanic rocks which are exposed in the Minnesota River valley in southwestern Minnesota are also very ancient. Uranium - lead concordia and whole-rock rubidium - strontium dating of granitic gneisses that cut these rocks indicate that they were intruded approximately 3.55 billion years ago (6). Gneisses, schists, and amphibolites of the Beartooth Mountains of Wyoming and the Bighorn Mountains of Montana are almost as old. Uranium - lead and whole-rock rubidium - strontium dates on these rocks indicate that they were metamorphosed between 3.2 and 3.4 billion years ago (7).

Similar rocks are found in the Superior Province of the Canadian Shield. Many of the gneisses are paragneisses, formed by the metamorphism of layered sedimentary rocks (Fig. 9.7). These paragneisses are chemically similar to graywacke, the probable parent rock. In some areas, the paragneisses grade into granitic rocks which contain a faint layering. The grain size and the composition of these granites are variable in contrast to the relative uniformity found in most granite batholiths. It is likely that such granites were formed from paragneisses by granitization, a process involving recrystallization of the parent rock and addition of silica, sodium, and potassium. Other Archean gneisses contain discontinuous lenses of biotite, hornblende, quartz, and feldspar. These are probably orthogneisses formed by the metamorphism of granites or related rocks.

A thick sequence of low-grade metamorphic rock overlies the more intensely metamorphosed gneisses, schists, and amphibolites of the Superior Province. The contrast in metamorphic grade suggests that a major unconformity or fault separates the two sequences. North of Lake Superior, the low-grade metasedimentary and metavolcanic rocks have been subdivided into the Keewatin and Coutchiching sequences. Keewatin

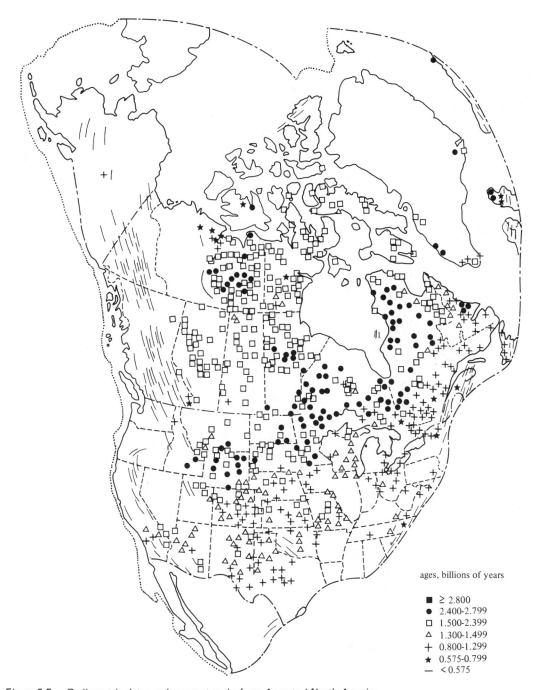

ages, billions of years

- ■ ≥ 2.800
- ● 2.400-2.799
- □ 1.500-2.399
- △ 1.300-1.499
- + 0.800-1.299
- ★ 0.575-0.799
- — < 0.575

Figure 9.5   Radiometric dates on basement rocks from Ancestral North America.

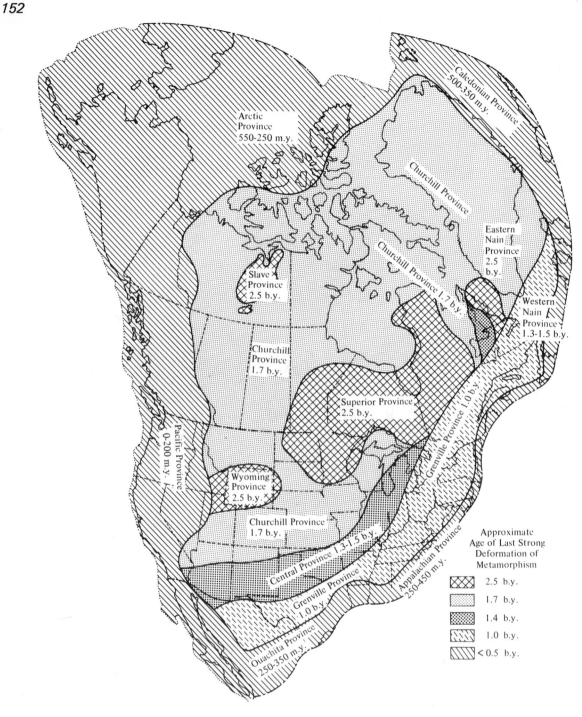

Figure 9.6   Geologic provinces of Ancestral North America.

Figure 9.7   Coarse-grained paragneiss from the Superior province near Nipigon, Ontario, Canada. The strongly layered character of the gneiss suggests that it was formed by the metamorphism of a sedimentary rock.

metavolcanic rocks are dominantly basalts and andesites that have been metamorphosed to greenstones.

Pillow structures, which are common in the greenstones, suggest submarine eruption of the lava (Fig. 9.8). Rubidium - strontium dating of these lavas indicates that they were erupted approximately 2.77 billion years ago (8). The Coutchiching Sequence consists of metamorphosed shale, graywacke, and iron formation. These deposits were derived from the weathering of adjacent basement uplifts and volcanic highlands. Uranium - lead dates on detrital zircons eroded from these areas demonstrate that at least part of the Coutchiching was deposited less than 2.75 billion years ago (8).

The Keewatin and Coutchiching were folded and intruded prior to the deposition of the overlying Timiskaming Sequence. A uranium - lead concordia plot from some of these granites gives an age of 2.73 billion years for their emplacement (8). The Timiskaming, which includes metaconglomerate and ripple-marked quartzites (Fig. 9.9), was folded, metamorphosed, and intruded approximately 2.55 billion years ago (9). Thus there were at least two and probably three phases of deformation during the Archean Era. The last of these disturbances affected much of Ancestral North America and is known as the Kenoran Orogeny in Canada.

The Scourian Sequence of northern Scotland is composed of metamorphosed sedimentary and

Figure 9.8   Pillow greenstone from the Superior province near Terrace Bay, Ontario, Canada. The outcrop has been striated and polished by the action of Pleistocene glaciers. The shape of the pillows indicates that the younger beds are on the left.

mafic volcanic rocks which were deformed and metamorphosed approximately 2.5 billion years ago (3). Northern Scotland may have been part of the continent of Ancestral North America at the end of the Archean Era.

The thickness and character of the Archean sedimentary and volcanic rocks of Ancestral North America would seem to indicate deposition in an island arc environment (10). However, these deposits occur over a very large area, rather than in a narrow belt typical of a single island arc (Fig. 9.10). Possibly several island arcs were joined together as a result of sea-floor spreading at the end of the Archean Era (Fig. 9.11). A modern analog would be the island arcs in the western part of the Pacific Ocean. One island arc extends southward from Japan through the Mariana Islands to Helen Island. A second arc extends southwestward from Japan through Taiwan to the Philippine Islands. Between the active island arcs, there are two linear chains of islands and seamounts that appear to be inactive island arcs (Fig. 9.12). Since

deep trenches and Benioff zones are associated with the active island arcs, presumably the sea-floor is moving under the arcs bringing them together. It is possible that the continent of Ancestral North America was formed in a similar manner.

Early Proterozoic Era
(2.5 - 1.7 billion years ago)

Deposits of Early Proterozoic age are found in many parts of the Canadian Shield, the Rocky Mountains, southwestern United States, Greenland, and northern Scotland. In the vicinity of Lake Superior and Lake Huron, Archean rocks are unconformably overlain by the Early Proterozoic Huronian Series. The Bruce Group, which is the oldest unit within the Huronian Series, consists predominantly of quartzite derived from sandstone of shallow-water origin. The Cobalt Group, which disconformably overlies the Bruce, contains a tillite with striated pebbles (Fig. 9.13). Striated pavement exposed below the tillite indicates ice

(a)

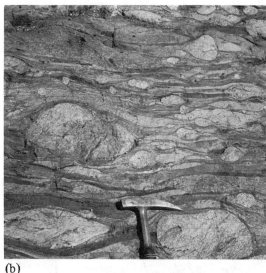

(b)

Figure 9.9   Timiskaming deposits in the Superior province of southern Ontario, Canada: (a) Ripple marks on steeply dipping quartzite near Bruce Mines. The variation of the ripple marks in different beds is indicative of a change in current direction. (b) Conglomerate in which the clasts have been deformed during the Kenoran Orogeny. Photographed along the Trans-Canada Highway near Wawa, Ontario.

movement toward the southwest (11). Poorly sorted conglomeratic deposits in northern Quebec, the northern peninsula of Michigan, southeastern Wyoming, and northern Utah may be glacial deposits laid down at the same time as the tillites of the Cobalt Group (12).

The Bruce and Cobalt groups were folded and intruded by mafic dikes and sills prior to the deposition of the overlying Anamikie Group. Samples from one of these intrusions have been dated by the whole-rock rubidium - strontium method at 2.16 billion years (13).

The Anamikie Group contains quartzite, silt-stone, limestone, iron formation, and minor amounts of coal. The Early Proterozoic deposits of iron formation in the Mesabi and Cuyuna districts of Minnesota and in the Gogebic, Marquette, and Menominee districts of Michigan are among the most important sources of iron ore in North America.

North of Lake Superior, the Huronian Series is only a few thousand feet thick, nearly horizontal, and almost unmetamorphosed. Near Lake Huron, it thickens to several tens of thousands of

feet and is intensely folded and metamorphosed. This marked thickening suggests a transition from platform to geosynclinal deposits. Most of the clastic deposits of the Huronian Series were derived from the Superior Province, but graywacke in the outer part of the geosyncline was supplied by volcanic and tectonic islands. These islands may have been part of an island arc bordering the continent. In this case, movement of the sea-floor beneath the island arc would have resulted in growth of the continent, as sedimentary and volcanic rocks that had been deposited on oceanic crust were carried beneath the island arc.

Thick sequences of Early Proterozoic volcanic and sedimentary rocks containing iron formation are found in a wide belt extending from the Great Lakes northeastward to Labrador, as well as in the Belcher Islands, in the Northwest Territories, and south of Hudson Bay (Fig. 9.14). Similar sequences of Early Proterozic age, but without iron formation, are found in several areas between southern Wisconsin and southern California.

Early Proterozoic deposits generally increase in thickness away from the Superior and Wyoming

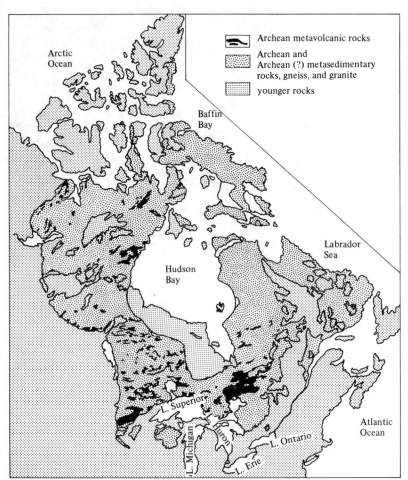

Figure 9.10   Geologic map showing the distribution of metavolcanic rocks in Archean sequences of the Canadian Shield. (From The Geological Map of Canada, Geological Survey of Canada, Ottawa.)

provinces and in many areas they are more than 15,000 ft (5000 m) thick (Fig. 9.14). The thickness and distribution of these deposits indicate that a narrow platform, and a wide geosyncline may have bordered the Superior and Wyoming provinces during the Early Proterozoic.

A major episode of folding, metamorphism, and intrusion known as the Huronian Orogeny occurred at the end of the Early Proterozoic (Fig. 9.15). This event created a mountain range approximately coincident with the Churchill Province. Whole-rock rubidium - strontium dating of

the Anamikie Group indicates that the metamorphism occurred 1.75 billion years ago (14).

Middle Proterozoic Era
(1.7 - 1.0 billion years ago)

In many areas, Early Proterozoic deposits are unconformably overlain by Middle Proterozoic sedimentary and volcanic rocks. These deposits vary from thin platform deposits, such as the Sioux Quartzite of South Dakota (15), to geosynclinal deposits tens of thousands of feet thick. The Belt Series of northern Idaho, western Montana, and southern British Columbia is a 35,000-ft (11,000-m) thick sequence of limestone, sandstone, and minor volcanic rocks. The basement on which the Belt Series rests was metamorphosed 1.7 billion years ago. Near Coeur d'Alene, Idaho, the Belt Series is intruded by uranium-bearing veins dated at between 1.0 and 1.2 billion years (16). Samples of volcanic rocks, shale, and glauconitic sandstone from the Belt have been dated at approximately 1.1 billion years (16). Evidently, deposition of the Belt Series occurred entirely during the Middle Proterozoic era. Except in the vicinity of granitic intrusions, these deposits have not been metamorphosed, but they were broadly folded prior to the deposition of the overlying Cambrian strata. In Glacier National Park, Pleisto-

cene glaciers have carved spectacular cliffs and pinnacles into the Belt Series (Fig. 9.16).

The Lower Purcell System of the Canadian Rockies is 50,000 ft (15,000 m) thick and is correlated with the Belt Series. These and similar deposits occur in a north - south trending zone extending from Arizona to northern Canada. The thickness and lithology of these sequences would indicate that they were laid down in a miogeosyncline.

During the early part of the Middle Proterozoic, between 25,000 and 50,000 ft (7500 and 15,000 m) of sandstone, shale, and volcanic rocks was deposited in northern Arizona. These deposits include the Vishnu Schist of the Grand Canyon region. The Vishnu, which rests on a gneissic

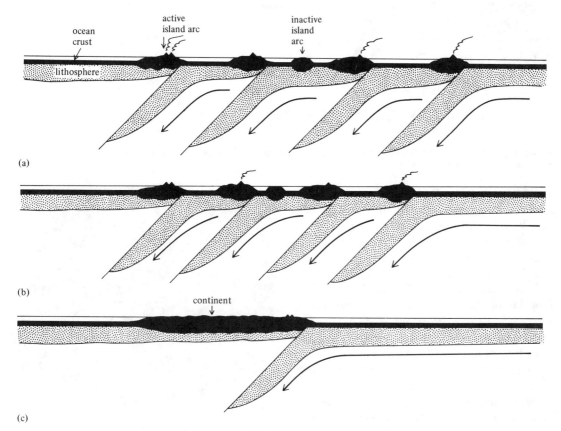

Figure 9.11 Diagrammatic sketch illustrating a possible method of formation of a continent. As the sea floor moves under several adjacent island arcs, they are brought together, forming a single continent.

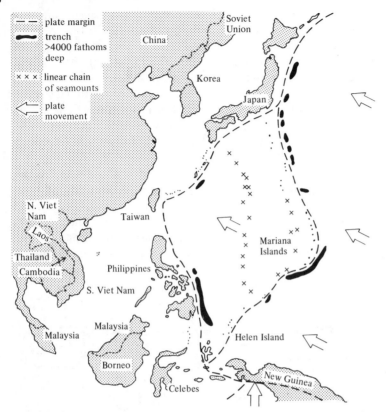

Figure 9.12 Modern island arcs in the western Pacific. The island arcs are moving toward one another as the sea floor descends beneath the arcs.

basement dated at 1.7 billion years, was folded, metamorphosed, and intruded for the first time 1.54 billion years ago; the process occurred again 1.3 billion years ago (17). The nature of the Vishnu Schist and correlative formations suggests deposition in a eugeosyncline (Fig. 9.16).

The Grand Canyon Series, which rests unconformably on the Vishnu Schist, is approximately 10,000 ft (3000 m) thick and contains abundant carbonates and few volcanics (Fig. 9.17). This series was intruded by diabase about 1.145 billion years ago (18). The lithology of the Grand Canyon Series seems to indicate a change from eugeosynclinal to miogeosynclinal conditions during the Middle Proterozoic. Such a change might have resulted from an outward growth of the continent during the Middle Proterozoic.

Figure 9.13 Early Proterozoic boulders from the Gowganda Tillite near Cobalt, Ontario show glacially produced striations. (Specimen collected by Dr. M.E. Wilson, photo courtesy of the Geological Survey of Canada, Ottawa.)

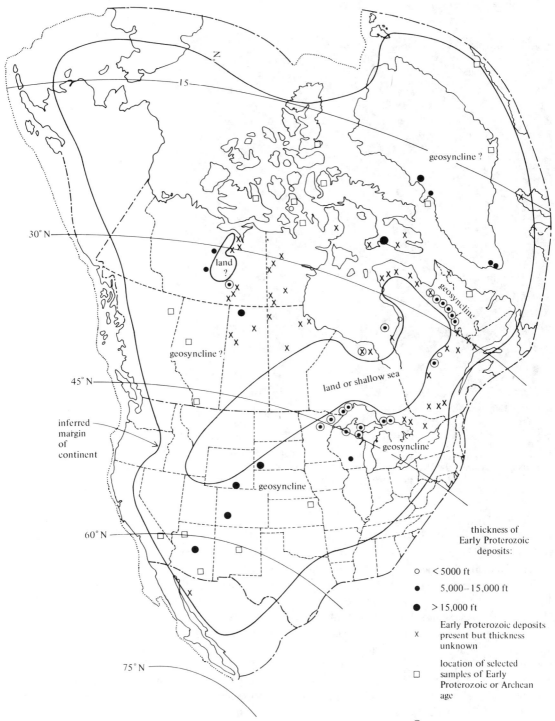

Figure 9.14 Paleogeographic map of Ancestral North America during the Early Proterozoic time, approximately 2.0 billion years ago, during the period of deposition of the Anamikie Group and related iron formations.

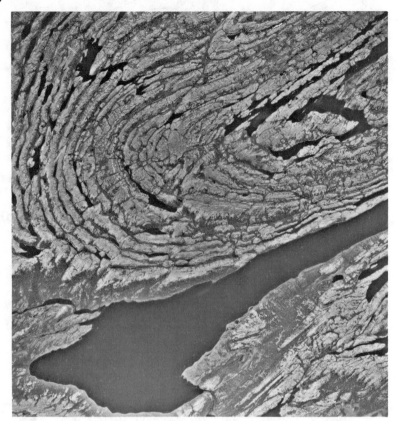

Figure 9.15 Aerial view of a plunging fold in eastern Quebec, Canada. These Early Proterozoic rocks were deformed during the Hudsonian Orogeny between 1.7 and 1.8 billion years ago. (Courtesy of the National Air Photo Library, Dept. of Energy, Mines and Resources (Canada)).

Tens of thousands of feet of metamorphosed carbonate, clastic, and volcanic rocks comprise the Grenville and Hastings series in the northern part of the Grenville Province. Volcanics interbedded with these deposits have been dated at between 1.2 and 1.3 billion years (19). Thick sequences of Middle Proterozoic sedimentary and volcanic rocks are also found in the states of Texas, Oklahoma, Missouri, and New York. Although these deposits are not continuous, they may have been part of a eugeosyncline extending along the southeastern border of the continent (Fig. 9.17). The Torridonian Series of northern Scotland is also a thick sequence of clastic sediments. This series unconformably overlies basement rocks that were meta-

morphosed 1.3 billion years ago (3) and may have been deposited in the same geosyncline as the Grenville Series and related sequences.

As much as 50,000 ft of shale, sandstone, conglomerate, and basaltic lava were deposited in the vicinity of Lake Superior during the Middle Proterozoic. These deposits are part of the Keweenawan Group and their areal distribution indicates that they were laid down in a basin whose center is located near the center of Lake Superior. Some of the lava flows and associated sedimentary rocks contain commercial quantities of native copper. These deposits were mined by the Indians long before the white men came to that part of the country. Rubidium - strontium dating of felsic volcanics from the Middle Keween-awan Sequence indicates that they were extruded approximately 1100 million years ago (20). This is about the same age as the Duluth Gabbro, a very large body of mafic intrusive rock exposed on the western end of Lake Superior (21).

The Grenville Orogeny ended deposition in the geosynclines that bordered Ancestral North America during the Middle Proterozoic. Potassium - argon and rubidium - strontium ages indicate that metamorphism occurred between 1.2 and 0.95 billion years ago (22). The presence of re-folded folds within the Middle Proterozoic

sequences shows that the Grenville Orogeny encompassed more than one episode of deformation. Large recumbent folds, the coarse texture, and the mineralogy of the rocks attest to the intensity of the deformation. Relatively unde-formed granitic intrusions in the Grenville Province have been dated at approximately 1.0 billion years. Presumably they were emplaced during the later stages of the orogeny (Fig. 9.18). The area of rocks deformed during the Grenville Orogeny is 3500 miles (5000 km) long and as much as 600 miles (1000 km) wide. This orogeny was one of the most intense ever in North America.

Figure 9.16 Gently dipping sedimentary rocks of the Belt Sequence of Middle Proterozoic age in Glacier National Park, Montana. (Courtesy of the Montana Highway Commission.)

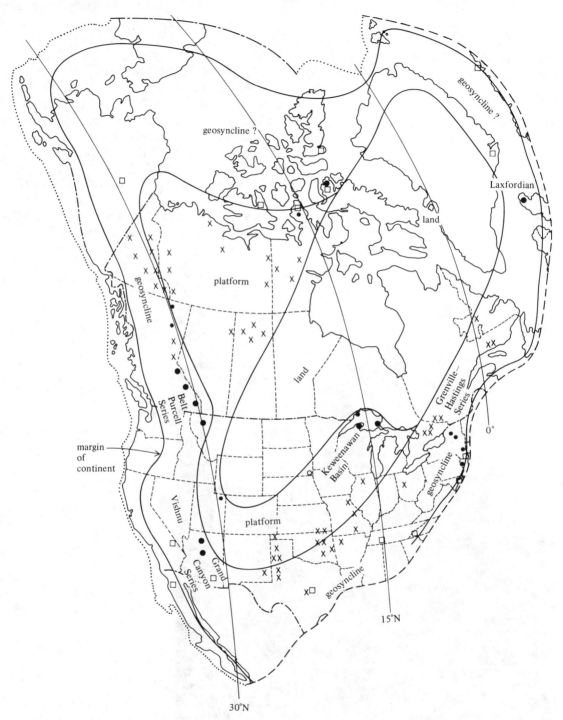

Figure 9.17 Paleogeographic map of Ancestral North America during the Middle Proterozoic approximately 1.3 billion years ago.

Figure 9.18   Granitic dikes of different ages in the Grenville Province near Parry Sound, Ontario, Canada.

Late Proterozoic Era
(1000 - 570 million years ago)

Sedimentary and volcanic rocks of Late Proterozoic age are found in both the Appalachian and
Cordilleran regions beneath thick sequences of
Paleozoic deposits. The Mount Rogers Group and
the Ocoee Series of the southern Appalachians
contain metavolcanics, conglomerate, sandstone,
and shale that have been subjected to varying
degrees of metamorphism. They were deposited on
a basement that was metamorphosed approximately 1.0 billion years ago during the Grenville
Orogeny. Rhyolites within the Mount Rogers Group
range in age from 850 to 950 million years (23).
This would place their time of deposition in the
early part of the Late Proterozoic.

In eastern Greenland and Spitsbergen, a thick
sequence of unmetamorphosed Precambrian sedimentary rocks of Middle and/or Late Proterozoic
age contains two beds of unsorted sedimentary
rocks which may be tillites (Figs. 9.19 and 9.20).
Striated boulders have been reported in these
deposits, but the deposits do not rest on a striated
pavement. The "tillites" occur 1200 to 2400 ft

(400- 800 m) below Cambrian beds (25). Proterozoic boulder beds from Scotland and Ireland may
correlate with these "tillites." Unfossiliferous
sequences of rocks in southern California and in
the Taconic Mountains of New York grade upward
into beds containing Cambrian fossils. These
deposits are sometimes referred to as Eocambrian
since they are apparently conformable with Cambrian strata.

ANCESTRAL EUROPE

The Precambrian history of Ancestral Europe is
similar to that of Ancestral North America. Rocks
of this age are exposed in the Baltic Shield, the
Ukrainian Shield, in southern Europe, and in a
belt extending from eastern Newfoundland to
eastern Massachusetts (Fig. 10.26). Throughout
most of eastern and central Europe, Precambrian
rocks are covered by flat-lying sedimentary rocks
of Paleozoic age.

Archean Era

The oldest known rocks in Ancestral Europe are
gneisses which crop out in the northeastern part of

Figure 9.19   Late Proterozoic deposits on the eastern margin of Ancestral North America include the thick sedimentary rocks of the Agardhsbjerg Formation in East Greenland. (Courtesy of H.R. Katz.)

the Baltic Shield and in the Ukrainian Shield (Fig. 9.21). Radiometric dates show that these rocks were metamorphosed between 3.0 and 3.1 billion years ago (27). Structural trends indicate that the two shields may be part of the same continental nucleus.

These ancient rocks are overlain unconformably by a sequence of metasedimentary and metavolcanic rocks that were deformed, metamorphosed, and intruded approximately 2.6 billion years ago (28). The chemical compositions and textures of these gneisses, schists, quartzites, and amphibolites suggest that they were probably shale, sandstone, graywackes, and volcanics before metamorphism. Thus the sequence would be eugeosynclinal in character.

Early Proterozoic Era

Early Proterozoic sedimentary and volcanic rocks crop out in the Baltic and Ukrainian shields and

are encountered in deep drilling in eastern Europe. Deposits of miogeosynclinal character include quartzite, dolomite marbles, iron formation, and mica schists that are more than 15,000 ft (5,000 m) thick (Fig. 9.22). These deposits occur discontinuously in a belt extending from northern Norway to the eastern Ukraine (Fig. 9.23). To the west is a belt containing tens of thousands of feet of metamorphosed Early Proterozoic shale and volcanics with few limestones and quartzites. These deposits are eugeosynclinal. Both eugeosynclinal and miogeosynclinal deposits rest on an Archean basement. The eugeosynclinal sequences may have been deposited in the vicinity of an island arc, but they do not appear to have been situated on the margin of the continent during the Early Proterozoic (Fig. 9.23).

The rocks in much of the area covered by the Baltic and Ukrainian shields were metamorphosed, deformed, and intruded at the end of the Early Proterozoic, 1.7 billion years ago (29). The moun-

GONDWANALAND
165

tain range that resulted from this deformation probably extended in a wide band from the Arctic Ocean to the Black Sea. It is possible that the orogenesis at the end of the Early Proterozoic may have been caused by an eastward movement of the sea-floor beneath the Baltic Shield. A general westward decrease in age within the Baltic and Ukrainian shields (Fig. 9.21) suggests a growth of the continent in that direction.

Middle Proterozoic Era
In Scandinavia, Middle Proterozoic quartzite, conglomerate, slate, and felsic volcanics rest unconformably on Early Proterozoic rocks. The thickness of these deposits generally increases toward the margins of the Baltic Shield. The deposits in the western part of the shield were deformed and metamorphosed 1.3 billion years ago (30). In southern Norway, still younger arkoses, carbonates, and volcanics were deformed and metamorphosed about 1.0 billion years ago (30). This deformation may have produced an extensive mountain range on the northwestern margin of Ancestral Europe at the end of the Middle Proterozoic.

Late Proterozoic Era
Sedimentary sequences with two interbedded "tillites" underlie Early Cambrian beds along the western margin of Scandinavia. These deposits rest unconformably on rocks dated at 1.0 billion years and are therefore Late Proterozoic in age. The "tillites" may be traced over 900 miles (1500 km) (25). Late Proterozoic deposits also occur in eastern Newfoundland, Nova Scotia, Maine, and eastern Massachusetts—all areas that are inferred to have been part of Ancestral Europe. These deposits consist of up to 15,000 ft of sedimentary and volcanic rocks which may have been laid down in the vicinity of an island arc on the margin of Ancestral Europe. They were deformed and intruded at the very end of the Late Proterozoic, approximately 575 million years ago (31).

GONDWANALAND
Precambrian rocks are exposed on all the continents that comprised Gondwanaland (Fig. 10.28). Rocks of this age are also present beneath a cover of Phanerozoic rocks in northern Florida, which

Figure 9.20 The Seveanor Tillite in Spitsbergen indicates glacial activity on the eastern margin of Ancestral North America at the end of the Late Proterozoic Era. (Photo by A. Hjelle, courtesy of S. Winsnes, Norsk Polarinstitutt, Ref. 24.)

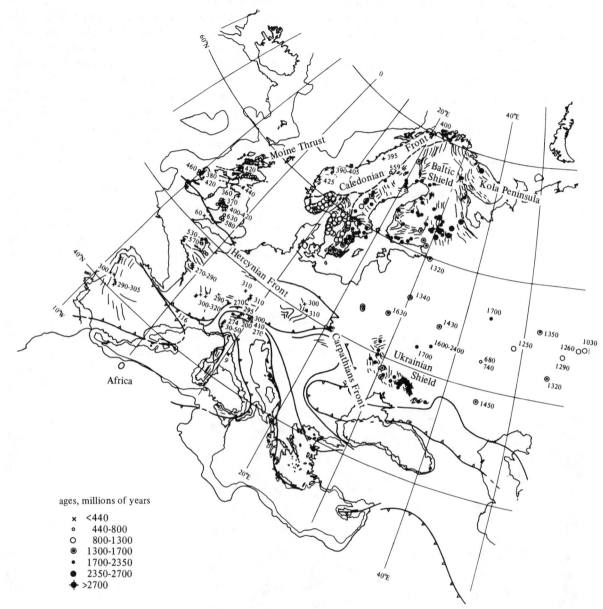

Figure 9.21   Radiometric dates for Europe. (From Hurley and Rand, Ref. 26. Copyright © 1969 by American Association for the Advancement of Science.)

was presumably part of Gondwanaland during the Precambrian.

Archean Era

The oldest known rocks in Gondwanaland are intensely metamorphosed gneisses, migmatites, and granulites, which crop out in southern Africa. Mica from a pegmatite cutting these deposits has been dated at 3.44 billion years (32). The oldest rocks are overlain by several Archean

(a)

(b)

Figure 9.22   Early Proterozoic rocks of the Karelian sequence in northern Norway: (a) Steeply dipping, ripple-marked quartzite, (b) pillow lavas. (Courtesy of Paul Reitan.)

sequences, separated from each other by unconformities (Table 9.1). The youngest Archean deposits are relatively undeformed and have been dated at as much as 3.0 billion years (33). Apparently the southern part of Africa has been stable since well before the end of the Archean (Fig. 9.24).

Gneisses in India (Fig. 9.25), Australia (Fig. 9.26), and Antarctica have been dated at more than 3.0 billion years (34). The ancient gneisses in India, Australia, and southern Africa are overlain by sequences containing greenstone and metasedimentary rocks that were deformed, metamorphosed, and intruded between 2.7 and 2.5 billion years ago (34). Similar radiometric ages are common throughout Gondwanaland. Evidently much of the continent was metamorphosed and intruded at the end of the Archean. At that time, large parts of the continent were transformed into mountain ranges.

Table 9.1   Precambrian systems in South Africa

System	Lithology	Age
Waterberg	Quartzites, shales, conglomerates, and volcanics	>1.42 b.y., <1.8
Loskop	Quartzites, shales, and conglomerates	>1.79
Transvaal	Dolomitic limestones, quartzites, and shales with mafic to felsic volcanic rocks and a thin tillite	A little greater than 1.95 b.y.
Ventersdorp	Mainly andesitic lavas	∿2.1
Witwatersrand	Quartzites, shales, and conglomerates with minor volcanics	2.15 b.y. or older
Dominion Reef	Lavas and quartzites	3.0 b.y.
Moodies System	Folded and metamorphosed conglomerate, arkose, quartzite, and graywacke	>3.0 b.y.

A plot of radiometric dates for Gondwanaland indicates a general decrease in age toward the east in Australia, toward the south in southern Africa, toward the west in South America, and toward the Pacific Ocean in Antarctica (Figs. 9.27 and 9.28). The general pattern is therefore one of decreasing age outward from the center of Gondwanaland. It is likely that this pattern resulted from the growth of the continent by accretion as the sea-floor moved under the margins of Gondwanaland.

Early Proterozoic Era

Early Proterozoic sedimentary and volcanic rocks are widely exposed in Gondwanaland. In central and western Australia, volcanics in a thick sequence of dolomite, sandstone, iron formation, chert, shale, lava, and tuff have been dated at between 2.0 and 2.2 billion years (36). Similar sequences occur in India, South Africa, and South America. Those in India have been dated at between 1.8 and 2.1 billion years, whereas those in southern Africa are between 1.95 and 2.3 billion years old (37). Early Proterozoic sequences in southern Africa include a tillite with striated pebbles which is exposed in an area of 11,500 square miles (32,000 km^2) (25).

Early Proterozoic deposits reach maximum thicknesses of 20,000 to 40,000 ft (6000 - 12,000 m). The distribution and thickness of these rocks would indicate that a geosyncline bordered Gond-wanaland at that time. Thinner platform deposits are widespread in the interior of the continent, although some areas in which Archean rocks are now exposed were probably above sea level and supplying sediments to the platform and geosyncline during the Early Proterozoic.

Whole-rock rubidium - strontium dating has indicated that widespread metamorphism took place 2.0 billion years ago in northeastern Brazil and western Africa (38). A somewhat younger episode of mountain building occurred near the margins of Gondwanaland at the end of the Early Proterozoic, approximately 1.7 billion years ago.

Middle Proterozoic Era

Thick sequences of Middle Proterozoic sedimentary and volcanic rocks occur in northern India, eastern Australia, southern Africa, northern Africa, Antarctica, and western South America. Sedimentary rocks of miogeosynclinal character grade into eugeosynclinal deposits near the margin of Gondwanaland. These rock units were metamorphosed and deformed approximately 1.0 billion years ago. The intensity of deformation and the grade of metamorphism increases toward the margin of the continent.

Late Proterozoic Era

Tillites are common within the Late Proterozoic deposits of Gondwanaland. In the Congo, two

tillite beds containing striated pebbles are interbedded with a relatively thin sequence of sedimentary and volcanic rocks (Fig. 9.29). The additional
presence of stromatolitic bioherms within this sequence (Fig. 9.30) is evidence of rapid climatic
fluctuations during the Late Proterozoic in central
Africa. These Late Proterozoic deposits were broadly folded, faulted, and intruded approximately 625
million years ago. The uranium mineralization of
the Katanga region also occurred at this time (40).

Tillites are found in South Australia, New
South Wales, Western Australia, the Northern
Territories, and Tasmania below fossiliferous beds
of Cambrian age (Fig. 9.31). Three glacial advances
have been postulated for the Late Proterozoic of
Australia. Two, and locally three, tillite horizons
contain striated pebbles. In one area, a glacial
pavement has been found below the tillite (41).
Varvelike shale deposits have also been identified.
Intensive deformation, metamorphism, and intru-

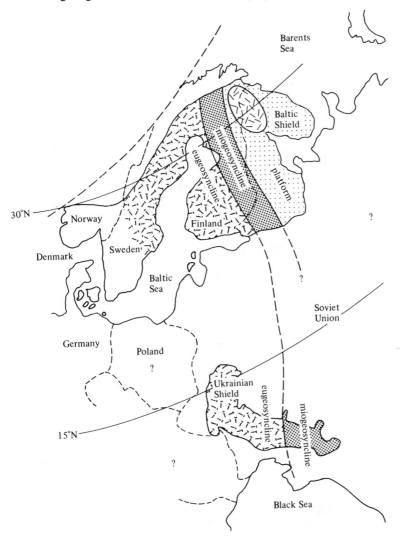

Figure 9.23   Paleogeographic map of part of Ancestral Europe during the Early Proterozoic, approximately 2.0 billion years ago.

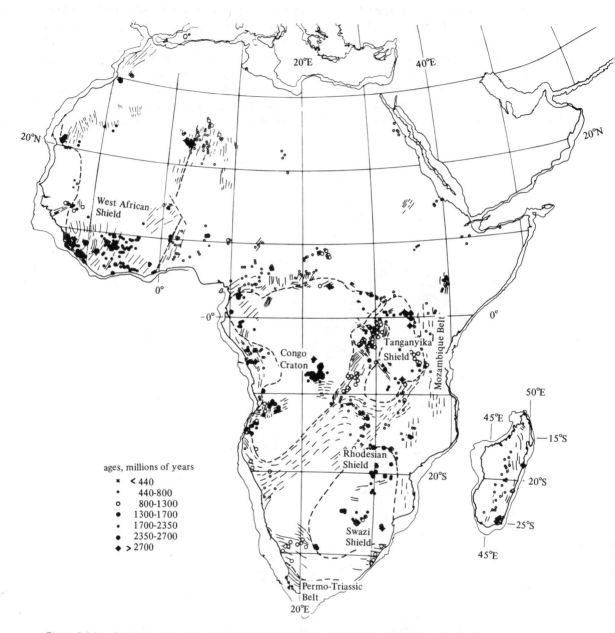

Figure 9.24   Radiometric dates for basement rocks in Africa. Data were obtained mainly by the potassium-argon method. (From Hurley and Rand, Refs. 26 and 34. Copyright © 1969 by American Association for the Advancement of Science.)

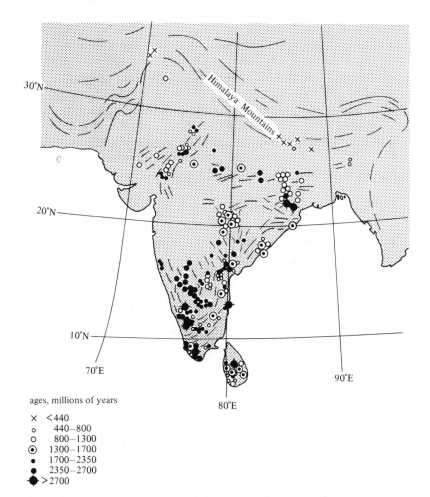

ages, millions of years

×	< 440
o	440–800
o	800–1300
⊙	1300–1700
•	1700–2350
●	2350–2700
◆	> 2700

Figure 9.25 Radiometric dates for basement rocks in India. (Data from Hurley and Rand, Refs. 26 and 34. Copyright © 1969 by American Association for the Advancement of Science.)

sion took place between 650 and 500 million years ago in some parts of Gondwanaland (Fig. 7.19). This widespread deformation affected not only the margins of Gondwanaland but the center of the continent as well.

ANCESTRAL CHINA

Precambrian rocks occur at the surface in eight relatively small shield areas in China as well as one in South Korea and another in Viet Nam (Fig. 10.39). Although few radiometric dates are available, the relative ages of these sequences are

known. The oldest rocks are complexly folded gneisses and mafic intrusives. This sequence, which may be Archean, is unconformably overlain by chlorite, mica, and hornblende schists. Relatively undeformed limestones, quartzites, and shales of Middle of Late Proterozoic age lie unconformably above the schists and gneisses.

ANCESTRAL SIBERIA

Crystalline rocks of Precambrian age are exposed in three relatively small shields in Siberia, and unmetamorphosed Proterozoic strata are found near the borders of these shields and in the geosynclines encircling the continent (Fig. 10.38). The oldest known rocks—gneisses of the Anabar Shield in northern Siberia—are dated at more than 3.55 billion years by the uranium - lead method (42). Somewhat younger gneisses, as well as schists of the Angara Shield were folded, metamorphosed, and intruded approximately 2.6 billion years ago and are therefore also Archean.

A still younger sequence of metasedimentary and metavolcanic rocks unconformably overlies the Archean sequence and was, in turn, intruded 1.8 billion years ago. This shows that these rocks were deposited during the Early Proterozoic.

The youngest sequence, the Sinian Complex, is composed of shale, sandstone, limestone, and dolomites. Since this sequence is conformably overlain by Lower Cambrian sedimentary rocks, it is in part Late Proterozoic.

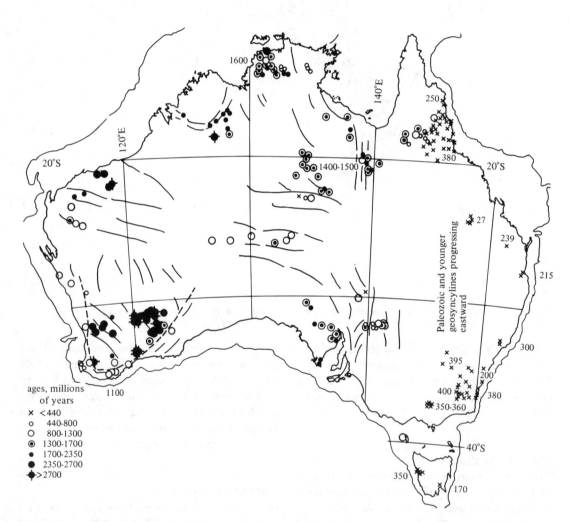

Figure 9.26 Radiometric dates for basement rocks in Australia. (Data from Hurley and Rand, Ref. 26. Copyright © 1969 by American Association for the Advancement of Science.)

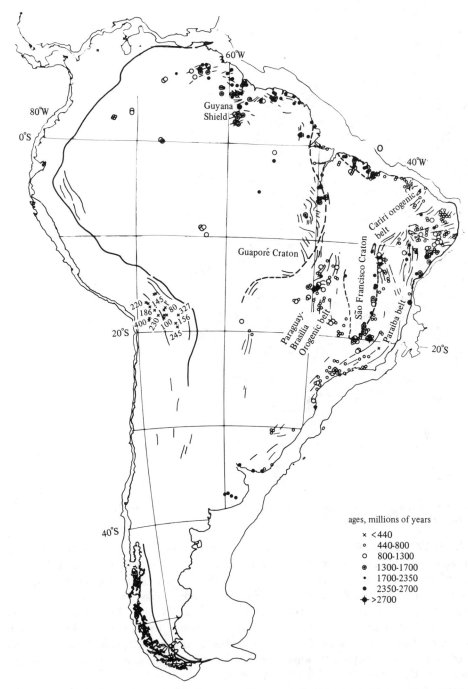

Figure 9.27   Radiometric dates for basement rocks in South America. (From Hurley and Rand, Ref. 26. Copyright © 1969 by American Association for the Advancement of Science.)

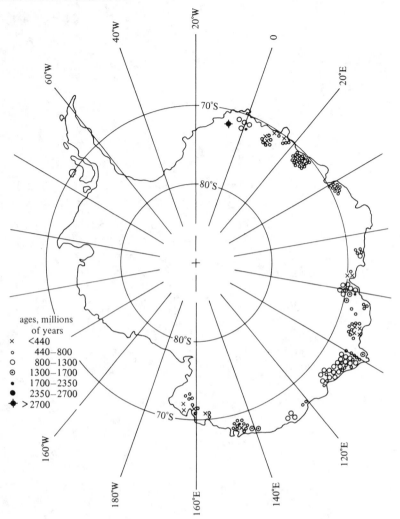

Figure 9.28 Radiometric dates for Precambrian basement rocks in Antarctica. (Data primarily from Piaciotto and Coppez, Ref. 35.)

RATE OF CONTINENTAL GROWTH

As we have indicated, the general decrease in age from the center of the continents toward the margins suggests the growth of the Precambrian continents by accretion of sedimentary and volcanic rocks onto older continental nuclei. However, it should be pointed out that some relatively ancient rocks occur near the margins of some continents. In southern California, rocks 1.0 billion years old crop out only 200 miles (330 km) from the margin of the continent, and 1.7 billion year old rocks crop out only 300 miles (500 km) from the continental margin. In India, rocks dated at 2.55 billion years occur 300 miles from the margin of Gondwanaland, and rocks dated at 3.2 billion years are found 350 miles from the continental margin. In Scotland, rocks 2.6 billion years old occur less than 200 miles from the margin of what is believed to have been Ancestral North America.

The occurrence of ancient rocks near the margins of the continents suggests that the rate of

Figure 9.29 Glacial varves with erratic blocks in Great Conglomerate Complex of Late Proterozoic age, central Katanga. (Photo by G. Mortelmans, from Cahen and Lepersonne, Ref. 39.)

continental growth has not been constant. The continents appear to have grown very rapidly during the early part of the Archean and at a decreasing rate since that time.

VARIATION IN TIME AND SPACE

The Precambrian histories of the various continents are remarkably similar. The general pattern is well illustrated by a comparison of the geologic cross sections of the Lake Superior and Grand Canyon regions (Figs. 9.32 and 9.33, respectively), which may serve as a model for the development of the Precambrian continents. A typical Precambrian history might be reconstructed as follows:

1. a. Deposition of graywacke, shale, and mafic volcanics.
   b. Very intense folding and metamorphism to form gneiss, schist, and amphibolite,

Figure 9.30 Stromatolitic reef of Late Proterozoic age near Kimpess in the Lower Congo. (Photo by J. Lepersonne, from Cahen and Lepersonne, Ref. 39.)

Figure 9.31   The Late Proterozoic Sturt tillite at Sturt Creek, South Australia. (Courtesy of Warren Hamilton, U.S. Geological Survey.)

and emplacement of granitic plutons, followed by uplift and erosion.

2. a. Deposition of shale and graywacke, with minor limestone and quartz sandstone, and mafic or intermediate volcanics.

   b. Intense folding and metamorphism of these rocks to phyllite or schist, quartzite, marble, and greenstone or amphibolite; intrusion of granitic plutons, uplift, and erosion.

3. a. Deposition of sandstone, shale, limestone, and dolomite with minor mafic to felsic volcanics.

   b. Moderate folding or faulting accompanied by mild metamorphism, uplift and erosion. Granitic intrusions are not common.

4. a. Deposition of limestone, dolostone, shale, and sandstone.

   b. Uplift and erosion.

The older sequences generally contain more graywacke and volcanic rocks, whereas the younger sequences contain more carbonates and quartz sandstone. Felsic and intermediate volcanic rocks are more abundant in younger sequences than in older sequences. Furthermore, the inten-sity of deformation and metamorphism within any one area decreases from the older to the younger sequences. These changes suggest an increasing stability of a given area with time and a transition in depositional environments from eugeosyncline to miogeosyncline to platform. Such changes could be associated with continental growth and outward migration of the margin of the continent.

There is also a lateral variation of rock type and intensity of deformation within the younger Precambrian sequences. Graywacke and volcanic rocks increase toward the continental margins, limestones and quartz sandstone being more common nearer the center of the continent. The younger Precambrian rocks near the margins are generally more highly metamorphosed and contain more volcanics and granitic intrusives than sequences deposited at the same time nearer the center of the continent. For example, the Middle Proterozoic rocks of the Grenville Province are intensely deformed and highly metamorphosed, but the Middle Proterozoic rocks of the Keweenawan Sequence to the west are only slightly deformed. Increasing deformation toward the continental margin may result from compression between tectonic plates and the effect of the sea-floor descending beneath the continent.

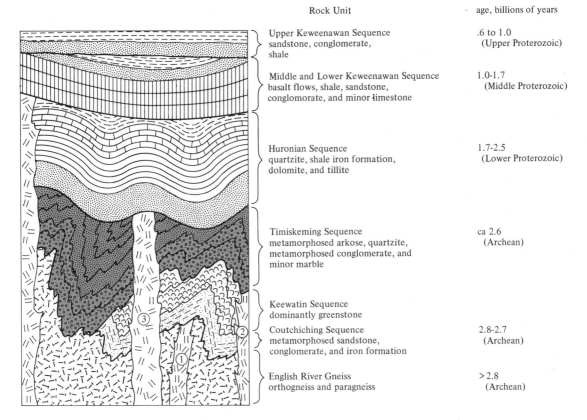

Rock Unit	age, billions of years
Upper Keweenawan Sequence sandstone, conglomerate, shale	.6 to 1.0 (Upper Proterozoic)
Middle and Lower Keweenawan Sequence basalt flows, shale, sandstone, conglomorate, and minor limestone	1.0-1.7 (Middle Proterozoic)
Huronian Sequence quartzite, shale iron formation, dolomite, and tillite	1.7-2.5 (Lower Proterozoic)
Timiskeming Sequence metamorphosed arkose, quartzite, metamorphosed conglomerate, and minor marble	ca 2.6 (Archean)
Keewatin Sequence dominantly greenstone	
Coutchiching Sequence metamorphosed sandstone, conglomerate, and iron formation	2.8-2.7 (Archean)
English River Gneiss orthogneiss and paragneiss	>2.8 (Archean)

Figure 9.32 Diagrammatic cross section of the Precambrian deposits in the Lake Superior region.

PRECAMBRIAN LIFE

There is no direct evidence from the fossil record to demonstrate just how life first evolved on the earth. Fossils of primitive algalike organisms have been found in rocks that have been radiometrically dated at more than 3.3 billion years, but life may have evolved long before that time.

Archean

The fossil record indicates that the Archean seas were occupied by bacterialike and algalike organisms. These organisms produced highly resistant cells resembling spores (Fig. 9.34). The oldest rocks known to contain possible evidence of life are carbonaceous cherts of the Onverwacht Group at the base of the Swaziland "System" in South Africa. These rocks are approximately 3.35 billion years old (43, 44). Although the organic nature of

these microstructures has not been conclusively established, their association with carbonaceous matter supports an organic origin. These structures may have been produced by algae or similar organisms. The overlying rocks of the Fig Tree Series are intruded by a pegmatite dike dated at 3.0 billion years, and they contain microstructures that are almost certainly fossils (44, 45)(Fig. 9.34b). These probably represent the remains of primative algae and bacteria.

The earliest known stromatolites occur in the 3.0 billion year old Bulawayan "System" in Rhodesia. The organisms that built the stromatolites were probably similar to the modern blue-green algae and if so, they were photosynthetic forms.

Analysis of the Archean Soudan Formation in Minnesota has shown the presence of petroleum-

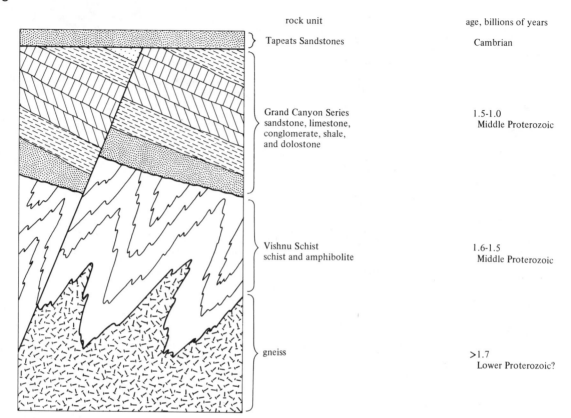

rock unit	age, billions of years
Tapeats Sandstones	Cambrian
Grand Canyon Series sandstone, limestone, conglomerate, shale, and dolostone	1.5-1.0 Middle Proterozoic
Vishnu Schist schist and amphibolite	1.6-1.5 Middle Proterozoic
gneiss	>1.7 Lower Proterozoic?

Figure 9.33   Diagrammatic cross section of the Grand Canyon region.

like substances (46). Carbon-rich chert from this 2.7 billion year old unit contains compounds such as paraffin, which are presumably of biological origin. Fossils representing blue-green algae have also been reported from this formation.

Early Proterozoic

Microfossils in the Gunflint Formation in southern Ontario, Canada, have been studied in detail (47). This unit, which is part of the Huronian System, is approximately 2 billion years old and contains algal cherts. Analysis has revealed the presence of hydrocarbons (paraffins), bacteria, algal filaments resembling modern blue-green algae, sporelike structures, and more complex organisms resembling protozoans. The state of preservation is such that the three-dimensional morphology of the

microfossils has been retained (Fig. 9.35). In South Africa, carbonaceous matter is also present in rocks of latest Archean or Early Proterozoic age (48). This carbonaceous matter contains hydrocarbons of probable organic origin.

Fossil stromatolite reefs are common within the Huronian System in the vicinity of the Great Lakes. They are also found near Great Slave Lake in northern Canada, and in the Rocky Mountains (Fig. 9.36). The reefs consist of concentrically laminated masses thought to have been formed by an algal form similar to modern blue-green algae. Fossil bacteria, fungi, and molds are present in the Gowganda Formation of the Huronian System in southern Canada (49).

Graphitic bodies and anthracitic coal of possible organic origin occur in Early Proterozoic

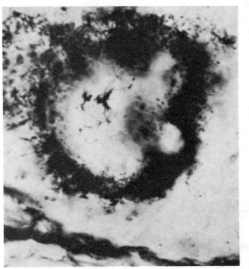
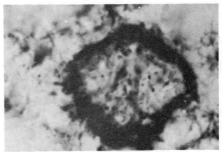
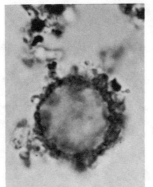

(a)

Figure 9.34 Archean microfossils: (a) Algalike forms in the Onverwacht Series from South Africa (From Engel et al., Ref. 43. Copyright © 1968 by American Association for the Advancement of Science.); (b) Spheroidal, algalike microfossils *(Archaeosphaeroides barbertonensis)* from the Fig Tree Series. (Courtesy of J.W. Schopf and E.S. Barghoorn, Ref. 45. Copyright © 1967 by American Association for the Advancement of Science.)

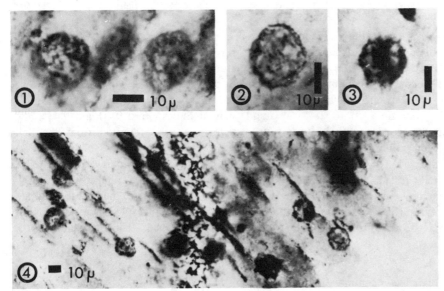

(b)

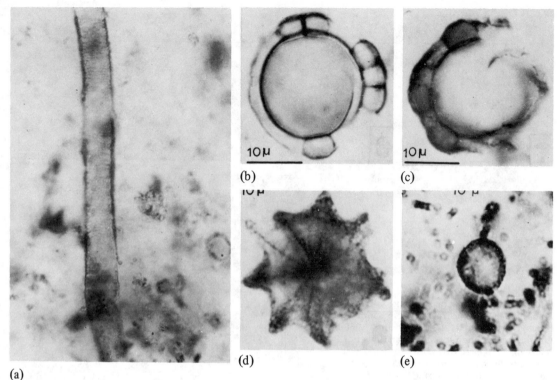

Figure 9.35 Early Proterozoic microfossils from the Gunflint Chert of the Anamikie Group:
(a) *Anamikiea septata*, an algal filament; (b) and (c) *Eosphaera tyleri*, an organism of uncertain
affinity; (d) *Kakabekia umbellata*, an organism of unknown affinity; (e) *Huroniospora microreticulata*.
(Courtesy of E. S. Barghoorn, Harvard University, Ref. 47. From Barghoorn and Tyler, copyright ©
1965 by American Association for the Advancement of Science.)

black shales in Michigan and in the Labrador Geosyncline (50). Chromatographic analysis of the coals indicates the presence of strongly pigmented organic compounds. The coal and graphite may have been derived from algal remains that were preserved under reducing conditions.

Curved tubular markings occur in Early Proterozoic quartzite of the Huronian Series in southern Ontario, Canada (51) (Fig. 9.37). These sand casts morphologically resemble worm trails, but most investigators now favor an inorganic origin for these structures (52). Other tubular markings in Early Proterozoic deposits may be worm trails.

Middle Proterozoic
Stromatolite reefs are relatively common in the Belt Series in the Rocky Mountains and the Grand Canyon Series in Arizona (Fig. 9.38). Spherical microfossils found within the Grand Canyon Series may be the remains of algae similar to those which produced the stromatolites (Fig. 9.39). Possible impressions of jellyfish have also been found in the Grand Canyon Series (Figs. 9.40 and 9.41), and worm trails have been reported from the Belt Series (Fig. 9.42). Although inorganic origins have been proposed for these structures, it is probable that they were produced by multicellular organisms. If the latter is the case, then these trails and impressions are the oldest known remains of organisms more advanced than one-celled plants.

Late Proterozoic
The scarcity of fossil remains of multicellular organisms in Late Proterozoic strata has been a major concern to paleontologists who are accus-

tomed to finding abundant fossils in Cambrian strata. The Cambrian fauna was so advanced and diversified that it must have been the result of a long and complex evolutionary history. However, there is little evidence of this development preserved in the fossil record.

Several examples of Late Proterozoic fossils of unicellular and multicellular organisms come from Australia. In central Australia, cherts of the Bitter Springs Limestone contain the remains of blue-green algae, the earliest recorded green algae, and algal stromatolites (55). The green algae take the form of unpartitioned filaments, threads, and spheres, similar to the resting zygote of modern green algae.

Perhaps the most complex fauna of Precambrian age is the Ediacara fauna from South Australia (56). This fauna, which is named after the Ediacaran Hills, is about 600 million years old and includes well-preserved impressions of jellyfish, worms, worm trails, sea pens, sponges, coral-like forms, arthropods, and algae (Fig. 9.43). Since

the strata are several hundred feet below the established base of the Cambrian, a Late Proterozoic age is suggested for this fauna.

In Esmeralda County, Nevada, possible annelid trails have been found 3000 feet below the base of the Lower Cambrian (57). In eastern Newfoundland, numerous fossils have been found in the Conception Group, which is probably Late Proterozoic in age. These fossils include spindle-shaped, leaf-shaped, and lobate forms (Fig. 9.44). The biological affinity of these organisms is somewhat uncertain, but they may be related to soft-bodied corals and jellyfish. Fossils similar to the spindle- and leaf-shaped forms are also found in Late Proterozoic rocks near Leicester, England (Fig. 9.45). Since both eastern Newfoundland and England were part of Ancestral Europe during the Late Proterozoic, the similarity in fossils is understandable.

Although fossils are generally not used in correlating Precambrian strata, the distinctive morphology of Late Proterozoic stromatolites makes them useful in correlating beds of this age (58).

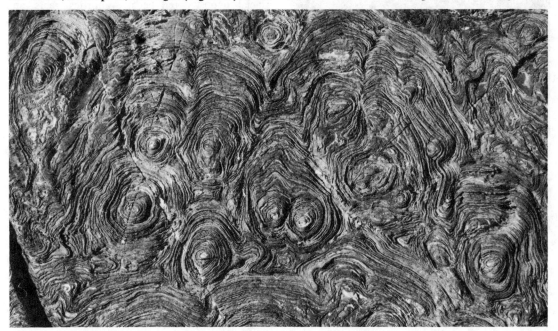

Figure 9.36 Early Proterozoic. Stromatolites viewed from the top on a surface parallel to bedding, Great Slave Lake area of northern Canada. (Photo by P. F. Hoffman, courtesy of the Geological Survey of Canada, Ottawa.)

Figure 9.37 Tubular markings on an Early Proterozoic quartzite north of Lake Huron. These structures resemble worm trails but are probably a type of mudcrack. (Photo by P. F. Hofmann, courtesy of the Geological Survey of Canada, Ottawa and the American Association for the Advancement of Science.)

Spores have been used in the determination of relative ages within the Late Proterozoic Sinian Complex of the Soviet Union and China. The assemblage of spores from these rocks is quite distinct from those of Early Cambrian age (59).

PRECAMBRIAN CLIMATES

The reconstruction of ancient climates should be made within the context of polar wandering and

Figure 9.38 Algal stromatolite, *Collenia multiflabella*, from the Middle Proterozoic rocks of Glacier National Park, Montana. (Courtesy of the U.S. Geological Survey.)

continental drift, since climatic indicators must be compared with paleolatitudes rather than with modern latitudes. Paleoclimates may be investigated by plotting such data as the locations of glacial deposits, red beds, evaporites, reefs, and thick limestones on reconstructions of the continents for each of the Precambrian eras.

Precambrian rock sequences provide evidence of a wide fluctuation in climate, but generally the ranges of climates are similar to that indicated for the Phanerozoic. Thick limestones are commonly located near the paleoequator, and therefore these regions were probably warm for the most part, just as they are today. However, at least twice during the Precambrian continental glaciation occurred in the middle and low latitudes.

Archean

Unfortunately, little can be determined about the climatic conditions of the Archean, an era covering almost half of geologic time. One reason for this lack of data is that most Archean deposits have been metamorphosed to such an extent that the original character of the rock is not evident. However, even if paleoclimatic indicators were preserved in these rocks, it would be difficult to make any inferences about Archean climates, owing to the scarcity of paleomagnetic data for rocks of Archean age.

Early Proterozoic

The tillite of the Cobalt Group north of Lake Huron indicates cold climatic conditions for this area approximately 2.3 billion years ago. A study of the striated pavement below the tillite indicates that ice moved toward the southeast on three separate occasions (11). Paleomagnetic measurements show that the North Pole was located at 22°N, 97°W at the time of the glaciation. This would mean that the area glaciated was at a latitude of 60°N during the Early Proterozoic and that the ice moved toward the North Pole. Such movement may be analogous to that of the Pleistocene ice sheets that moved from the Canadian and Baltic Shields northward toward the Arctic Ocean. Much of the Superior province was presumably at a relatively high elevation following

Figure 9.39 Spherical microfossil of uncertain biological affinity from the Chuaria Shale, Grand Canyon Series, magnification, 750X. (Courtesy of C. Downie.)

Figure 9.40 Probable impression of a fossil jellyfish from the Nankoweap Group, Grand Canyon Series. Although the organic nature of this structure has been questioned, it has been designated as *Brooksella canyonensis.* (From Bassler, Ref. 54. Courtesy of the Smithsonian Institution.)

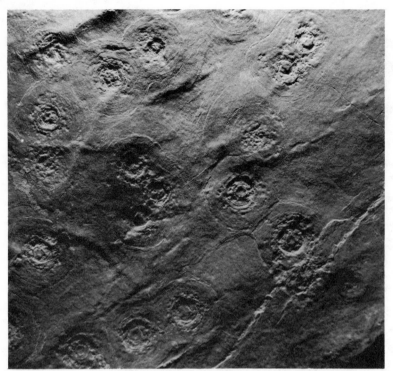

Figure 9.41 Possible jellyfish from the Unkar Group south of Bright Angel Creek, Arizona. (Courtesy of Raymond M. Alf Museum, Webb School of California, Claremont.)

the Kenoran Orogeny, so that the ice movement would have been influenced by the paleoslope from the Superior Province toward the geosyncline. The glacial deposits of Early Proterozoic age in southern Africa may have been deposited at the same time as those of the Cobalt Group.

The thick limestones, dolomites, and iron formation of late Early Proterozoic age in North America, Africa, Australia, and India are indicative of deposition under warm climates. The iron concentrations in these deposits may have resulted from deep weathering in a tropical environment. During such lateritic weathering, large amounts of iron and silica would have been carried in solution from the continents to the oceans, where they may have been precipitated as iron oxide and chert. Paleomagnetic data indicate that most deposits of iron formation were laid down within 45° of the equator. That latitude is consistent with the suggestion that these deposits were the result of lateritic weathering (Figs. 9.13 and 9.22).

Figure 9.42 Probable worm trails from the Middle Proterozoic rocks of the Belt Series, Dawson Pass, Glacier National Park, Montana. (Courtesy of Fenton and Fenton, Ref. 53.)

(a)　　　　　　　　　　　　　　(b)

(c)　　　　　　　　　　　　　　(d)

Figure 9.43　Late Proterozoic fossils from the Ediacaran Hills of Australia: (a) *Dickinsonia costata*; (b) *Spriggina floundersi*, a segmented worm; (c) *Ranges grandis*, a sea pen; (d) *Medusinites asteroides*, a possible jellyfish. (Courtesy of Prof. M. F. Glaessner, Ref. 56. Copyright © 1961 by Scientific American, Inc., all rights reserved.)

The iron formation and carbonates of the Anamikie Group are significantly younger than the tillites of the Cobalt Group. The change in climate in the Great Lakes region between the times of deposition of these two sequences may have been due in part to movement of the North Pole away from this area between the beginning and the end of the Early Proterozoic.

Middle Proterozoic
The thick limestones of the Belt Series and the thick marbles of the Grenville Series suggest deposition in a warm climate. The presence of salt and gypsum in Grenville deposits suggests the climate may also have been somewhat dry (61). Paleomagnetic data indicate that both units were deposited within 30° of the Middle Proterozoic equator (Fig. 9.17).

Late Proterozoic
Tillites of Late Proterozoic age have been reported from all the continents except Antarctica. How-

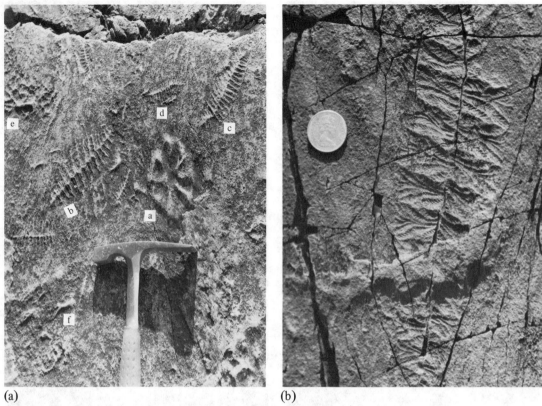

(a)   (b)

Figure 9.44   Late Proterozoic fossils from southeastern Newfoundland: (a) Round lobate organisms probably jellyfish [a, f]; spindle-shaped organisms [b, c, e]; leaf-shaped organism [d]; stalk [g]; and base [h] of the leaf-shaped organisms. (b) Enlargement of spindle-shaped organism. (Courtesy of S. B. Misra, Ref. 60.)

ever, only those from Greenland, Scandinavia, central Africa, Australia, and China have been well substantiated in age and lithologic character (Fig. 9.46). It is not certain that even these tillites were deposited contemporaneously, since the Late Proterozoic covers more than 400 million years. However, most of these deposits occur within several hundred meters of Cambrian beds and most of them contain two tillite horizons. This evidence suggests that there were two worldwide glacial advances during the Late Proterozoic.

When paleolatitudes are plotted for the areas affected by these glaciations, it is found that the Late Proterozoic "tillites" of Greenland, Scandinavia, and Australia lay on or very close to the equator during the Late Proterozoic, whereas the

"tillites" of China and central Africa were within 40° of the equator (Fig. 9.46). This indicates that the climatic cooling just prior to the Cambrian Period was even more severe than that which occurred during the Late Tertiary and Pleistocene. The mystery of why the earth cooled to such a degree invites further study and speculation.

An evaporite series of Late Proterozoic age from Australia was deposited near the paleo-equator according to paleomagnetic data.

SUMMARY

The oldest known rocks have been found in Greenland, central North America, southern Africa, and Siberia. Radiometric dating indicates that

Figure 9.45 *Charnia masoni*, a spindle-shaped organism from Late Proterozoic deposits of England, three-quarters natural size. This impression is possibly of a primitive pennatulid type of coelenterate or of a fucoid seaweed. It was found in the Charnian rocks of Charnwood Forest near Leicester. (Courtesy of Trevor D. Ford.)

these rocks are more than 3.5 billion years old. Rock sequences more than 2.5 billion years old have been found on all continents. These sequences contain abundant volcanic rocks and graywacke which may have been laid down in the vicinity of island arcs. Intense and very widespread deformations occurred on all continents at the end of the Archean, 2.5 billion years ago. Possibly these deformations were associated with the joining of several separate island arcs to form the five continental nuclei.

During the Early Proterozoic, geosynclines formed on the margins of the five continental nuclei. Volcanism and sea-floor spreading at the outermost margin of the continents resulted in the accretion of new material to the continents. Especially intense deformations at the end of the Early Proterozoic affected all five ancestral continents. Folding of geosynclinal deposits before and during these orogenies helped to increase the area of the continents.

Middle Proterozoic geosynclines developed on the margins of the continents, but since the continents were larger, deposition occurred further from their centers. The cycle of deformation and resulting growth of the continents was repeated, and by the end of the Precambrian, the continents had nearly reached their present size.

The earliest evidences of life are microscopic spherical-shaped bodies which are found in rocks approximately 3.4 billion years old. These bodies were probably produced by algae or similar organisms. Numerous stromatolites in Early and Middle Proterozoic deposits indicate that algae were abundant in the seas. The oldest well-documented fossils of multicellular organisms are probable worm trails and jellyfish from Middle Proterozoic rocks. Numerous fossils of soft-bodied multicellular organisms have been found in rocks of Late Proterozoic age.

We have few indications of severe climates for most of the Precambrian. It is likely that the equatorial zones were generally warm, and the middle latitudes were for the most part free of glaciers. However, the climate was relatively cool during the early part of the Early Proterozoic when glaciers apparently moved into the middle latitudes. A worldwide episode of cold climate occurred near the end of the Late Proterozoic, and during this time glaciers occupied the middle and low latitudes.

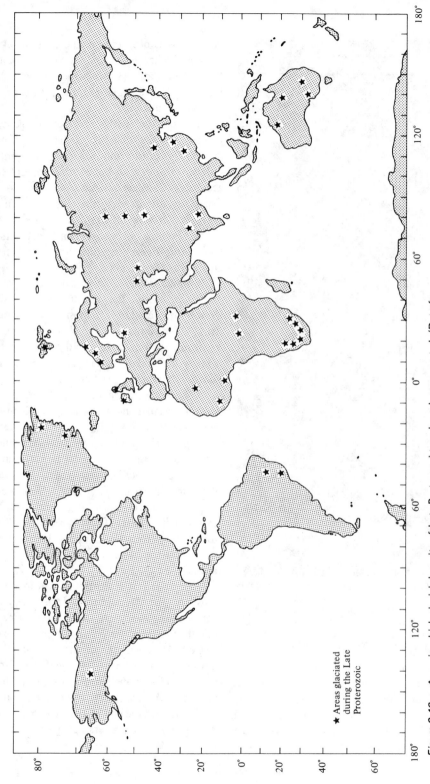

Figure 9.46 Areas in which glacial deposits of Late Proterozoic age have been reported. (Data from Harland, Ref. 25.)

★ Areas glaciated during the Late Proterozoic

REFERENCES CITED

1. G. Gastil, 1960, The distribution of mineral dates in time and space: *Amer. Jr. Sci.*, v. 258, p. 1.

2. R. K. Wanless, R. D. Stephens, G. R. Lachance, and C. M. Edmonds, 1968, *Age Determinations and Geological Studies—K - Ar Isotopic Ages, Report 8:* Geological Survey of Canada paper 67-2 part A, Ottawa, Canada.

3. B. J. Giletti, R. St. J. Lambert, and S. Moorbath, 1961, The basement rocks of Scotland and Ireland, *in* J. L. Kulp, ed., *Geochronology of Rock Systems*: New York Academy of Sciences, v. 91, p. 464.

4. J. A. Grant, 1964, Rubidium - Strontium Isochron study of the Grenville Front near Lake Timagami, Ontario: *Science*, v. 146, p. 1049.

   T. E. Krough and G. L. Davis, 1971, Paragneiss studies in the Georgian Bay area 90 km southeast of the Grenville Front: *Annual Report of the Geophysical Laboratory—1969 - 1970*, p. 339.

5. L. P. Black, et al., 1971, Isotopic dating of very early Precambrian amphibolite facies gneisses from the Godthaab district, West Greenland; *Earth Planet. Sci. Lett.*, v. 12, p. 245.

6. S. S. Goldich, C. E. Hedge, and T. W. Stern, 1970, Age of the Morton and Montevideo Gneiss and related rocks, southwestern Minnesota: *Geol. Soc. Amer. Bull.*, v. 81, p. 3671.

7. E. J. Catazaro and J. L. Kulp, 1964, Discordant zircons from Little Belt (Montana), Beartooth (Montana), and Santa Catalina (Arizona) Mountains: *Geochim. Cosmochim. Acta*, v. 28, p. 87.

   R. A. Heimlich and P. O. Banks, 1968, Radiometric age determinations, Bighorn Mountains, Wyoming: *Amer. Jr. Sci.*, v. 266, p. 180.

8. S. R. Hart and G. L. Davis, 1969, Zircon U - Pb and whole-rock Rb - Sr ages and early crustal development near Rainy Lake, Ontario: *Geol. Soc. Amer. Bull.*, v. 80, p. 595.

9. C. E. Hedge and F. G. Walthall, 1963, Variations in radiogenic strontium found in volcanic rocks: *Science*, v. 140, p. 1214.

10. R. E. Folinsbee et al., 1968, A very ancient island arc, *in* L. Knopoff, C. L. Drake, and P. J. Hart, eds., *The Crust and Upper Mantle of the Pacific Area*: American Geophysical Union Monograph 12, p. 441.

11. D. A. Lindsey, 1966, Sediment transport in a Precambrian ice age: The Huronian Gowganda Formation: *Science*, v. 154, p. 1442.

12. K. C. Condie, 1967, Petrology of the late Precambrian tillite (?) association in northern Utah: *Geol. Soc. Amer. Bull.*, v. 78, p. 1317.

    G. M. Young, 1970, An extensive Early Proterozoic glaciation in North America?: *Palaeogeography, Palaeoclimatology, Palaeoecology*, v. 7, p. 85.

13. R. Van Schmus, 1965, The geochronology of the Blind River - Bruce mines area, Ontario, Canada: *Jr. Geol.*, v. 73, p. 755.

14. Z. E. Peterman, 1966, Rb - Sr dating of middle Precambrian metasedimentary rocks of Minnesota: *Geol. Soc. Amer. Bull.*, v. 77, p. 1031.

15. E. G. Lidiak, 1971, Buried Precambrian rocks of South Dakota: *Geol. Soc. Amer. Bull.*, v. 82, p. 1411.

16. J. D. Obradovich and Z. E. Peterman, 1968, Geochronology of the Belt Series, Montana: *Can. Jr. Earth Sci.*, v. 5, p. 737.

17. C. A. Anderson, 1963, Simplicity in structural geology, *in* C. C. Albritton, Jr., ed., *The Fabric of Geology:* Freeman-Cooper, San Francisco, Calif., p. 175.

    B. J. Giletti and P. E. Damon, 1961, Rubidium - strontium ages of some basement rocks from Arizona and northwestern Mexico: *Geol. Soc. Amer. Bull.*, v. 72, p. 639.

18. L. T. Silver, 1960, Age determinations on Precambrian diabase differentiates in the Sierra Andra, Gila County, Arizona (abstr.): *Geol. Soc. Amer. Bull.*, v 71, p. 1973.

    P. E. Damon, D. E. Livingston, and R. C. Erickson, 1962, New K - Ar dates for the Precambrian of Pinal, Gila, Yavapai, and Coconimo Counties, Arizona: *N. Mex. Geol. Soc. Field Conf. Guidebook*, v. 13, p. 56.

    L. T. Silver, 1963, The use of cogenic uranium-lead isotope systems in zircons in geochronology, *in Radioactive Dating*: International Atomic Energy Agency, Vienna, Austria, p. 279.

19. L. T. Silver and S. B. Lumbers, 1966, *Geochronologic Studies in the Bancroft - Madoc area of the Grenville Province, Ontario, Canada*: Geological Society of America Special Paper 87, p. 156.

20. W. R. Van Schmus, 1971, Rb - Sr age of Middle Keweenawan Rocks, Mamainse Point and vicinity, Ontario, Canada: *Geol. Soc. Amer. Bull.*, v. 82, p. 3221.

21. G. Faure, S. Chandhuri, and M. D. Fenton, 1969, Ages of the Duluth gabbro complex and of the Endion Sill, Duluth, Minnesota: *Jr. Geophys. Res.*, v. 74, p. 720.

22. G. R. Tilton, G. W. Wetherill, G. L. Davis, and M. N. Bass, 1960, 1000 million-year-old minerals from the eastern United States and Canada: *Jr. Geophys. Res.*, v. 65, p. 4173.

23. D. W. Rankin, T. W. Stern, J. C. Reed, Jr., and M. F. Newell, 1969, Zircon ages of felsic volcanic rocks in the upper Precambrian of the Blue Ridge, Appalachian Mountains: *Science*, v. 166, p. 741.

24. T. S. Winsnes, 1965, The Precambrian of Spitsbergen and Bjørnøya, *in* K. Rankama, ed., *The Precambrian*: Wiley-Interscience, New York, v. 2, p. 1.

25. M. Schwarzback, 1963, *Climates of the Past:* Van Nostrand-Reinhold, New York.

    W. B. Harland, 1964, Evidence of late Precambrian glaciation and its significance, *in* A. E. M. Nairn, ed., *Problems in Paleoclimatology*: Wiley-Interscience, New York, p. 119.

26. P. M. Hurley and J. R. Rand, 1969, Pre-drift continental nuclei: *Science*, v. 164, p. 1229, June 13.

27. S. I. Zykov, A. I. Tugarinov, I. V. Belkov, and E. V. Bilikova, 1964, The age of the oldest formations of the Kola Peninsula: *Geochemistry International*, no. 2, p. 262.

    N. P. Semenenko, 1967, Deep crustal structure in Ukrainian crystalline shield: *Int. Geol. Rev.*, v. 9, p. 49.

    E. V. Sobotovich, S. M. Grasheheenko, V. M. Aleksandruk, and M. M. Shats, 1965, Determining the age of the oldest rocks by the lead - isochron method and spectroscopic Sr - isotope methods: *Int. Geol. Rev.*, v. 7, no. 11, p. 1907.

28. O. Kouvo and G. R. Tilton, 1966, Mineral ages from the Finnish Precambrian: *Jr. Geol.*, v. 74, p. 421.

29. K. Rankama, ed., 1963, *The Precambrian*: Wiley-Interscience, New York, v. 1.

    N. P. Semenenko, S. P. Rodionov, I. S. Usen-ko, I. L. Lichak, and I. D. Tsarovsky, 1960, Stratigraphy of the Pre-Cambrian of the Ukrainian Shield: Pre-Cambrian Stratigraphy and correlations, *21st International Geological Congress Proceedings,* Copenhagen, p. 108.

30. J. L. Kulp and H. Neumann, 1961, Some potassium - argon ages on rocks from the Norwegian basement, *in* J. L. Kulp, ed., *Geochronology of Rock Systems*: New York Academy of Sciences. v. 91, p. 469.

31. H. W. Fairbairn and P. M. Hurley, 1969, Northern Appalachian geochronology as a model for interpreting ages in older orogens, *in Variations in Isotopic Abundances of Strontium, Calcium, and Argon and Related Topics*: Department of Geology and Geophysics, M.I.T., Cambridge, Mass., p. 11.

32. H. Allsopp, H. Roberts, G. Schreiner, and D. Hunter, 1962, Rb - Sr age measurements on various Swaziland granites: *Jr. Geophys. Res.*, v. 67, p. 5307.

33. A. L. Hales, 1961, An upper limit to the age of the Witwatersrand System, *in* J. L. Kulp, ed., *Geochronology of Rock Systems*: New York Academy of Sciences, v. 91, p. 524.

34. A. R. Crawford, 1969, India, Ceylon, and Pakistan: New age data and comparisons with Australia: *Nature*, v. 223, p. 380.

    M. Halpern, 1970, Rubidium - strontium date of possibly 3 billion years for a granitic rock from Antarctica: *Science*, v. 169, p. 977.

    A. R. Crawford, 1969, Reconnaissance Rb - Sr dating of the Precambrian rocks of southern peninsular India: *Jr. Geol. Soc. India*, v. 10, p. 117.

    A. R. Crawford, 1970, The Precambrian geochronology of Rajasthan and Bundelkhand, northern India: *Can. Jr. Earth Sci.*, v. 7, p. 91.

    S. N. Sarkar, E. K. Gerling, A. A. Polkanov, and F. V. Chukrov, 1967, Pre-Cambrian geochronology of Nagpur-Bhandara-Drug, India: *Geological Magazine*, v. 104, no. 6, p. 37.

    S. N. Sarkar, A. K. Saha and J. Miller, 1967, Potassium-argon ages from the oldest metamorphic belt in India: *Nature*, v. 215, p. 946.

35. E. Picciotto and A. Coppez, 1961, Bibliographie des mesures d'ages absolus en Antarctique: Extrait des *Annales de la Société Geologique de Belgique*, v. 85, p. 263.

36. A. F. Trendall, 1968, Three great basins of Precambrian banded iron formation deposition: A systematic comparison: *Geol. Soc. Amer. Bull.*, v. 79, p. 1527.

37. U. Aswathanarayana, 1964, Age determination of rocks and geochronology of India, *in* B. C. Roy, ed., *22nd International Geological Congress Proceedings*: New Delhi, p. 1.

L. O. Nicolaysen et al., 1958, New measurements relating to the absolute age of the Transvaal System and of the Bushveld Igneous Complex: *Geol. Soc. S. Afr. Trans.*, v. 48, p. 161.

C. B. van Niekerk and A. J. Burger, 1964, The age of the Ventersdorp System: *Union of South Afr. Geol. Survey Annals*, v. 3, p. 75.

38. P. M. Hurley et al., 1967, Test of continental drift by comparison of radiometric ages: *Science*, v. 157, p. 495.

39. L. Cahen and J. Lepersonne, 1967, The Precambrian of the Congo, Rwanda, and Burundi, *in* K. Rankama ed., *The Precambrian*, v. 3, p. 143.

40. L. Cahen, 1961, Review of geochronological knowledge in middle and northern Africa, *in* J. L. Kulp, ed., *Geochronology of Rock Systems*: New York Academy of Sciences, v. 91, p. 535.

41. D. A. Brown, K. S. W. Campbell, and K. A. W. Crook, 1968, *The Geological Evolution of Australia and New Zealand*: Pergamon Press, New York.

42. A. Ya. Krylov et al., 1963, Absolute ages of rocks of the Anabar Shield: *Geochemistry*, p. 1193.

43. B. Nagy and L. A. Nagy, 1969, Early Precambrian Onverwacht microstructures: Possibly the oldest fossils on earth?: *Nature*, v. 223, p. 1226.

A. E. J. Engel, B. Nagy, and L. A. Nagy, 1968, Alga-like forms in Onverwacht Series, South Africa: Oldest recognized lifelike forms on earth: *Science*, v. 161, p. 1005, Sept. 6.

P. M. Hurley et al., 1972, Ancient age of the Middle Marker horizon, Onverwacht Group, Swaziland sequence, South Africa: *Earth Planet. Sci. Lett.*, v. 14, p. 360.

44. H. L. Allsopp, T. J. Ulrych, and L. O. Nicolaysen, 1968, Dating some significant events in the history of the Swaziland System by the Rb-Sr isochron method: *Can. Jr. Earth. Sci.*, v. 5, p. 605.

45. E. S. Barghoorn and J. M. Schopf, 1966, Microfossils three billion years old from the Precambrian of South Africa: *Science*, v. 152, p. 758.

J. W. Schopf and E. S. Barghoorn, 1967, Algae-like fossils from the Early Precambrian of South Africa: *Science*, v. 156, p. 508, April 28.

K. C. Condie, J. E. Macke, and T. O. Reimer, 1970, Petrology and geochemistry of early Precambrian graywackes from the Fig Tree Group, South Africa: *Geol. Soc. Amer. Bull.*, v. 81, p. 2759.

46. W. G. Meinschein, 1965, Soudan Formation: Organic extracts of Early Precambrian rocks: *Science*, v. 150, p. 601.

47. E. S. Barghoorn and S. A. Tyler, 1965, Microorganisms from the Gunflint Chert: *Science*, v. 147, p. 563, Feb. 5.

48. J. Hoefs, 1967, Carbon isotope composition of carbonaceous matter from the Precambrian of the Witwatersrand System: *Science*, v. 155, p. 1096.

49. T. A. Jackson, 1967, Fossil actinomycetes in Middle Precambrian glacial varves: *Science*, v. 155, p. 1003.

50. S. A. Tyler et al., 1957, Anthracite coal from Precambrian upper Huronian black shale of the Iron River district, northern Michigan: *Geol. Soc. Amer. Bull.*, v. 68, p. 1293.

51. H. J. Hofmann, 1967, Precambrian fossils (?) near Elliot Lake, Ontario: *Science*, v. 156, p. 500, April 28.

52. G. M. Young, 1969, Inorganic origin of corrugated vermiform structures in the Huronian Gordon Lake Formation near Flack Lake, Ontario: *Can. Jr. Earth Sci.*, v. 6, p. 795.

53. C. L. Fenton and M. A. Fenton, 1937, Belt Series of the North: *Geol. Soc. Amer. Bull.*, v. 48, p. 1873.

54. R. S. Bassler, 1941, A supposed jellyfish from the Pre-Cambrian of the Grand Canyon: *Proc. U. S. Nat. Hist. Mus.*, v. 48, p. 519.

55. J. W. Schopf, 1968, Microflora of the Bitter Springs Formation, Late Precambrian, central Australia: *Jr. Paleontol.*, v. 42, p. 651.

56. M. F. Glaessner, 1961, Pre-Cambrian animals: *Sci. Amer.,* v. 204, no. 3, p. 72.

M. F. Glaessner, 1971, Geographic distribution and time range of the Ediacara fauna: *Geol. Soc. Amer. Bull.,* v. 82, p. 509.

57. S. A. Kirsch, 1971, Chaos structure and Turtleback Dome, Mineral Ridge, Esmeralda County, Nevada: *Geo. Soc. Amer. Bull.,* v. 82, p. 3169.

58. M. F. Glaessner, W. V. Preiss, and M. R. Walter, 1969, Precambrian columnar stromatolites in Australia: Morphological and stratigraphic analysis: *Science,* v. 164, p. 1056.

M. E. Raaben, 1969, Columnar stromatolites and late Precambrian stratigraphy: *Amer. Jr. Sci.,* v. 267, p. 1.

59. D. V. Nalivkin, 1960, *The Geology of the U.S.S.R.,* transl. S. I. Tomkeieff: Pergamon Press, New York.

60. S. B. Misra, 1969, Late Precambrian (?) fossils from southeastern Newfoundland: *Geol. Soc. Amer. Bull.,* v. 80, p. 2133.

61. M. J. de Wit, 1970, Evidence for salt deposits in the Appalachian/Caledonian Orogen: *Nature,* v. 227, p. 829.

62. A. J. Froeloch and E. A. Drieg, 1969, Geophysical-geologic study of the northern Amadeus Trough, Australia: *Amer. Assoc. Petrol. Geol. Bull.,* v. 53, p. 1978.

GENERAL REFERENCES FOR PRECAMBRIAN AND PHANEOZOIC ERAS

North America

E. H. Bailey, ed., 1966, *Geology of Northern California:* California Division of Mines and Geology Bull. 190, San Francisco.

M. P. Billings, 1956, *The Geology of New Hampshire:* New Hampshire State Planning and Development Commission, Concord.

O. E. Childs and B. W. Beebe, 1963, *Backbone of the Americas—Tectonic history from pole to pole:* American Association of Petroleum Geologists, Mem. 2.

T. H. Clark and C. W. Stearn, 1968, *Geological Evolution of North America:* Ronald Press, New York.

A. J. Eardley, 1962, *Structural Geology of North America:* Harper & Row, New York.

G. W. Fisher et al., eds., 1970, *Studies of Appalachian Geology: Central and Southern:* Wiley-Interscience, New York.

J. Gilluly, 1965, *Volcanism, tectonism, and plutonism in the western United States:* Geological Society of America Special Paper 80.

——, 1967, Chronology of tectonic movements in the western United States: *Amer. Jr. Sci.,* v. 265, p. 306.

W. E. Ham and J. L. Wilson, 1967, Paleozoic epeirogeny and orogeny in the central United States: *Amer. Jr. Sci.,* v. 265, p. 332.

P. B. King, 1951, *The Tectonics of Middle North America:* Princeton University Press, Princeton, N. J.

——, 1959, *The Evolution of North America:* Princeton University Press, Princeton, N. J.

G. E. Murray, 1961, *Geology of the Atlantic and Gulf Coastal Province of North America:* Harper & Row, New York.

G. O. Raasch, ed., 1960, *Geology of the Arctic,* Proceedings of the First International Symposium on Arctic Geology: University of Toronto Press, vols. I and II.

J. Rodgers, 1967, Chronology of tectonic movements in the Appalachian region of eastern North America: *Amer. Jr. Sci.,* v. 265, p. 408.

E-an Zen et al., eds., 1968, *Studies of Appalachian Geology: Northern and Maritime:* Wiley-Interscience, New York.

Central America, South America, and the Antilles

H. J. Harrington, 1962, Paleogeographic development of South America: *Amer. Assoc. Petrol. Geol. Bull.,* v. 46, p. 1773.

W. F. Jenks, ed., 1956, *Handbook of South American geology: An explanation of the geologic map of South America:* Geological Society of America, Mem. 65.

C. Schuchert, 1935, *Historical Geology of the Antillean Caribbean Region:* Wiley, New York.

Europe

G. Abrard, 1948, *Geologie de la France:* Payot, Paris.

J. G. C. Anderson and T. R. Owen, 1968, *The Structure of the British Isles:* Pergamon Press, New York.

J. Aubouin, 1965, *Geosynclines:* Elsevier, New York.

G. M. Bennison and A. E. Wright, 1969, *The Geological History of the British Isles:* St. Martin's Press, New York.

R. Brinkmann, 1969, *Geologic Evolution of Europe,* 2nd ed., transl. J. E. Sanders: Hafner, New York.

S. von Bubnoff, 1926 - 1936, *Geologie von Europa,* 2 vol.: Gebrüder Borntraeger, Berlin.

J. K. Charlesworth, 1963, *Historical Geology of Ireland:* Oliver & Boyd, Edinburgh.

C. Y. Craig, 1965, *The Geology of Scotland:* Archon Books, Hamden, Conn.

J. Goguel, 1965, *Geologie de la France:* Presses Universitaires de France, Paris.

A. Heim, 1922, *Geologie der Schweiz:* Chr. Herm. Touchnitz Verlag, Leipzig.

O. Holtedahl, 1960, Geology of Norway: *Norg. Geol. Unders.,* no. 208.

M. R. W. Johnson and F. H. Stewart, 1963, *The British Caledonides:* Oliver Boyd, Edinburgh.

G. Knetsch, 1963, *Geologie von Deutschland:* Enke, Stuttgart.

D. V. Nalivin, 1960, *The Geology of the U.S.S.R.,* transl. S. I. Tomkeieff: Pergamon Press, New York.

D. H. Rayner, 1967, *The Stratigraphy of the British Isles:* Cambridge University Press, New York.

M. G. Rutten, 1969, *The Geology of Western Europe:* Elsevier, Amsterdam.

T. Strand, 1971, *Scandinavian Caledonides:* Wiley-Interscience, New York.

L. J. Wills, 1951, *A Paleogeographical Atlas of the British Isles and Adjacent Parts of Europe:* Blackie, London.

Africa and the Middle East

A. L. Du Toit, 1954, *The Geology of South Africa,* 3rd ed.: Hafner, New York.

F. Bender, 1968, *Geologie von Jordanien:* Gebrüder Borntraeger, Berlin.

R. Furon, 1963, *Geology of Africa,* transl. A. Hallam and L. A. Stevens: Hafner, New York.

S. H. Haughton, 1963, *The Stratigraphic History of Africa South of the Sahara:* Hafner, New York

R. Said, 1962, *The Geology of Egypt:* Elsevier, New York.

R. Wolfart, 1967, *Geologie von Syrien und dem Libanon:* Gebrüder Borntraeger, Berlin.

A. J. Whiteman, 1971, *The Geology of the Sudan Republic:* Clarendon Press, Oxford.

Asia, Southeast Asia, and Indonesia

M. G. Audley-Charles, 1968, The Geology of Portuguese Timor: *Mem. Geol. Soc. London,* No. 4.

Ta Ch'ang, 1963, *The Geology of China: Communist China:* U. S. Department Commerce, Office of Technical Service, Joint Publication Research Service, Washington, D.C.

H. L. Chhibber, 1934, *The Geology of Burma:* Macmillan, London.

A. Gansser, 1964, *Geology of the Himalayas:* Wiley, New York.

M. S. Kirschnan, 1949, *Geology of India and Burma:* Madras Law Journal Office, Madras.

J. S. Lee, 1939, *The Geology of China:* T. Murby, London.

H. Y. Liu, 1959, *Palaeogeographic Maps of China* (in Chinese): Scientific Press, Peking.

J. B. Scrivenor, 1931, *The Geology of Malaya:* Macmillan, London.

W. D. Smith, 1924, *Geology and Mineral Resources of the Philippine Islands:* Bureau of Printing, Manila.

J. H. F. Umbgrove, 1938, Geological History of the East Indies: *Amer. Assoc. Petrol. Geol. Bull.,* v. 22, p. 1.

R. W. van Bemmelen, 1949, *The Geology of Indonesia:* Government Printing Office, The Hague.

D. N. Wadia, 1966, *Geology of India;* St. Martin's Press, New York.

Australia, New Zealand, and Japan

D. A. Brown, K. S. W. Campbell, and K. A. W. Crook, 1968, *The Geological Evolution of Australia and New Zealand:* Pergamon Press, New York.

T. W. E. David, 1950, *The Geology of the Commonwealth of Australia*: Arnold, London.

M. Minato, M. Gorai, and M. Hunahashi, eds., 1965, *The Geologic Development of the Japanese Islands*: Tsukiji Shokan, Tokyo.

F. Takai, T. Matsumoto, and R. Toriyama, eds., 1963, *Geology of Japan*: University of California Press, Berkeley.

Antarctica

R. J. Adie, ed., 1962, *Antarctic Research*: American Geophysical Union, Geophysical Monograph 7, p. 26.

_____ ed., 1964, *Antarctic Geology:* Wiley, New York.

J. B. Hadley, ed., 1965, *Geology and Paleontology of the Antarctic*: Antarctic Research Series, v. 6, American Geophysical Union, Publ. 1299.

W. Hamilton, 1963, *Tectonics of Antarctica*: American Association of Petroleum Geologists, Mem. 2.

World

B. Kummel, 1961, *History of the Earth*: Freeman, San Francisco and London.

Geologic History by Systems

W. J. Arkell, 1956, *Jurassic Geology of the World:* Oliver & Boyd, Edinburgh.

J. K. Charlesworth, 1957, *The Quaternary Era,* 2 vols.: Arnold, London.

E. D. McKee, S. S. Oriel, et al., 1967, *Paleotectonic Maps of the Permain Systems, maps 1 - 450*: Government Printing Office, Washington, D. C.

_____1967, *Paleotectonic Investigations of the Permian System in the United States: Geological Survey Professional Paper 515*: Washington, D. C.

D. H. Oswald, ed., 1967, *International Symposium on the Devonian System:* Alberta Society of Petroleum Geologists, Calgary.

K. Rankama, 1965, *The Quaternary:* Wiley-Interscience, New York, v. 1 - 2.

_____, ed., 1963, *The Precambrian:* Wiley-Interscience, New York, v. 1 - 4.

J. Rogers, ed., 1956, *El Sistema Cambrinco, su Paleogeografia y el Problema de su Base; Part 1: Europa, Africa, Asia; Part 2: Australia, America* (20th Congreso Geologico Internacional, Mexico).

R. L. Sherlock, 1948, *The Permo-Triassic Formations, A World Review:* Hutchinson Scientific and Technical, London.

F. E. Zeuner, 1959, *The Pleistocene Period:* Hutchinson Scientific and Technical, London.

The Oceans

Initial Reports of the Deep Sea Drilling Project: U.S. Government Printing Office, v. 1 - 12.

10
The Early Paleozoic Era

The abundant and complex life represented in the early Paleozoic fossil record contrasts markedly with the almost complete absence of advanced organisms in Precambrian deposits. Where sufficient paleontological data are available, early Paleozoic deposits may be correlated not only within a continent, but between continents as well. Such correlations permit the realignment of major structural features, such as geosynclines and orogenic belts, and assist in the reconstruction of ancient continents.

In many parts of the world the boundary between fossiliferous Cambrian sedimentary rocks and sparsely fossiliferous Precambrian deposits is marked by an unconformity (Fig. 10.1). In other areas, thick, apparently conformable sequences underlie the lowest fossiliferous beds. Here the base of the Cambrian is placed either just below the oldest fossiliferous deposits or at the first unconformity below these deposits. The latter criterion is somewhat unsatisfactory, since beds that are conformable in one area may be unconformable in another. Thus strata designated as Cambrian could be the time-stratigraphic equivalents of Precambrian deposits elsewhere (Fig. 10.2).

Figure 10.1   The unconformity between the Cambrian Flathead Sandstone and the Archean gneiss is plainly visible on the highway between Cody, Wyoming, and Yellowstone National Park. This unconformity represents 2 billion years of erosion. The tilting of the Cambrian sandstone occurred during the Laramide Orogeny in the early Tertiary.

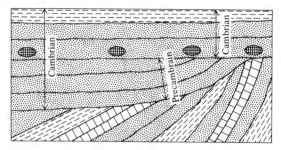

Figure 10.2 Diagrammatic sketch illustrating how beds that are conformable in one area may be unconformable in an adjacent area. If the Cambrian is placed at the first unconformity below the lowest beds containing fossils with hard parts, beds that were deposited contemporaneously would be Cambrian in the left-hand side of the diagram and Precambrian in the right-hand side.

EARLY PALEOZOIC GEOGRAPHY

Paleomagnetic and structural data are useful in reconstructing movements of the Paleozoic continents. If based on the present position of the continents, the early Paleozoic paleomagnetic poles are widely divergent (Fig. 7.17). Even when recalculated for the Pangaea reconstruction of the continents, these poles do not coincide (Fig. 7.41). For example:

1. The recalculated pre-Carboniferous polar wandering curve for Gondwanaland diverges significantly from that of both North America and Europe (Fig. 7.41). This indicates that Gondwanaland was probably separated from North America and Europe by a wide ocean prior to the Carboniferous (Fig. 10.3). However, if it is assumed that North America moved progressively closer to Gondwanaland during the early Paleozoic, the paleomagnetic poles for North America and Gondwanaland agree reasonably well. This movement may have occurred through relative rotation about a pole of spreading located several thousand miles west of South America (Fig. 10.3b).

2. The recalculated Precambrian and Cambrian paleomagnetic poles for North America and Europe during the Precambrian and Cambrian do not agree, but for Ordovician through Triassic times they are in reasonable agreement (Fig. 7.41).

Thus it may be inferred that these two continents were separated during the Precambrian and Cambrian and were joined from the Ordovician through the Triassic.

3. The early Paleozoic polar wandering curve of Ancestral Siberia does not agree with that of Ancestral Europe. However, the Silurian through Triassic paleomagnetic poles agree for these two continents (Fig. 7.17). Thus it is probable that Ancestral Siberia and Ancestral Europe were not joined until the Silurian.

4. The Cambrian paleomagnetic pole of Ancestral China is located relatively close to that of Ancestral Siberia, which suggests that these two continents were rather close to each other at that time (Fig. 10.3).

ANCESTRAL NORTH AMERICA

During the early Paleozoic, thick sequences of sedimentary and volcanic rock were deposited in the geosynclines that encircled Ancestral North America. These include the Appalachian, Ouachita, Cordilleran, Franklin, East Greenland, and Caledonian Geosynclines (Fig. 10.4). In each, a eugeosyncline, characterized by deeper water deposits, bordered a miogeosyncline with shallow-water deposits. The early Paleozoic seas repeatedly inundated the continental interior and deposited a relatively thin cover of sedimentary rock (Fig. 10.5).

Appalachian Geosyncline

The thickness of the lower Paleozoic deposits in the continental interior of Ancestral North America averages approximately 5000 ft (1500 m), but the deposits are several times thicker in the Appalachian Mountains. This increase in thickness is well illustrated by an isopach map for the Ordovician System (Fig. 10.6). Ordovician deposits, which accumulated at a rate of more than 40 ft per million years, occur in an elongate belt extending from Newfoundland to northern Georgia and Alabama. The Appalachian Geosyncline lies along the junction of Ancestral North America with Ancestral Europe and Gondwanaland.

(a)

Figure 10.3. Reconstruction of the continents during the early Paleozoic with paleoclimatic indicators. Paleolatitudes are based on paleomagnetic data from samples which were tested for magnetic stability. The average paleomagnetic pole for each continent (Table 7.1) was used to compute an average pole position for the continents as a whole. (a) Cambrian, (b) Ordovician (directions of ice movement are from R. W. Fairbridge, Ref. 1), (c) Silurian.

(b)

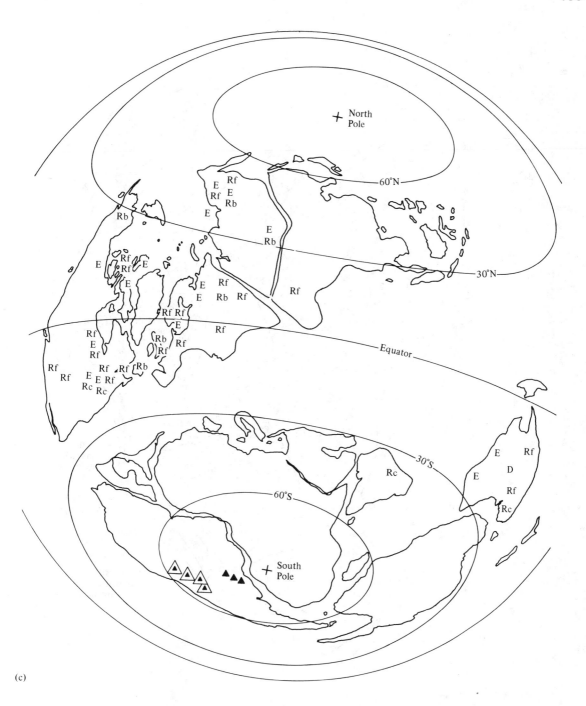

(c)

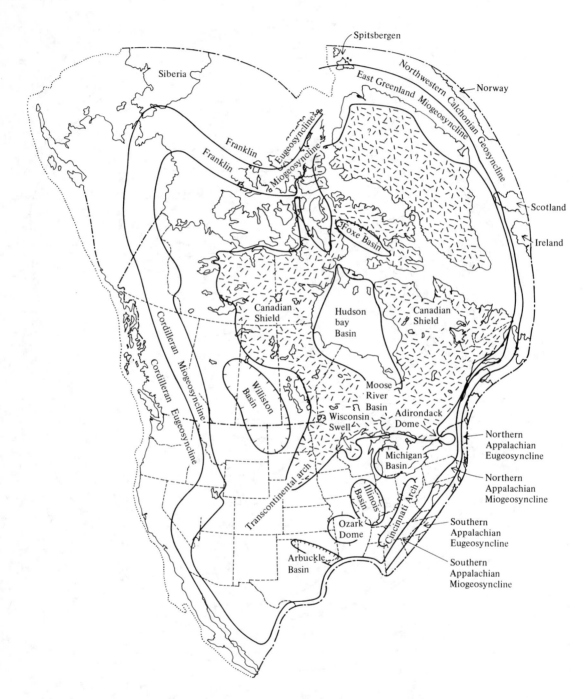

Figure 10.4 Tectonic elements of Ancestral North America during the early Paleozoic.

The boundary between Ancestral North America and the other continents was probably located along the belt of highly metamorphosed and intensely deformed Paleozoic rocks near the center of the Appalachian fold belt. The Appalachian Geosyncline probably consists of geosynclines that were originally marginal to these continents and were brought together during their joining. It can be divided into northern and southern segments (Fig. 10.4).

Northern Appalachian Geosyncline

During the early Paleozoic, a thick layer of limestone, dolostone, shale, and sandstone was deposited in the shallow seas of the miogeosyncline. At the same time, shales and graywackes were being deposited in the deeper waters of the eugeosyncline to the east of the miogeosyncline (Fig. 10.4).

Miogeosyncline. At the beginning of the Cambrian, the seas were largely confined to the eugeosyncline. As they spread westward into the miogeosyncline, a basal sandstone was deposited unconformably on the metamorphic rocks of the Grenville Province. The near-shore depositional environment was replaced by deeper water as the sea moved westward, and carbonates were deposited on the sandstone. In southern Pennsylvania, the carbonate sequence, which ranges from Early Cambrian through Middle Ordovician, is approximately 10,000 ft (3000 m) thick.

Early Paleozoic deposits of the miogeosyncline are generally fossiliferous carbonates which are assigned to the *shelly facies*. The presence of abundant cross-bedding, oolites, algal stromatolites, mud cracks, and erosional disconformities suggests deposition in shallow, well-aerated water. John Rodgers has proposed that this sequence was deposited on an immense carbonate bank, analogous to the modern Bahama Banks (3). In this case, the boundary between the miogeosyncline and the eugeosyncline might have been a reef front that separated the miogeosyncline from the adjacent eugeosyncline (Fig. 10.7). Dolostones, which

are abundant within the Cambrian and Early Ordovician deposits of the miogeosyncline, are indicative of deposition in an arid climate. Paleomagnetic data show that the miogeosyncline was between 25 and 30° south latitude during the Late Cambrian (Fig. 10.5b).

The abundance of carbonates and the absence of angular unconformities in Cambrian through early Middle Ordovician deposits indicate that little deformation occurred in or adjacent to the miogeosyncline during this interval. However, midway through the Middle Ordovician, uplift and erosion in the miogeosyncline was followed by deposition of clastic sediments from the east. The source of the clastics was a broad *geanticline* which developed within the eugeosyncline.

Continued uplift of the geanticline resulted in the sliding of great masses of rock from the geanticline onto the sediments of the miogeosyncline (Fig. 10.8). The slide masses are collectively referred to as the Taconic Allochthon. E-an Zen has postulated that the sediments of the allochthon were originally deposited along what is now the Green Mountain Anticlinorium (4) (Fig. 10.9). Boulders, cobbles, and pebbles in the deposits beneath the allochthon can be matched with rocks found in the allochthon (Fig. 10.10). Thus deposition of the sediments beneath the allochthon was essentially contemporaneous with its emplacement. Since Middle Ordovician deposits are found beneath, within, and on top of the allochthon, it must have been emplaced entirely within the Middle Ordovician (6). The emplacement of the allochthon occurred during the Vermontian phase of the Taconic Orogeny, which was the most widespread deformation to affect the northern Appalachian Geosyncline since its formation. The geanticline contributed sediments to the miogeosyncline during the later part of the Middle Ordovician and the early part of the Late Ordovician. These deposits, which include the Martinsburg Formation, form a clastic wedge that generally increases in thickness toward the southeast.

The sediments of the miogeosyncline were folded, faulted, and metamorphosed during the

Late Ordovician (3). This deformation was the terminal phase of the Taconic Orogeny. The age of this deformation is clearly discernible in eastern Pennsylvania, where deformed Martinsburg strata are unconformably overlain by the Early Silurian Tuscarora Sandstone. In southeastern New York and northern New Jersey, Precambrian gneisses have been thrust over shales equivalent in age to the Martinsburg Formation. A study of gravity and magnetic anomalies in the vicinity of the thrust sheet indicates that the thrust has a displacement of many tens of miles (7). The involvement of basement rocks shows that thrusting was not a result of gravity sliding as in the Taconic Allochthon. Since the youngest rocks involved are Late Ordovician shales, thrusting probably occurred during the terminal phase of the Taconic Orogeny. Uplifting during this time resulted in the deposition of a clastic wedge that is thickest in eastern Pennsylvania (Fig. 10.11). This wedge includes red shales, sandstones, and conglomerates of the Queenston Shale and the Juniata Formation.

Following the terminal phase of the Taconic Orogeny, a thick sequence of conglomerate and sandstone was deposited in the geosyncline by rapidly moving streams and rivers (9). These strata are time transgressive, and they vary from Early Silurian to Middle Silurian in age. Eastward thick-

ening of Late Silurian red beds indicates that deformation and uplift occurred east of the miogeosyncline. This orogeny, the Salinic Disturbance, was relatively minor compared with the Late Ordovician and late Paleozoic deformations in the Appalachian region. It occurred just before the terminal phase of the Caledonian Orogeny in Europe.

Eugeosyncline. The sedimentary rocks of the miogeosyncline grade eastward into eugeosynclinal sequences of shale, sandstone, and volcanics which are of Cambrian and Ordovician age. These deposits include the Taconic Sequence of eastern New York and western Vermont and the East Vermont Sequence of eastern Vermont (3). Graptolites are among the most abundant fossils in the black shales of the eugeosyncline, and hence such deposits are referred to as the *graptolite facies.* The black shales may have been deposited below wave base in relatively deep, poorly aerated water. One probable indication of different depositional environments is that trilobites in the graptolite facies seldom resemble those in the shelly facies. The Ordovician fossils of the graptolite facies are rather abundant and varied, which may indicate deposition in moderately deep water rather than in abyssal depths (10).

Figure 10.5 Paleogeographic maps of Ancestral North America: (a) Lowermost Cambrian, (b) Late Cambrian (Middle Trempealcauan) maximum transgression, (c) Uppermost Early Ordovician, (d) Late Ordovician, Richmond Stage.

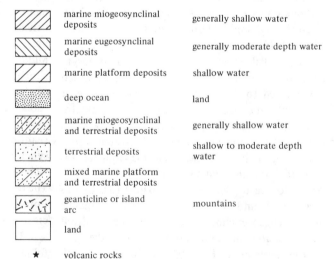

marine miogeosynclinal deposits	generally shallow water
marine eugeosynclinal deposits	generally moderate depth water
marine platform deposits	shallow water
deep ocean	land
marine miogeosynclinal and terrestrial deposits	generally shallow water
terrestrial deposits	shallow to moderate depth water
mixed marine platform and terrestrial deposits	
geanticline or island arc	mountains
land	
★ volcanic rocks	

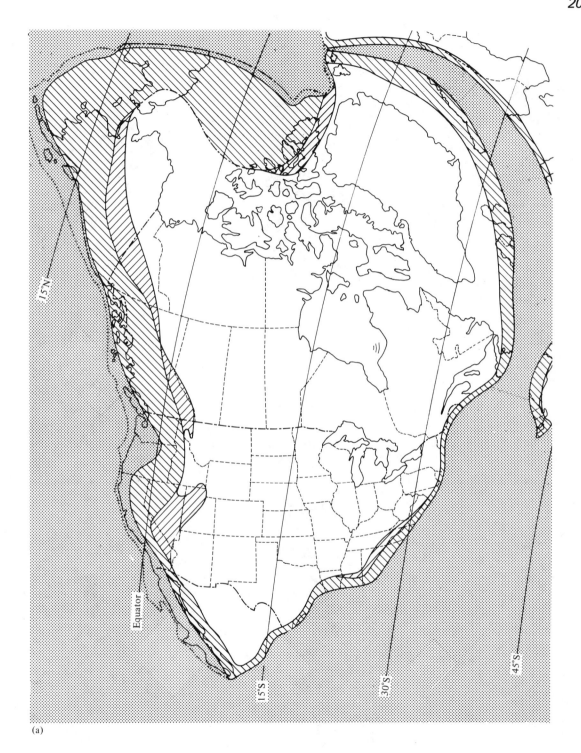

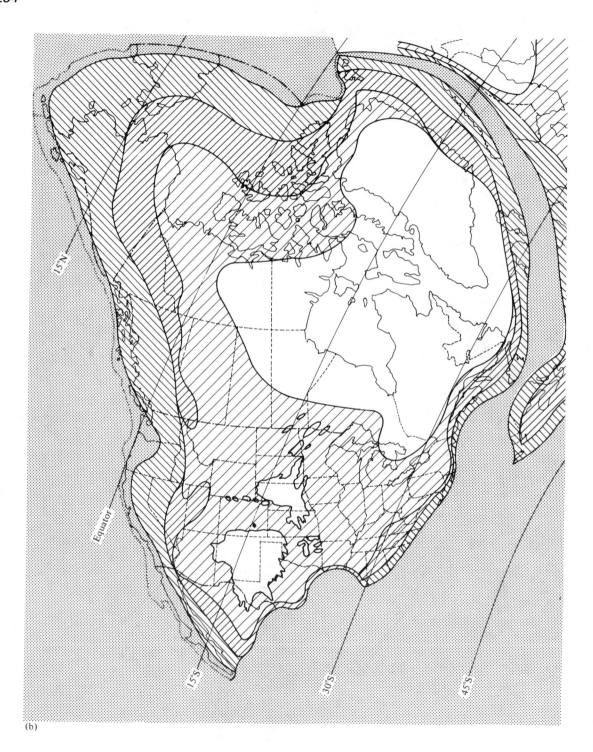

(b)

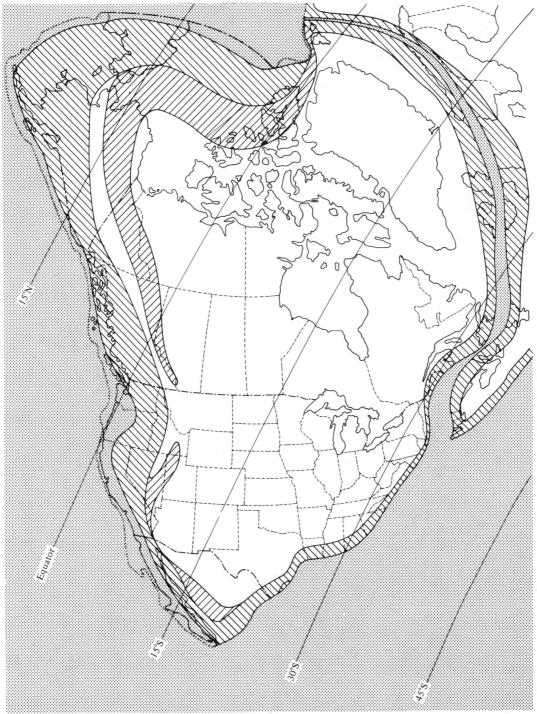

(c)

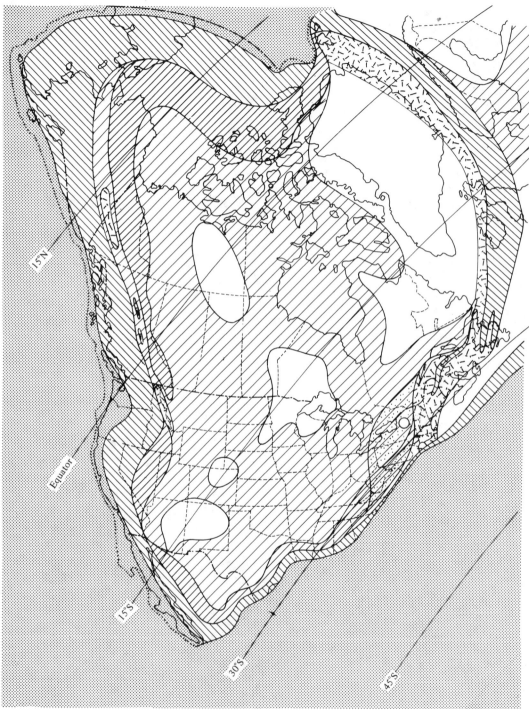

(d)

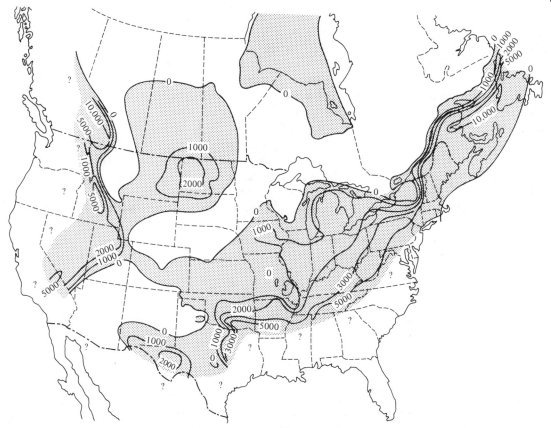

Figure 10.6   Isopach map showing the thickness of Ordovician deposits in the United States and southern Canada. (Compiled by R. Macomber, *in* Sloss, Dapples, and Krumbein, Ref. 2.)

Large blocks of limestone, found in the Cambrian and Ordovician deposits near the western margin of the eugeosyncline, may be the result of submarine landslides (Fig. 10.12). It is likely that these blocks were part of a reef complex bordering a large carbonate bank. Loosened by wave action, they may have slid down the reef front into the deeper waters of the eugeosyncline.

In eastern Vermont, the eugeosynclinal deposits rapidly increase in thickness eastward. Many of the Paleozoic deposits of the Appalachian Geosyncline display similar eastward thickening. Some geologists have suggested that a broad landmass named Appalachia, consisting primarily of crystalline rocks, lay to the east of the geosyncline. More recently, the existence of such a borderland has been questioned. Marshal Kay (8) proposed that much of the clastic sediment of the Appalachian geosyncline originated in volcanic and tectonic islands within the geosyncline. He also suggested that during the Paleozoic, the Appalachian Geosyncline may have been similar to the Indonesian Island Arc. The sedimentary and volcanic rocks of the eugeosyncline are in fact quite similar to those which have been deposited in the vicinity of modern island arcs. Furthermore, the volcanic rocks of the East Vermont Sequence were probably rhyolites, andesites, and basalts before metamorphism. Since andesites of Quaternary age are almost entirely restricted to island arcs, the East Vermont Sequence may have been deposited in a similar environment.

In modern island arcs, an inner volcanic arc is generally bordered by an outer tectonic arc. In the volcanic arc, relatively young volcanic rocks commonly overlie a basement composed of deformed metasedimentary and metavolcanic rocks that have been intruded by granitic rocks. The tectonic arc is made up of a chain of islands containing deformed sedimentary and volcanic rocks. The deposits of the eugeosyncline may have been derived largely from the erosion of such an island arc. However, the scarcity of volcanic fragments in the eugeosynclinal deposits indicates that the erosion of volcanoes did not produce the bulk of the clastics.

Some of the older deposits in the center of the Appalachian Geosyncline have no known basement and may have been deposited on oceanic

crust (3, 6). However, the deposits of the East Vermont Sequence rest unconformably on Precambrian gneisses in the vicinity of the Green Mountain Anticlinorium (4) and were not, therefore, deposited on oceanic crust.

The earliest deformation in the eugeosyncline occurred during the early Middle Ordovician in western Newfoundland. The Taconic Allochthon was emplaced during this deformation, the Bonnian phase of the Taconic Orogeny. During the latter Middle Ordovician and the early Late Ordovician, a thick sequence of clastic sediments, beginning with a basal conglomerate, was deposited in the eugeosyncline (3).

During the terminal phase of the Taconic Orogeny, the eugeosynclinal deposits were

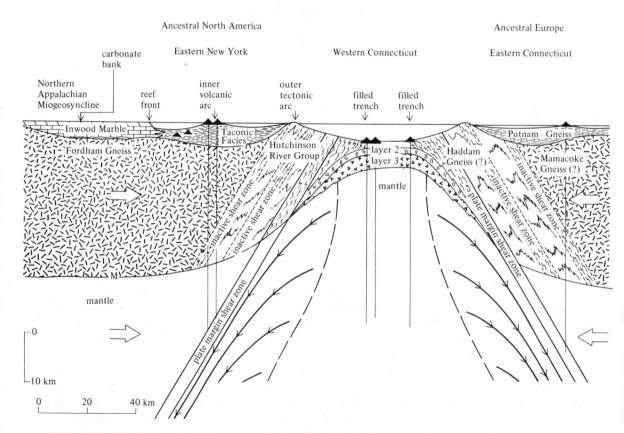

Figure 10.7 Restored cross section across the Northern Appalachian Geosyncline from Ancestral North America to Ancestral Europe during the Cambrian. (After Seyfert and Leveson, Ref. 5).

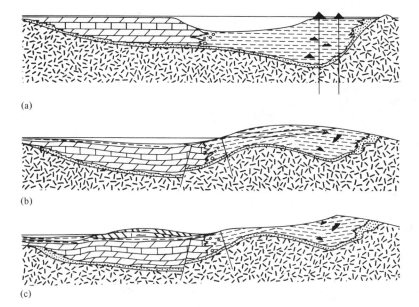

(a)

(b)

(c)

Figure 10.8 Diagrammatic sketches illustrating the formation of the Taconic Allochthon: (a) Deposition of eugeosynclinal and miogeosynclinal sequences during the Cambrian, (b) uplift within the eugeosyncline with clastic sediment deposited over the carbonates of the miogeosyncline, (c) gravity sliding of the allochthon onto the deposits of the miogeosyncline.

strongly folded. Large recumbent folds or *nappes* were transported westward toward the miogeosyncline (11). The resulting highlands supplied sediments for a basal conglomerate of Early Silurian age. This conglomerate is overlain, in turn, by a thick sequence of marine clastic sediments, volcanic rocks, and fossiliferous limestone.

In central Maine, southeastern Quebec, and central Newfoundland, an episode of folding, metamorphism, and intrusion occurred during the Late Silurian or Early Devonian (13). The intrusions have been dated at approximately 400 million years. Pebbles of Silurian slates from Early Devonian deposits in central Maine indicate that the metamorphism was relatively weak (13). This orogenic episode correlates with the Salinic Disturbance. Figure 10.13 presents radiometrically determined dates of igneous and metamorphic rocks from the Appalachian Geosyncline.

The orogenic activity within the Northern Appalachian Geosyncline may have been caused by collisions of large crustal plates during the early Paleozoic. In particular, the volcanism and deformation may have been associated with the joining of Ancestral North America and Ancestral Europe and the resulting movement of the sea-floor under Ancestral North America. The Middle Ordovician deformations might have resulted from the initial joining of these two continents, and the Late Ordovician and Late Silurian orogenic episodes might have been caused by continuing compression between these continents.

Southern Appalachian Geosyncline

The northern half of the Southern Appalachian Geosyncline developed on the margin of Ancestral North America adjacent to Gondwanaland. The boundary between these two continents is probably located within the belt of intensely metamorphosed rocks of the Piedmont Province, which extends from central Virginia to central Georgia.

Miogeosyncline. The Chilhowee Group has 6000 ft (2000 m) of quartzite, shale, and conglomerate. This group, which contains Cambrian trilobites

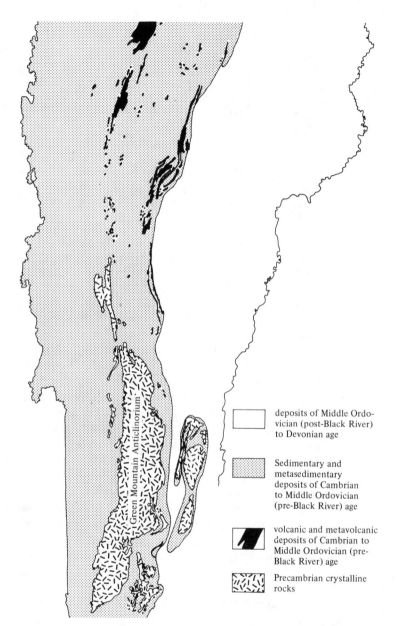

Figure 10.9 Geologic map of Vermont showing the distribution of volcanic rocks of Cambrian through early Middle Ordovician age. The concentration of volcanics indicates that this area was a geosyncline at the time. (Modified from the Centennial Geological map of Vermont, published by the Vermont Geological Survey.)

Figure 10.10 Conglomerate beneath the Taconic Allochthon east of Albany, New York. The large blocks within this deposit are thought to have slid or have been eroded from the advancing thrust sheet. (Photo by E-an Zen, U.S. Geological Survey.)

near the top, is partly or completely Cambrian in age, depending on how the base of the Cambrian is defined. Studies of paleocurrent directions indicate that the sediments were derived from the west or northwest, presumably from Precambrian crystalline rocks of the craton. Overlying the Chilhowee Group is a sequence of limestone and dolostone 9000 ft (3000 m) thick; it ranges in age from Early Cambrian to Middle Ordovician.

In the Middle Ordovician, large amounts of clastic sediment were deposited in the miogeosyncline. These clastics thicken toward the east and grade westward into limestones (Fig. 10.14). The source of the sediments may have been highlands within an island arc on the outermost margin of the continent. This uplift was formed during the Blountian phase of the Taconic Orogeny.

Eugeosyncline. Some of the metamorphosed sedimentary and volcanic rocks in the Piedmont Province may be equivalent to the Chilhowee Group (Fig. 10.15). According to radiometric dating, the rocks of the Piedmont were metamorphosed more than 350 million years ago (Late Devonian) (15). Some of these deposits may have been laid down on oceanic crust (3).

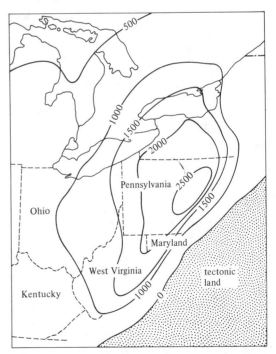

Figure 10.11 Isopach map showing the variation in thickness of Upper Ordovician sediments in Pennsylvania and adjacent states. (After Kay, Ref. 8.)

Figure 10.12 Cow Head conglomerate in western Newfoundland. (Photo by E-an Zen, U.S. Geological Survey.)

Ouachita Geosyncline

Lower Paleozoic sedimentary rocks in the Ouachita Mountains consist of 3000 ft (1000 m) of platform deposits. Magnetic surveying and deep drilling indicate that the axis of the Ouachita Geosyncline lay south of the present Ouachita Mountains during the early Paleozoic. The geosynclinal deposits are now concealed beneath Mesozoic and Cenozoic sediments in the Gulf Coastal Plain.

Cordilleran Geosyncline

The Cordilleran Geosyncline extends in a wide belt from Alaska through the western United States and Canada into Mexico (Fig. 10.4). It joins the Franklin Geosyncline in the north and the Ouachita Geosyncline in the south, and consists of a miogeosyncline and a eugeosyncline. The miogeosyncline generally received carbonates during the early Paleozoic; the eugeosyncline accumulated a great thickness of clastic and volcanic rocks.

Miogeosyncline. Lower Paleozoic deposits in the miogeosyncline include limestones and dolostones with some interbedded sandstones and shales (Fig. 10.16). These deposits often overlie sedimentary rocks of Late Proterozoic age with apparent conformity. In the Canadian Rockies, the direction of

dip of crossbedding in Cambrian sandstones indicates that the sand came from the craton to the east. In general, there was little deformation within the miogeosyncline during the early Paleozoic. However, an angular unconformity has been reported from Middle Cambrian deposits in Montana (16), and disconformities occur between Lower and Middle Ordovician and between Middle Silurian and Middle Devonian beds in many parts of the miogeosyncline (17).

Eugeosyncline. During the early Paleozoic, the miogeosyncline was bordered on the west by a eugeosyncline (Fig. 10.4). In the Klamath Mountains of northern California, W. Porter Irwin has divided the eugeosynclinal deposits into five elongated belts which more or less parallel the Pacific shoreline (17) (Fig. 10.17). These belts are separated by a series of low-angle thrust faults. Radiometric dating of schists and amphibolites from the Central Metamorphic Belt indicates that metamorphism occurred during the Devonian and Carboniferous (18). Thus the parent sediments and volcanics were probably deposited in early Paleozoic time. Fossils in the Western Paleozoic and Triassic Belt indicate that deposition occurred from the Silurian or Devonian to the Late Triassic

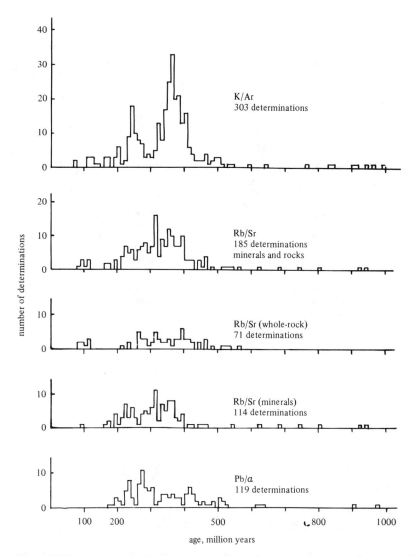

Figure 10.13 Frequency distribution of radiometric dates on igneous and metamorphic rocks from the Appalachian Geosyncline. Peaks of igneous and metamorphic nativity occur during: Early, Middle, and Late Ordovician, Late Silurian, Middle and Late Devonian, Late Mississippian, Late Pennsylvanian, Middle Permian, and Late Triassic or Early Jurassic. (From Lyons and Faul, Ref. 12.)

(18). Deposits in the Eastern Klamath Belt include thick sequences of shale, sandstone, conglomerate, chert, volcanic rocks, and minor limestone (Fig. 10.18).

Although some geologists believe that all the rocks of the eugeosyncline were deposited in deep water on oceanic crust (19), several factors suggest that the eugeosynclinal deposits of the Eastern Klamath Belt and those of western Nevada were deposited in relatively shallow water in the vicinity of an island arc:

1. The widespread unconformities occurring in these deposits indicate extensive subaerial erosion (18 and 20). It is not likely that eugeosynclinal deposits would have been repeatedly

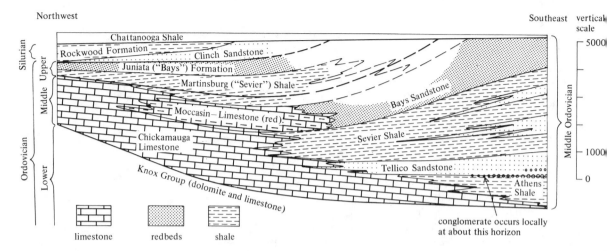

Eastern Tennessee

Figure 10.14 Restored cross section of the Ordovician and Silurian rocks in the Valley and Ridge Province of eastern Tennessee. The thickening of the clastic sediments eastward indicates that their source lay in that direction. The unconformity at the base of the Chattanooga Shale indicates a Devonian deformation affected this area. (After P. B. King, Ref. 14. Reproduced with permission of The American Association of Petroleum Geologists.)

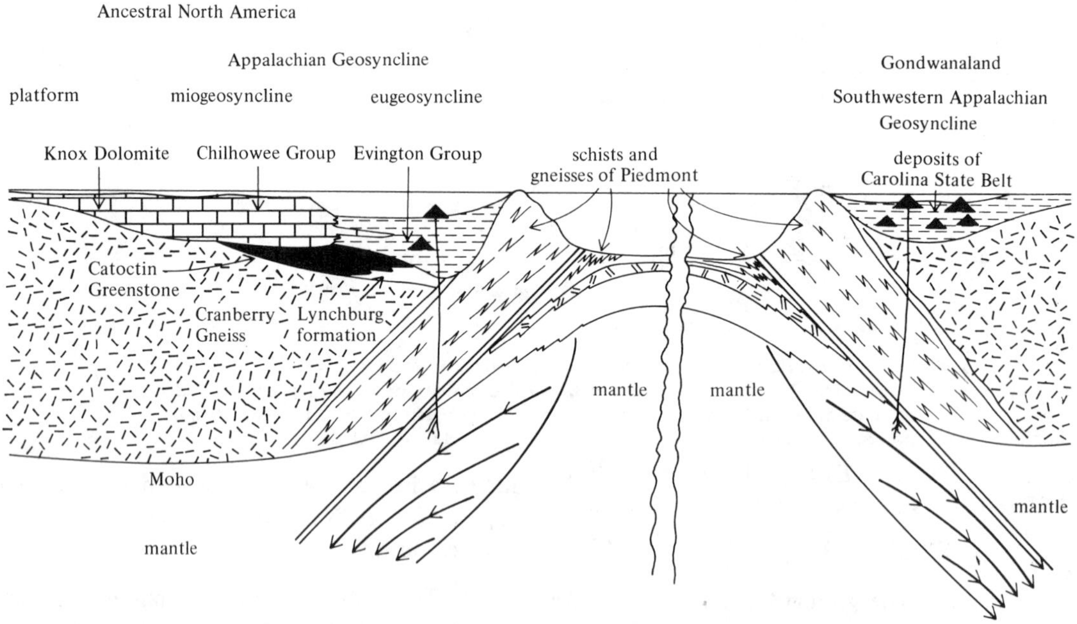

Ancestral North America

Appalachian Geosyncline

Gondwanaland

platform   miogeosyncline   eugeosyncline   Southwestern Appalachian
                                                                  Geosyncline

Knox Dolomite   Chilhowee Group   Evington Group   schists and   deposits of
                                                          gneisses of Piedmont   Carolina State Belt

Catoctin
Greenstone

Cranberry   Lynchburg
Gneiss   formation

mantle   mantle

Moho

mantle   mantle

mantle

Figure 10.15 Restored cross section across the Southern Appalachian Geosyncline during the Cambrian. The section extends southeastward from eastern Kentucky through Virginia to eastern North Carolina.

Figure 10.16   Breached anticline in Banff National Park, Canada. The cliff at the left is composed of Cambrian quartzite overlain by Cambrian limestone. The great thickness of these deposits indicates deposition in a miogeosyncline.

uplifted above sea level and then submerged to great depths.

2. Andesites are common in the eugeosyncline. At the present time, such volcanics are restricted to island arcs.

3. Shallow-water fossils are locally abundant, especially in the limestones (18).

The Central Metamorphic Belt and the Western Paleozoic and Triassic Belt (Fig. 10.17) contain metamorphosed sequences of rhythmically bedded cherts, black shales, graywackes, mafic volcanic rocks, and minor limestone (Fig. 10.19). In general, volcanic flows in these two belts are more common and tuffs less common than in the eastern part of the eugeosyncline. Several lines of evidence suggest that the rocks within the Central Metamorphic Belt and the Western Paleozoic and

Triassic Belt were deposited in deep water on oceanic rather than on continental crust:

1. The rhythmically bedded cherts contain fossil radiolarians and may therefore be lithified radiolarian oozes, similar to those found today in the deep oceans.

2. There is no known granitic basement in either belt.

3. The deposits are more intensely deformed than correlative sequences to the east. A granitic basement would tend to minimize the effects of compressive stress.

The clastics of the eugeosyncline were probably derived in part from volcanic and tectonic islands in an island arc. Some of the graywackes and shales may have been carried down the continental slope by turbidity currents and deposited

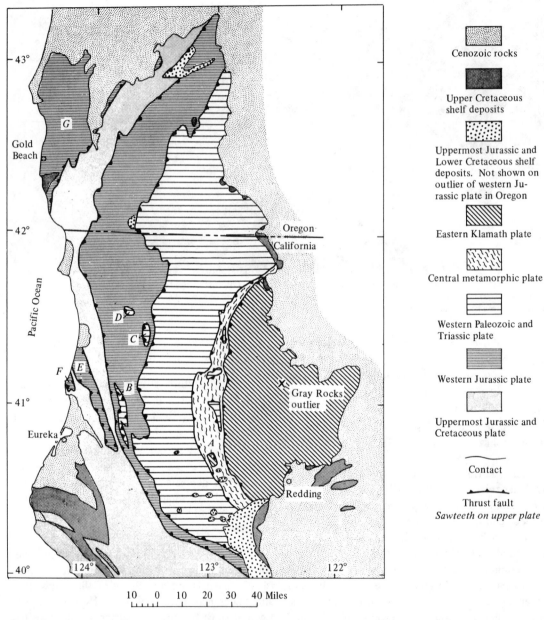

Figure 10.17 Geologic map of northern California and southern Oregon showing the distribution of Irwin's five belts and the thrust faults forming their boundaries. (From Irwin, Ref. 18.)

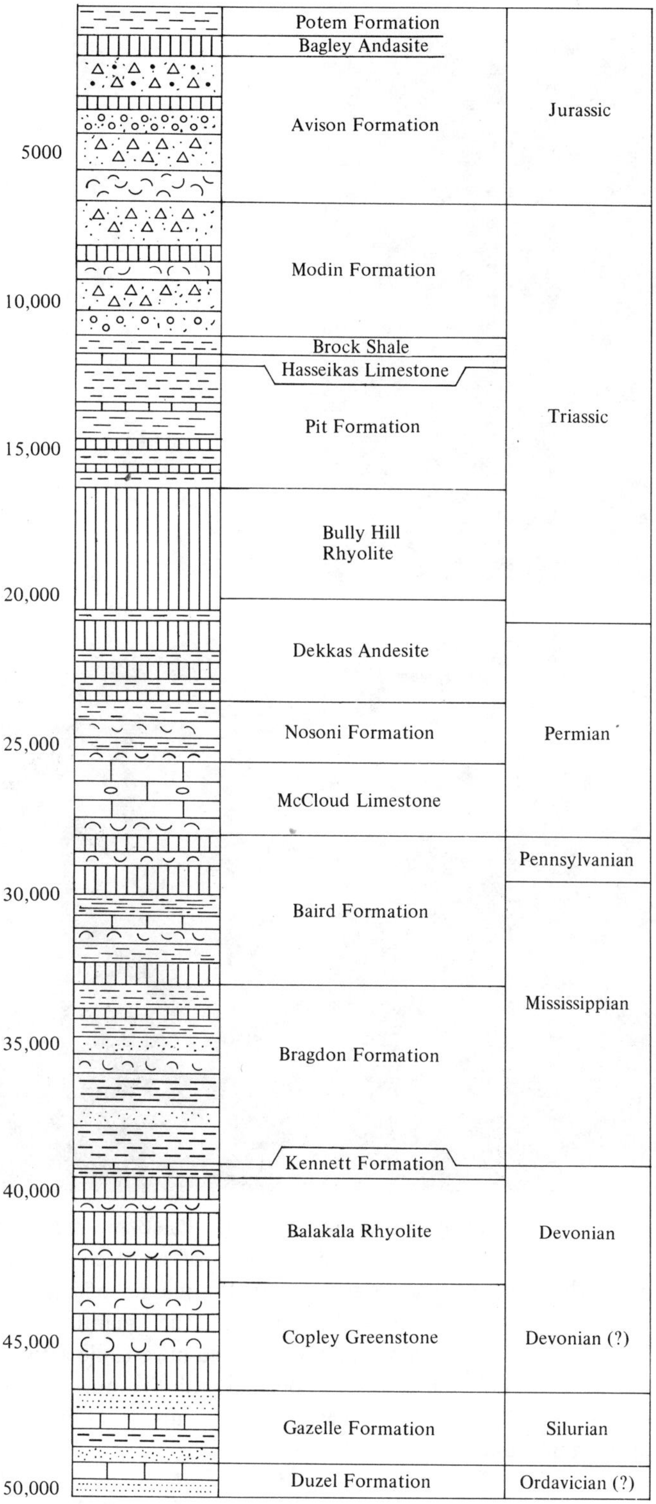

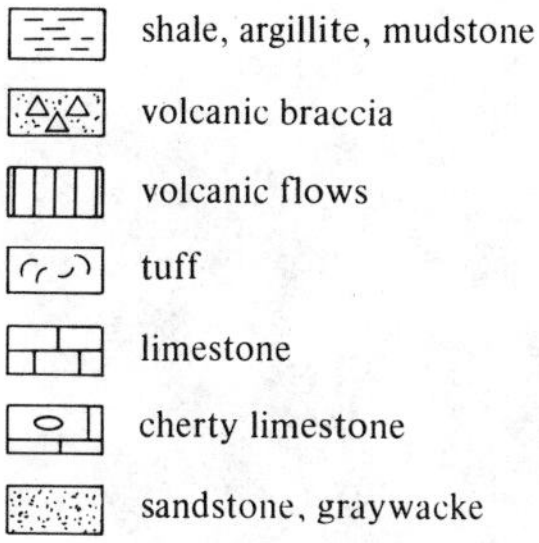

Figure 10.18 Stratigraphic section illustrating the deposits of the Eastern Klamath Belt. The abundance of felsic and intermediate volcanic rocks suggests deposition in the vicinity of an island arc. (Data from Irwin, Ref. 18.)

(a)

Figure 10.19 Rocks of the Western Paleozoic and Triassic Belt in Northern California: (a) Rhythmically bedded chert near Sawyers Bar, California; (b) pillow greenstone near Cecilville, California. (b)

in a trench adjacent to such an island arc. The presence of a great thickness of early Paleozoic clastics in the eugeosyncline suggests that nearby highlands were being uplifted during deposition of the sediments.

Cause of Volcanism, Deformation, and Metamorphism. Tectonic activity in the Cordilleran Geosyncline during the early Paleozoic may have been caused by movement of an oceanic plate beneath an island arc on the margin of the continent. Partial melting of the mantle as it descended would have resulted in volcanic activity, and compression associated with the movement of the sea-floor may have produced folding and uplift.

Some of the early Paleozoic rocks of the Central Metamorphic Belt and the Western Paleozoic and Triassic Belt contain glaucophane and lawsonite, which are minerals that form under conditions of high pressure and low temperature. Continental areas have a geothermal gradient too high for these minerals to form, but high pressures and low temperatures could occur as sediments were carried rapidly beneath a continent. In such an area, sediment deformation and continental growth would occur more or less continuously.

Franklin Geosyncline

The Franklin Geosyncline extends from the Arctic Islands in northernmost Canada to northeastern Greenland (Fig. 10.4). In the west, it joins the Cordilleran Geosyncline and in the east, the East Greenland Geosyncline. Cambrian sequences of geosynclinal thickness are not exposed, but Ordovician and Silurian sedimentary rocks in the southern part of the geosyncline belong to the shelly facies and are dominantly limestone and dolostone. Deep drilling on Bathurst Island has revealed the presence of about 4000 ft (1200 m) of Ordovician evaporites (Fig. 10.20). The Silurian section, which is 10,000 ft (3000 m) thick is one of the thickest Silurian carbonate sequences in the world. Paleomagnetic data indicate that in Ordovician and Silurian time the Franklin Geosyncline was between 10 and 15° north of the equator. Deposits of the shelly facies grade northward into thin black shales of the graptolite facies. Isolated carbonate bodies, which are found with the shales of the graptolite facies, may represent reefs that were built up above the muddy bottom of the sea to form shoals and low islands (Fig. 10.20).

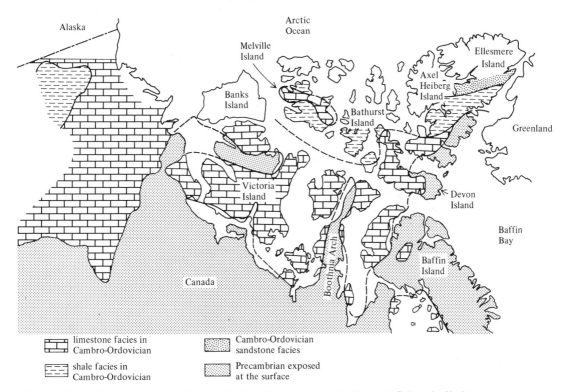

limestone facies in
Cambro-Ordovician

shale facies in
Cambro-Ordovician

Cambro-Ordovician
sandstone facies

Precambrian exposed
at the surface

Figure 10.20 Changes in facies within the Franklin Geosyncline during the early Paleozoic. Notice the carbonate bank on Melville Island and the coarsening of sediments toward the north end of Ellesmere Island. (After Douglas et al., Ref. 21.)

Except for the absence of volcanic rocks, these deposits resemble the eugeosynclinal rocks of the Northern Appalachian Geosyncline. A eugeosynclinal sequence of graywacke, shale, sandstone, and volcanic rocks that is exposed in northern Ellesmere Island and northwestern Axel Heiberg Island contains Ordovician fossils. A mild deformation near the end of the Silurian created the Boothia Arch (Fig. 10.20) at about the same time as the terminal phase of the Caledonian Orogeny in Europe.

East Greenland Geosyncline

The East Greenland Geosyncline extends from northeastern Greenland southward along the coast

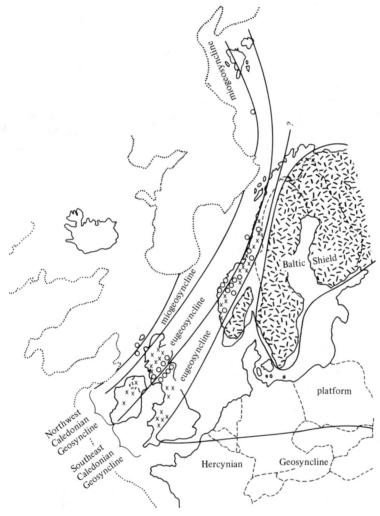

x Early Paleozoic volcanic rocks
o serpentinites and other ophiolitic rocks
·········· 1000 m depth contour

Figure 10.21 Distribution of early Paleozoic volcanics, serpentinites, and related rocks within the Caledonian Geosyncline. The boundary between Ancestral North America and Ancestral Europe is inferred to have been along the belt of serpentinites and other ophiolitic rocks.

to Scoresby Sound. It joins the Franklin Geosyncline on the north, and it was connected to the Northwestern Caledonian Geosyncline before separation of the continents. A basal sandstone of Cambrian age was deposited disconformably on Late Proterozoic strata. These sandstones are overlain by sequences of carbonates between 6000 and 9000 ft (2000-3000 m) thick. The absence of Late Cambrian fossils in this sequence may indicate a disconformity between Middle Cambrian and Early Ordovician beds. These deposits are quite similar in lithology and fossil content to Cambrian and Ordovician miogeosynclinal deposits in Scotland, Spitsbergen, and eastern North America (21, 22). Since at least nine invertebrate species are common to the East Greenland, Appalachian, and Caledonian geosynclines, it is likely that shallow-water connections existed between these geosynclines.

Intense deformation occurred near the end of the Silurian. Great blocks were displaced westward along major thrust faults. This deformation was probably associated with the joining of Ancestral North America and Ancestral Europe. It is interesting to note that the zone along which the continents were joined in early Paleozoic time coincides approximately with the zone along which Europe and Greenland split apart during the Early Tertiary.

Caledonian Geosyncline

The Caledonian Geosyncline is a complex system of depositional troughs which may be divided into two separate geosynclines (Fig. 10.21). During the early Paleozoic the Northwest Caledonian Geosyncline bordered Ancestral North America, and the Southeast Caledonian Geosyncline bordered Ancestral Europe. The northwest segment extends from northwestern Ireland through Scotland and western Norway to Vest Spitsbergen. Part of this geosyncline fills the gap between the East Greenland Geosyncline and the Northern Appalachian Geosyncline. The Northwest Caledonian Geosyncline contains both miogeosynclinal and eugeosynclinal deposits. The miogeosynclinal and some of the eugeosynclinal deposits are known to rest on a sialic basement. However, the eugeosynclinal deposits near the boundary with the Southeast Caledonian Geosyncline may have been deposited on oceanic crust (Fig. 10.22).

Miogeosyncline. In the Northwest Highlands of Scotland and in Vest Spitsbergen, a sequence of Cambrian and Ordovician deposits 4000 to 8000 ft (1300-2600 m) thick unconformably overlies Late Proterozoic sedimentary rocks. As in the East Greenland Geosyncline, these deposits are dominantly carbonates, and a disconformity probably exists between Middle Cambrian and Early Ordovician beds. Silurian deposits are absent from the miogeosyncline. Between the Late Silurian and the Middle Devonian, miogeosynclinal deposits were thrust northward several tens of miles along the Moine Thrust (23).

Eugeosyncline. A belt of schists, gneisses, amphibolites, and marbles extends from northwestern Ireland through the central highlands of Scotland to western Norway. A trilobite of Early or Middle Cambrian age has been found near the top of the 13,000 ft (4000 m) thick Dalradian Sequence within this belt in central Scotland (24). A probable tillite located at the base of the Middle Dalradian suggests that the Lower and the Middle Dalradian are of Late Proterozoic age. The Upper Dalradian consists of graywacke, shale, limestone, and a few volcanic rocks. Algal stromatolites and mud cracks indicate that the Upper Dalradian was deposited in relatively shallow water (24). Considering the similarities of these deposits to the eugeosynclinal deposits of the Appalachian Geosyncline, it is likely that the Northwestern Caledonian and the Northern Appalachian geosynclines were continuous during the early Paleozoic.

The Dalradian Sequence was metamorphosed and folded at least three times prior to the deposition of Early Devonian strata. Potassium-argon dating of biotite and muscovite from the Dalradian show that metamorphism occurred at least 450 million years ago (Middle Ordovician) (25). It is

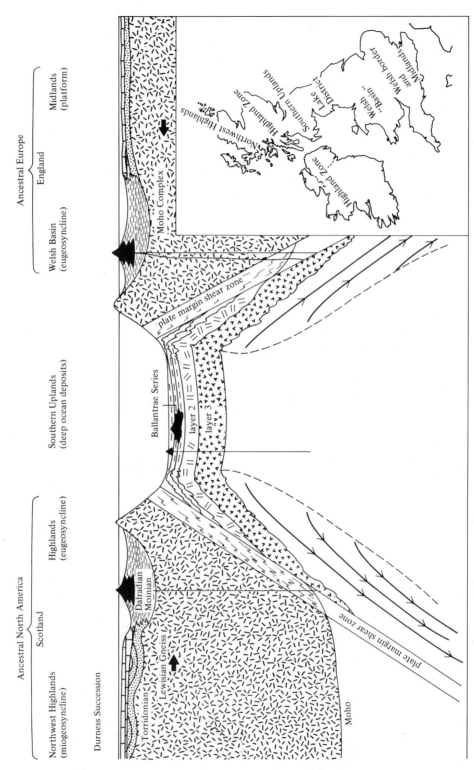

Figure 10.22 Restored cross section across the Caledonian Geosyncline during the Early Ordovician. Insert shows the major geographic divisions of the United Kingdom and Ireland.

222

likely that at least one episode of folding and metamorphism corresponds to an early phase of the Taconic Orogeny in the northern Appalachians.

The Ballantrae Series was deposited during the Early Ordovician and is the oldest known sequence in the Southern Uplands of Scotland. The Ballantrae includes basaltic pillow lavas, volcanic breccias, and tuffs interbedded with graptolitic shales and radiolarian cherts; all of these have been intruded by large serpentinites. The lithologic character of the Ballantrae Series, and the lack of a known basement suggest deposition on oceanic rather than on continental crust.

The Ballantrae Series was deformed prior to deposition of the overlying Middle Ordovician Barr Group and at approximately the same time as the Vermontian or Bonnian Phase of the Taconic Orogeny in the Appalachians. The Barr Group is part of a very thick sequence of clastic sedimentary and volcanic rocks ranging in age from Middle Ordovician to Upper Silurian. Similar deposits are also found in southern Ireland and western Norway (Fig. 10.22).

The Middle Ordovician deposits of the Southern Uplands of Scotland contain a fauna which is remarkably similar to that in the Middle Ordovician deposits of the Southern Appalachian Geosyncline. Approximately 45% of the brachiopod species are common to both regions and many of these species have not been found elsewhere (26). These relationships suggest that a shallow seaway connected the two areas during the Middle Ordovician. However, the Middle Ordovician fauna of the Southern Uplands is only remotely similar to that of the Welsh "Basin" which is now only a little

Figure 10.23   Folded Silurian graywackes and shales in southeastern Scotland. These rocks were folded during the Caledonian Orogeny at the end of the Silurian. (Crown Copyright Geological Survey Photo. Reproduced by permission of the Controller, Her Britannic Majesty's Stationery Office.)

over 100 miles to the southeast (26) (Fig. 10.22) Apparently some sort of barrier, such as a land-mass or a deep sea, prevented faunal migration between these two regions.

A widespread unconformity in the southern part of the eugeosyncline separates folded Late Silurian strata from deposits of latest Silurian to Early Devonian age (Fig. 10.23). This deformation, which was the terminal phase of the Caledonian Orogeny, ended extensive marine deposition in the geosyncline. Younger upper Paleozoic deposits in this region are confined to local basins.

Cause of the Deformation. The first deformation and metamorphism in the northwestern Caledonian Geosyncline occurred between Early Cambrian and Middle Ordovician time. Deformation also occurred during the Middle Ordovician, the Late Ordovician, and the Late Silurian. These orogenic episodes were probably associated with plate collisions. The movement of Ancestral North America relative to Ancestral Europe probably caused much of the early Paleozoic folding in this region. If the Lower Ordovician Ballantrae Series was deposited on oceanic crust, the time of its first deformation and subaerial erosion (Middle Ordovician) may mark the initial contact of Ancestral North America and Ancestral Europe. Subsequent deformations may have been caused by continued compression between these continents.

Continental Interior

The continental interior of Ancestral North America includes that part of the continent which lies between the geosynclines. It consists mainly of lowlands except in the western region, where Cenozoic folding and faulting has produced several mountain ranges. As referred to here, the continental interior is a more extensive region than the stable interior of the continent, or craton.

At the beginning of the Cambrian, the seas were largely confined to the ocean basins and the geosynclines, but two major transgressions of the seas onto the continental interior occurred during the early Paleozoic (Fig. 10.24). The first trans-gression began in Middle Cambrian time and reach-ed its maximum extent during the Late Cambrian (Fig. 10.5b). Basal sandstones of this transgression, such as the Potsdam Sandstone of northern New York, the Flathead Sandstone of Wyoming, and the Tapeats Sandstone in the Grand Canyon, were deposited in near-shore environments. These sand-stones range in age from Middle Cambrian to Early Ordovician, a period covering about 75 million years. Minor regressions during the overall trans-gression produced a complex interfingering of rock units (Fig. 10.24).

As the seas moved toward the center of the continent, basal sandstones were covered by shale or carbonates. During the maximum transgression of the seas, much of the continental interior was inundated, and only higher areas of the Canadian Shield and portions along the Transcontinental Arch remained above sea level (Fig. 10.44). Thus sedimentation was continuous over much of the continental interior between Cambrian and Ordovician time. The warm, shallow seas of the continental interior were favorable to the development of marine life, and fossils are abundant in early Paleozoic deposits. The seas retreated from the continental interior during the Early Ordovician (Fig. 10.5c).

In the Middle Ordovician, the seas again returned to the continental interior. A brief regression in mid-Late Ordovician time may have been associated with Late Ordovician glaciation in the southern continents. Another advance and withdrawal of the seas in latest Ordovician was followed by another transgression, which reached its maximum in the Middle Silurian.

During times of crustal unrest in the Appalachian and Cordilleran geosynclines, clastic sediments were carried across the geosynclines onto the continental interior. For example, extensive marine shales, such as the Utica Shale in central New York, were deposited in the eastern part of the continental interior during and after the Vermontian Phase of the Taconic Orogeny (Middle Ordovician), and continental sandstones and shales were deposited in the eastern continental interior

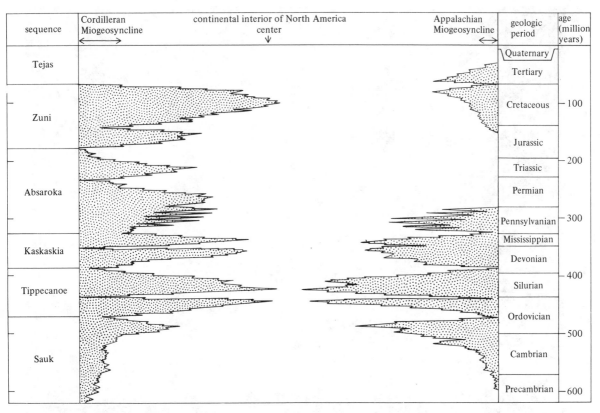

sequence	Cordilleran Miogeosyncline ⟵⟶	continental interior of North America center ↓	Appalachian Miogeosyncline ⟷	geologic period	age (million years)
Tejas				Quaternary	
				Tertiary	
Zuni				Cretaceous	100
				Jurassic	
				Triassic	200
Absaroka				Permian	
				Pennsylvanian	300
Kaskaskia				Mississippian	
				Devonian	
Tippecanoe				Silurian	400
				Ordovician	
Sauk				Cambrian	500
				Precambrian	600

Figure 10.24 Transgressions and regressions of the seas across the continental interior of North America during the Phanerozoic. (Modified considerably after Sloss, Ref. 27.)

as highlands were uplifted during the terminal phase of the Taconic Orogeny (Late Ordovician) (Fig. 10.25). Dolostones of Early and Middle Silurian age form prominent escarpments in the northeastern and north-central United States. One of these, the Lockport Dolostone, is the cap rock of Niagara Falls (Fig. 10.25).

During the Late Silurian, extensive red beds and evaporites were deposited in several basins in the eastern continental interior. The circulation within these basins had been restricted by the development of coral reefs and only limited amounts of seawater of normal salinity entered the basins. The presence of thick evaporites, red beds, and dolomites suggests that deposition occurred in an arid climate, and the abundant reefs prove that the climate was rather warm. Paleomagnetic studies indicate that the northeastern and north-central areas of the United States were at a latitude of about 20°S during the Silurian. A gradual withdrawal of the seas from the continent began during the Middle Silurian and continued to the end of the period.

ANCESTRAL EUROPE

The geosynclines that encircled Ancestral Europe during the early Paleozoic included the Caledonian, Uralian, and Hercynian geosynclines, as well as the eastern half of the Northern Appalachian Geosyncline (Fig. 10.26). Nova Scotia, eastern Newfoundland, New Brunswick, and eastern New England were left on the North American side of the Atlantic Ocean when Europe and North America separated during the late Mesozoic.

Figure 10.25 Silurian marine and continental deposits at the Whirlpool below Niagara Falls. These deposits being much thinner than those in the Appalachian Geosyncline to the east, are therefore platform deposits. (Photo by G. P. Morehead, Jr.)

Southeastern Caledonian Geosyncline

The Southeastern Caledonian Geosyncline extends from southern Ireland and northwestern England through western Norway to Spitsbergen (Fig. 10.26). It contains as much as 30,000 ft (10,000 m) of shale, quartz sandstone, and graywacke, as well as minor amounts of limestone, conglomerate, and andesitic to basaltic volcanic rocks. These eugeosynclinal deposits grade eastward into relatively thin platform deposits. The Cambrian trilobite fauna of the Southern Caledonian Geosyncline is remarkably similar to that found in eastern Newfoundland, Nova Scotia, New Brunswick, and eastern Massachusetts (28). This similarity suggests that there was a relatively shallow seaway between

this geosyncline and the eastern half of the Northern Appalachian Geosyncline.

Unconformities, which are common in the early Paleozoic sequences, were the result of minor episodes of folding and uplift within the geosyncline. In Wales, there are two unconformities in the Cambrian, three in the Ordovician, and at least three in the Silurian sequence. The deformations culminated at the end of the Silurian with the terminal phase of the Caledonian Orogeny, which ended marine sedimentation in the geosyncline (Fig. 10.27). Reconstruction of the continents based on paleomagnetic data indicates that early Paleozoic deformation and volcanism may have been associated with the movement of the sea-

floor under Ancestral Europe as plates containing
these two continents moved toward each other
(Fig. 10.22).

Northern Appalachian Geosyncline

The eastern half of the Northern Appalachian Geosyncline extends from eastern Newfoundland
through eastern New Brunswick and Nova Scotia
to eastern New England (Fig. 10.26). The deposits
in the geosyncline include tens of thousands of
feet of deformed, metamorphosed clastics and
volcanics, which grade eastward into thinner
sequences of mildly deformed limestone, shale,
and sandstone.

Miogeosyncline. In eastern Massachusetts, Cambrian quartzite, red and green shales, and red limestones overlie granitic rocks that have been dated
at between 580 and 610 million years (29). Sequences with similar lithologies and similar fossils
are also found in southeastern New Brunswick,
Nova Scotia, and eastern Newfoundland (28, 30).
These deposits, which are between 500 and 11,000
ft (170 and 3000 m) thick, are of geosynclinal
thickness.

Marine beds younger than Middle Cambrian
have not been found along the Atlantic border of
New England, which may have been above sea
level during most of the Paleozoic. A relatively
continuous sequence of marine strata ranging in

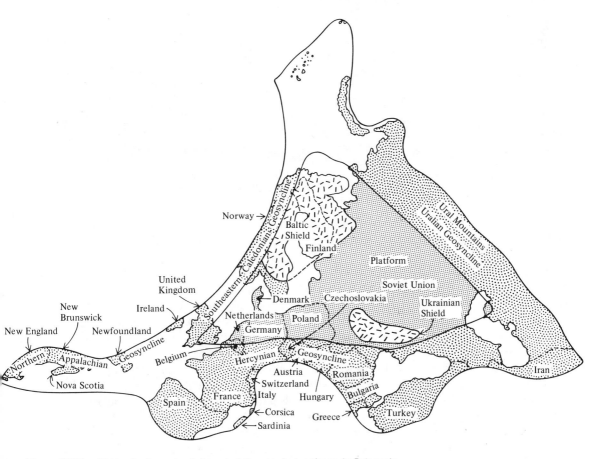

Figure 10.26 Tectonic elements of Ancestral Europe during the early Paleozoic.

Figure 10.27   The Aberystwyth Grits of Silurian age were tilted during the Caledonian Orogeny at the end of the Silurian. Western coast of Wales, England. (Crown Copyright Geological Survey photo. Reproduced by permission of the Controller, Her Britannic Majesty's Stationary Office.)

age from Early Cambrian to Middle Ordovician occurs in western Newfoundland and southeastern New Brunswick. However, except for a narrow trough in western Nova Scotia, most of the miogeosyncline was probably emergent after the Late Ordovician (31).

Eugeosyncline. The eugeosyncline, extending from eastern Connecticut and western Rhode Island through eastern Massachusetts, central New Hampshire, eastern Maine, and central New Brunswick to central Newfoundland, contains more than 40,000 ft (13,000 m) of early Paleozoic deposits (32). These deposits have been subjected to deformation and metamorphism ranging from broad folding and slight recrystallization to intense fold-

ing and high-grade metamorphism. Before metamorphism, the eugeosynclinal deposits were graywackes, shales, siltstones, conglomerates, volcanic rocks, and limestones.

The absence of latest Ordovician and earliest Silurian deposits from much of the eugeosyncline and the presence of angular unconformities at the base of most Silurian deposits indicates that an episode of folding and erosion occurred near the end of the Ordovician (33). Furthermore, an absence of latest Silurian and earliest Devonian deposits in the eugeosyncline indicates that the early Paleozoic deposits were uplifted at about the same time as the terminal phase of the Caledonian Orogeny (31). Silurian eugeosynclinal deposits presumably accumulated on continental crust, since widespread subaerial erosion could occur only in areas underlain by crust thick enough to stand above sea level. If this is correct, Ancestral North America and Ancestral Europe were joined before the beginning of the Silurian.

Hercynian Geosyncline

Extending eastward across southern Europe from Spain to the Caspian Sea, the Hercynian Geosyncline forms a link between the Caledonian and Uralian geosynclines (Fig. 10.26). Deposits of the Hercynian Geosyncline are dominantly clastics and volcanics with only minor limestone. There are no miogeosynclinal deposits associated with these eugeosynclinal deposits. Volcanic activity was especially intense during the Late Cambrian and Early and Middle Ordovician.

In Sardinia, on the southernmost margin of Ancestral Europe, Ordovician shale rests with angular unconformity on Cambrian strata (34). In southern France, an angular unconformity has been reported between Middle and Lower Ordovician beds (35). These unconformities provide evidence that this region was tectonically active during the early Paleozoic. The cause of the deformation and volcanism may have been movement of oceanic plates under Ancestral Europe as it moved toward Gondwanaland.

Uralian Geosyncline

The Uralian Geosyncline is located on the eastern margin of Ancestral Europe and extends the length of the Ural Mountains (Fig. 10.26). This geosyncline consists of a eugeosyncline bordered by miogeosynclines both to the east and to the west (36). Thick sequences of limestone, dolomite, shale, and quartzite in the western miogeosyncline grade eastward into even thicker sequences of shale, arkosic sandstone, and felsic volcanics. Little is known about the eastern miogeosyncline, since it is largely covered by thick Cenozoic sediments.

Unconformities in the early Paleozoic sequences provide evidence of several episodes of folding, crustal uplift, and withdrawal of the seas from the geosyncline. Large mafic and ultramafic

intrusions, which contain important deposits of chromite, asbestos, and platinum, occur on the east flank of the Urals. These intrusions were probably emplaced during an episode of folding in Late Ordovician or Early Silurian time. Folding also occurred in the geosyncline at the end of the Silurian. Early Paleozoic deformations may have been related to the joining of Ancestral Europe and Ancestral Siberia. Paleomagnetic data for these two continents indicates that they were separated until the Silurian. Ordovician deformation may have been caused by the movement of the sea-floor under Ancestral Europe as the continents approached each other. The late Silurian phase of deformation may have resulted from the collision of continental plates.

Continental Interior

The lower Paleozoic strata of the continental interior include sandstone, shale, limestone, and blue clay. The clay, which is of Cambrian age, is remarkable in that it has remained unconsolidated for more than 500 million years. Apparently it was never covered by strata thick enough to cause compaction during all that time. The lower Paleozoic rocks of the continental interior are quite fossiliferous. On the island of Gotland, in the Baltic Sea, limestones contain corals remarkably similar to those of North America. Evidently shallow seas connected these areas during the Silurian. Several disconformities in the lower Paleozoic sequences provide evidence of transgressions and regressions of the seas into the interior of the continent.

GONDWANALAND

Paleomagnetic data indicate that Gondwanaland was separate from the other continents during the early Paleozoic (Fig. 10.3). At that time, geosynclines formed a continuous belt around the continent (Fig. 10.28). These geosynclines included the Tasman Geosyncline in eastern Australia, the Buller Geosyncline in New Zealand, the Southern Tethyan Geosyncline extending

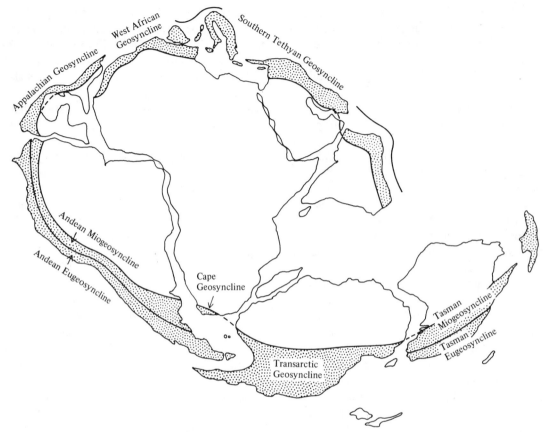

Figure 10.28   Tectonic elements of Gondwanaland during the early Paleozoic.

from northern India to northern Africa, the West African Geosyncline, the eastern part of the Southern Appalachian Geosyncline, the Andean Geosyncline, the Cape Geosyncline in southern Africa, and the Transantarctic Geosyncline.

Tasman Geosyncline

During the Cambrian, the Tasman Geosyncline was divided into a western miogeosyncline with a shallow-water, shelly facies and an eastern eugeosyncline with a deeper water, graptolitic facies (Fig. 10.29). A thick sequence of Lower and Middle Cambrian carbonates in the miogeosyncline contains algal stromatolites, oolites, and intraformational breccias – all features of shallow-water deposition. Sedimentation ceased during the Middle Cambrian, and the deposits were folded in Middle or Late Cambrian time.

During the early Paleozoic, geanticlines were uplifted within the eugeosyncline (Fig. 10.30), and these furnished clastics to adjacent troughs. Basaltic and andesitic volcanic rocks are common along the geanticlines, which may have been part of an island arc on the continental margin (Fig. 10.29). The deformation and volcanism may have been associated with movement of the sea-floor under the eugeosyncline.

Buller Geosyncline

The Buller Geosyncline, which underlies most of New Zealand, was the site of rapid deposition of clastic sediments and volcanic rocks throughout the early Paleozoic (37). Andesites and basalts in the northwestern part of the geosyncline may have been deposited in an island arc environment. The fauna in the associated sedimentary rocks is similar to that in the early Paleozoic deposits of Australia. It is therefore probable that New Zealand and Australia were relatively close during the early Paleozoic.

In the southeastern part of the geosyncline, a sequence of metamorphosed graywackes, shales, and basalts contains fossils as old as the Permian. These deposits are underlain by a great thickness of unfossiliferous rocks, including some that may have been deposited on ocean crust during the early Paleozoic.

Southern Tethyan Geosyncline

The Southern Tethyan Geosyncline extends westward from the Himalayas across West Pakistan, Iran, Turkey, Greece, Yugoslavia, and Italy to North Africa (Fig. 10.28). From Cambrian through Devonian time, this geosyncline was continuous with the West African Geosyncline. Until deposition ceased in the Southern Tethyan Geosyncline in the early Tertiary, it was separated from the Northern Tethyan Geosyncline by a deep ocean. The two geosynclines were brought to-

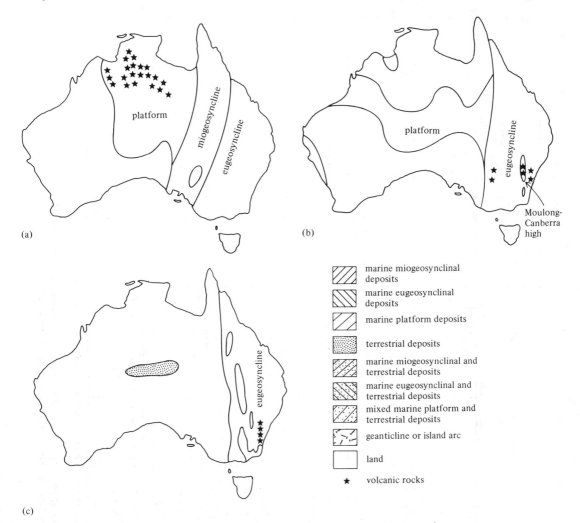

Figure 10.29 Paleogeographic map of the Tasman Geosyncline of eastern Australia (a) Middle Cambrian, (b) Early Ordovician, (c) Early Silurian ages. (Modified after Brown et al. Ref. 37. Used with permission of Pergamon Press, Oxford.)

Geosynclines

Age		Transantarctic, Antarctica	Ref.	Tasman, Australia	Ref.	Buller, New Zealand	Ref.	Andean, South America	Ref.	West African	Ref.	Cape, South Africa	Ref.	Southern Tethyan, India	Ref.
Permian	Upper	intrusion		deformation and intrusion	21			deformation and uplift	25	←---?-----deformation-----?---→		←---?---→ deformation			
	Middle														
	Lower		28*												
Carboniferous	Upper							intrusion	26						
	Lower			intrusion and deformation	21	uplift and intrusion	21	deformation	25	←---?---→ deformation					
Devonian	Upper	intrusion	28	deformation and intrusion	21			deformation	25						
	Middle													uplift	23
	Lower														
Silurian	Upper			deformation and intrusion	37			deformation and uplift	42 43						
	Middle									uplift					
	Lower										39				
Ordovician	Upper			deformation and intrusion	37			deformation and uplift	43						
	Middle														
	Lower														
Cambrian	Upper	deformation and intrusion	46	←---? uplift and deformation ---→	37							intrusion and deformation	44		
	Middle														
	Lower														

*References numbers for Late Paleozoic data refer to Chapter 11

Figure 10.30 Comparison of the history of deformation of Australia, New Zealand, Antarctica, South America, Africa, and India, during the Paleozoic.

gether as India and Africa were separated from the other continents of Gondwanaland and collided with Eurasia.

The lower Paleozoic deposits of the Southern Tethyan Geosyncline include between 5000 and 17,000 ft (1500 and 5500 m) of sandstone, shale, limestone, and minor conglomerate. The scarcity of volcanic rocks here indicates that these sequences are miogeosynclinal in character. During the Cambrian, a thick sequence of salt was deposited in the Middle East, which according to paleomagnetic data was then at a latitude of 20° south. No major breaks in sedimentation occur in the lower Paleozoic sequences of the geosyncline, although there is a possible disconformity at the top of the Silurian in the Himalayas (Fig. 10.31). The lack of deformation and volcanism may indicate that there was no subduction zone near the geosyncline during the early Paleozoic.

West African Geosyncline

A Paleozoic orogenic belt extends from the western end of the Southern Tethyan Geosyncline in Morocco along the west coast of Africa to Portuguese Guinea (Fig. 10.28). Although the geology of this area is not well known, it was probably a geosyncline during the early Paleozoic. Geosynclinal deposits include dolomite, shale, sandstone, and limestone. Since rocks as young as Devonian are involved in the folding of this belt, the final deformation occurred after that time (39).

Southern Appalachian Geosyncline

Before separation of the continents, the West African Geosyncline was continuous with the eastern half of the Southern Appalachian Geosyncline (Fig. 10.28). Early Paleozoic sedimentary rocks in this geosyncline are largely covered by younger deposits, but they crop out in the Carolina Slate Belt of central North Carolina and South Carolina, as well as in adjacent parts of Virginia and Georgia. It is probable that these deposits were originally laid down in a eugeosyncline; they are at least 30,000 ft (10,000 m) thick and include volcanic flows, breccias, and tuffs interbedded with sand-

stone and slate. Fossils are rare, but it is likely that a Cambrian trilobite found in a stream bed in central North Carolina was derived from this sequence (40). Radiometric dating of rhyolites from the Carolina Slate Belt indicates that volcanism occurred during the Cambrian (41). The felsic composition of some of the volcanic rocks suggests that the eugeosyncline may have been part of an island arc (Fig. 10.15).

Andean Geosyncline

The Andean Geosyncline occupied the western margin of South America more or less along the present trend of the Andean Mountains (Fig. 10.28). Although the early Paleozoic deposits in the eastern part of the geosyncline do not contain volcanic rocks, those to the west contain appreciable volcanics. Thus the Andean Geosyncline consisted of both a miogeosyncline and eugeosyncline (Fig. 10.32).

The miogeosynclinal deposits are 5000 to 10,000 ft (1500-3000 m) thick and are dominantly clastics with only minor carbonates. Anhydrite of probable Cambrian age in western Bolivia was probably deposited under arid or semi-arid conditions (42). According to paleomagnetic studies, this area was at a latitude of approximately 45° south during the Cambrian. Glacial deposits of Early Ordovician and Middle Silurian age have been reported from western Bolivia and northwestern Argentina (42). This region was at approximately 60° south latitude during the Ordovician. It has been suggested that lower Paleozoic basalts and associated sedimentary rocks in western Colombia were deposited in the deep ocean on an oceanic rather than a continental crust (43).

Folding and metamorphism occurred in the geosyncline during the Late Ordovician and the later half of the Silurian. The early Paleozoic orogeny may have been related to the movement of an oceanic plate under an island arc on the margin of South America.

Cape Geosyncline

The Cape Geosyncline, located on the southern tip

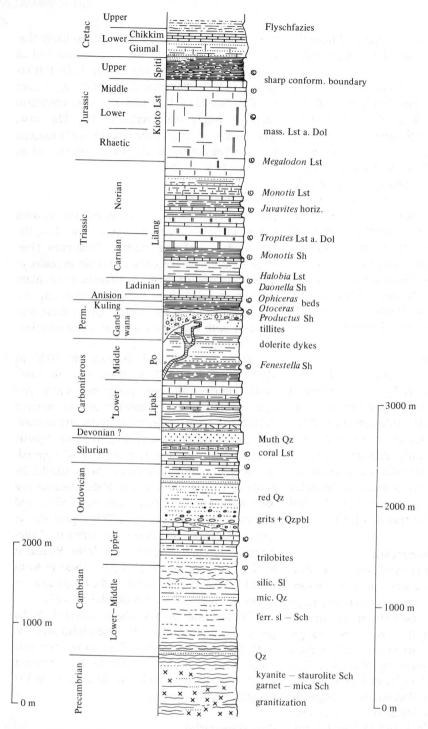

Figure 10.31   Columnar section of the deposits in the Spiti region in the western Himalayas. (After A. Gansser, Ref. 38.)

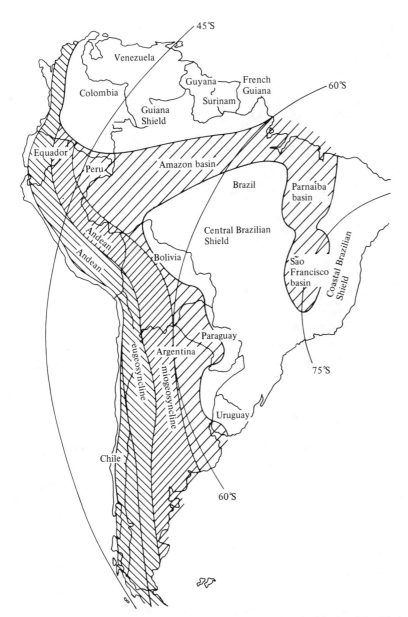

45°S
Venezuela
Guyana French
Guiana
Colombia
Surinam 60°S
Guiana
Shield
Equador
Peru
Amazon basin
Brazil Parnaíba
basin
Central Brazilian
Shield
Bolivia
São
Francisco
basin
Coastal Brazilian Shield
Paraguay
Argentina
75°S
eugeosyncline
miogeosyncline
Uruguay
Chile
60°S

Figure 10.32 Paleogeographic map of South America during the Silurian. (Modified after Harrington, Ref. 42. Used with permission of The American Association of Petroleum Geologists.)

of South Africa, is a small fragment of a geosyncline that extended from South America to Antarctica during the early Paleozoic (Fig. 10.28). The Table Mountain Series, a sequence of sandstone and shale 5000 ft (1500 m) thick, unconformably overlies the Cape Granites (Fig. 10.33).

The granites have been dated at approximately 510 million years (latest Cambrian) by the uranium-lead method (44), and fossils of late Ordovician age have recently been found near the top of the series. One of the shale horizons in the Table Mountain Series contains randomly distribu-

Figure 10.33 Unconformity between the Cape Granite (Ordovician) and the Table Mountain Series (Late Cambrian) at Cape of Good Hope, South Africa. The road is just above the unconformity. (Photo by Warren Hamilton, U.S. Geological Survey.)

ted pebbles and boulders, some having striated and faceted surfaces (Fig. 10.34). Although these markings suggest glacial transportation, the deposit does not rest on a grooved and striated basement. The lower part of the pebble-bearing shale is undeformed, but the upper part is folded and may have been overridden by a glacier shortly after deposition (45).

Transantarctic Geosyncline

The Transantarctic Geosyncline parallels the Transantarctic Mountains from the Wedell Sea to Victoria Land (Fig. 10.28). During the Early and Middle Cambrian, clastics, limestones, and volcanics were deposited with apparent conformity on Late Proterozoic deposits. The entire sequence was deformed, metamorphosed, and intruded by granites in Late Cambrian or Early Ordovician time between 450 and 520 million years ago (46) (Fig. 10.35). This deformation occurred at approximately the same time as the intrusion of the Cape Granites in South Africa and the deformation of the Cambrian deposits in the southern part of the Tasman Geosyncline.

The axis of the Transantarctic Geosyncline shifted toward the Pacific Ocean during Ordovician and Silurian time. A thick sequence of clastics unconformably overlies metamorphosed Cambrian deposits in northeast Victoria Land on the Pacific end of the Transantarctic Geosyncline. These deposits were intruded by granitic rocks in the Late Devonian, and therefore some are probably of early Paleozoic age (Fig. 10.36). A thick early Paleozoic sequence is also present in the Ellsworth Mountains near the center of the Transantarctic Geosyncline (47).

Continental Interior

Marine sedimentary rocks of early Paleozoic age are rather restricted in the continental interior of Gondwanaland. Widespread deformation and uplift in Gondwanaland during the Late Proterozoic and the early Paleozoic may have limited marine incursions into the continental interior.

Figure 10.34   Striated pebbles from the Table Mountain Series, South Africa. (Photo from DuToit, Ref. 45. Reproduced with permission of Oliver and Boyd, Edinburgh.)

Australia. During the early Paleozoic, a large depositional basin was located in central Australia, and several smaller basins lay along the western and northern borders of the continent (Fig. 10.29). Deposits in these basins include sandstone, shale, limestone, dolomite, gypsum, and anhydrite. Evidently, Australia was in an arid or semiarid climate during the early Paleozoic. Paleomagnetic studies show that Australia was within 30° of the equator at that time (Fig. 10.3).

India and Pakistan. In the Salt Range of northern West Pakistan, a 1200 ft (400 m) thick sequence of Lower Cambrian clastics contains salt pseudomorphs and gypsum. The evaporites indicate that deposition occurred under somewhat arid conditions. At the time of deposition, this area was located at approximately 25° south latitude according to paleomagnetic data (Fig. 10.3a). No lower Paleozoic sedimentary rocks are known from the continental interior of India.

Africa. Lower Paleozoic sedimentary rocks have been found in several basins along the west coast of Africa. The scarcity of Cambrian marine deposits suggests that the seas were largely confined to the geosynclines at that time. In the Sahara Desert, shales and sandstones of Late Ordovician age are overlain by tillites that contain striated and faceted boulders and rest on grooved and striated pavements (Fig. 10.37). The tillites, which are overlain by graptolitic shales of Early Silurian age, interfinger with marine deposits of Late Ordovician age near the northern margin of the continental interior (1,49). Late Ordovician glacial deposits cover a very large part of northern Africa

Figure 10.35   Large anticline in early Paleozoic carbonate and clastic rocks in the Neptune Range, Pensacola Mountains, Antarctica. (Photo by J. R. Ege.)

and were probably deposited by continental glaciers (Fig. 10.3b). The direction of ice movement, as determined by the orientation of striations, *roches moutonnés,* glacial valleys, and chatter marks, is from south to the north, *away* from the present equator (49). Paleomagnetic data indicate that the South Pole was located in west central Africa during the Late Ordovician. This is approximately the region from which the Late Ordovician ice sheets appear to have radiated. The Ordovician glacial deposits all occur within 50° of the Late Ordovician South Pole (Fig. 10.3b).

South America. Lower Paleozoic sedimentary rocks have been identified in the Amazon, Parnaiba, and Paraná Basins (Fig. 10.32). Although Cambrian and Ordovician fossils are rare, sedimentary rocks of these ages are thought to underlie fossiliferous Silurian sequences. The Iapo Formation of southern Brazil contains striated boulders, presumably of glacial origin. The exact age of these deposits is not known, but they un-

derlie Early Devonian strata. They may be the same age (Late Ordovician) as the Saharan Glacial deposits.

Antarctica. Early Paleozoic rocks have not yet been found in the continental interior of Antarctica, where a thick ice sheet covers the bedrock.

Southeastern United States and Southern Mexico. The original boundary between Ancestral North America and Gondwanaland probably lies along the belt of strongly metamorphosed rocks in the center of the Appalachian and Ouachita Orogenic Belts. Therefore, the southeastern United States and southern Mexico have been included in Gondwanaland, rather than in Ancestral North America. Wells in southeastern Alabama, southern Georgia, and northern Florida have penetrated relatively undisturbed shales of Ordovician and Silurian age (50). In central Florida, volcanic rocks possibly of early Paleozoic age overlie Late Proterozoic or

Cambrian granitic and metamorphic rocks. The latter rocks have been dated at between 480 and 530 million years (Late Cambrian to Early Ordovician) (51). Radiometric dates in this range are common elsewhere in Gondwanaland.

Relatively thin early Paleozoic sequences have been found in the states of Tamaulipas and Hidalgo, Mexico, approximately 250 miles (400 km) south of Texas. These deposits include conglomerate, sandstone, shale, and limestone which range from Early Cambrian to Late Silurian (52). A very thin sequence of limestone and shale of Cambrian and Ordovician age has been found in Oaxaca, approximately 200 miles (300 km) southeast of Mexico City (53).

ANCESTRAL SIBERIA

The geosynclines that bordered the continental interior of Ancestral Siberia include the Uralian Geosyncline in the west, the Angara Geosyncline in the south, the Northern Pacific Ocean Geosyncline in the east, and the Taimyr Geosyncline in the

north (Fig. 10.38). The lithologies of the geosynclinal sequences and their deformational histories are summarized in Table 10.1.

Volcanic and associated sedimentary rocks in the Angara Geosyncline were probably deposited in an island arc environment on the margin of Ancestral Siberia. If this is correct, the present location of these geosynclinal deposits near the center of the Asian continent would be the result of the joining of Ancestral Siberia and Ancestral China. The initial contact of these two plates may have occurred during the Silurian.

It is likely that the volcanic and associated sedimentary rocks of the Northern Pacific Ocean Geosyncline were deposited in the vicinity of an island arc. Since this region is presently near the margin of the continent, early Paleozoic deformation and volcanism was probably due to the movement of an oceanic plate under the continent.

The Taimyr Geosyncline is also located on the margin of the Asian continent. The folding and metamorphism that occurred within the geosyncline at the end of the Silurian may have been

Figure 10.36   Folded graywacke and siltstone of Ordovician (?) age, northern Victoria Land, Antarctica. (Photo by Warren Hamilton, U.S. Geological Survey.)

Figure 10.37 Ordovician glacial features in the Sahara Desert: (a) Tillite, (b) faceted boulder, (c) striated pavement, (d) *roche moutonnée.* [(a), (c) and (d) from S. Beuf et al., Ref. 48; (b) from R. W. Fairbridge.)

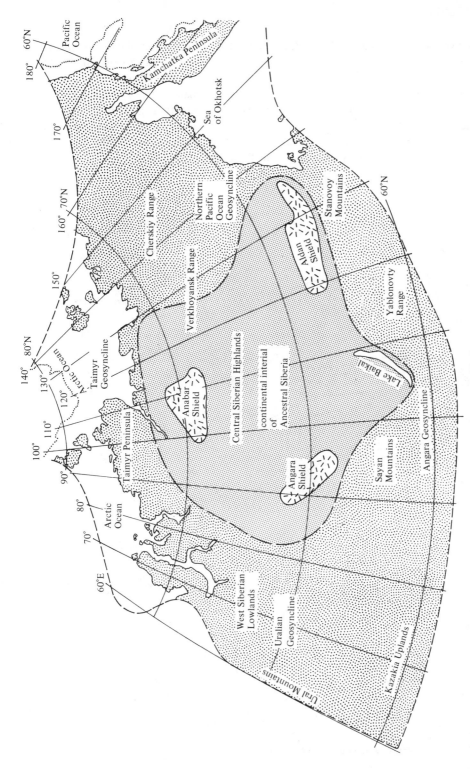

Figure 10.38 Tectonic elements of Ancestral Siberia during the early Paleozoic. Modern latitudes and mountains are included for the purposes of location.

caused by the movement of an oceanic plate under the continent.

ANCESTRAL CHINA

During the early Paleozoic, Ancestral China was bordered on the north by the Central Asia Geosyncline, on the southwest by the Northern Tethyan Geosyncline, on the south by the Indo- nesian Island Arc, and on the east by the Southern Pacific Ocean Geosyncline (Fig. 10.39). The continental interior of Ancestral China contains a number of small, isolated shields, separated by platform deposits (Fig. 10.39). A summary of the history of deposition and deformation within the geosynclines and the continental interior is given in Table 10.1.

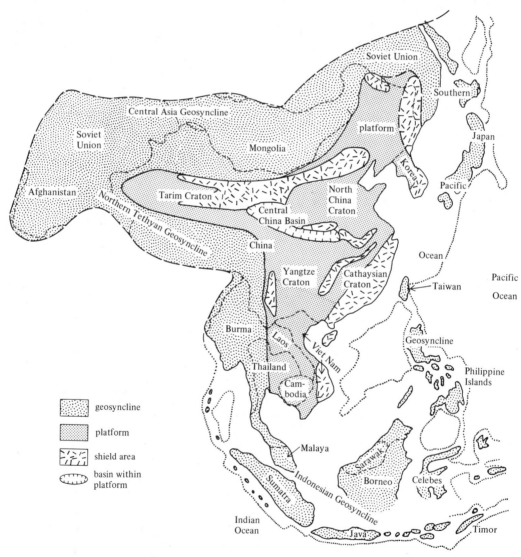

Figure 10.39   Tectonic elements of Ancestral China during the early Paleozoic.

Table 10.1   Summary of the depositional and deformational history of Ancestral Siberia and Ancestral China during the early Paleozoic.

	Deposits	Orogenic activity	Inferred climate
Ancestral Siberia			
Angara Geosyncline	Thick limestones and clastics in the north, mafic and felsic volcanics in the south	Folding at the end of the Silurian	Warm
Northern Pacific Ocean Geosyncline	Thick limestones, shale, sandstone, and volcanics—Silurian deposits are more than 10,000 ft (3000 m) thick	Uplift during the Ordovician; folding near the close of the Silurian	Warm, perhaps tropical—area was near paleoequator
Taimyr Geosyncline	Thick limestones, and also shale, sandstone, and felsic volcanics; sedimentary rocks dominantly marine; red beds and evaporites also present	Clastic wedge indicates deformation during the Late Silurian	Warm and dry
Continental Interior	Cambrian: sandstone, clay, dolostone, limestone, and evaporites—Ordovician: abundant limestone, clay, sandstone, evaporites and red beds—Silurian: argillaceous limestones, clays, and red beds	Local uplift during the Silurian	Warm and dry—paleolatitude 10°S during Cambrian
Ancestral China			
Southern Pacific Ocean Geosyncline	Very thick Cambrian limestones in southeast; felsic volcanics interbedded with Silurian shale and limestone in Japan	No major orogenic activity	Warm
Central Asia Geosyncline	Great thicknesses of clastics and volcanics; some limestones	Early Cambrian uplift and late Silurian folding and metamorphism	?
Northern Tethyan Geosyncline	Conglomerate, sandstone, shale, limestone, and volcanics; sequence is 15,000 ft (5000 m) thick in Tibet	None	?
Continental Interior	Cambrian shales grading upward into limestones—Early and Middle Ordovician limestones—Late Ordovician to Late Silurian shale, sandstone, and limestone	Middle Ordovician and Late Silurian uplift	Warm—paleolatitude between 30 and 60°N during Ordovician

Volcanic rocks are present in all the geosynclines bordering Ancestral China, and it is possible that the continent was encircled by island arcs during the early Paleozoic. If so, oceanic plates probably would have been descending on most if not all margins of the continent. The seas spread from the geosynclines onto the continental interior during the Early Cambrian. The abundance of the Cambrian limestones in both the geosynclines and the platforms is of interest in view of the relatively high paleolatitude of this region (Table 10.1). This suggests that during the Cambrian, the climate may have been relatively mild at high latitudes in Ancestral China. The seas retreated from most parts of the continental interior in Middle Ordovician time. During the late Ordovician, seas invaded only the southern part of the continental interior where they remained until the end of the Silurian.

EARLY PALEOZOIC LIFE

Early Cambrian fossil assemblages differ considerably from those of the Precambrian. Even where Cambrian and Precambrian sequences are apparently conformable, invertebrates with calcareous shells are not found in the latest Precambrian rocks. On the other hand, nearly all the important invertebrate phyla have been found in Lower Cambrian strata. Of the important index forms, only bryozoans, cephalopods, conodonts, stony corals, echinoids, pelecypods, graptolites, and vertebrates are missing from the Early Cambrian fossil record. All these groups appear in the fossil record by Middle Ordovician time.

Origin of Calcareous Shells

The reason for the sudden appearance of invertebrates with calcareous shells at the beginning of the Cambrian remains one of the great mysteries of geology. How could such a varied, advanced, and abundant fauna evolve with almost no fossil evidence of its development? As discussed in Chapter 9, impressions of a variety of advanced, soft-bodied invertebrates have been found in Late Proterozoic deposits in many parts of the world. Thus the lack of a record of the development of

Figure 10.40 Algal stromatolites of Cambrian age from Saratoga, New York. (New York State Museum and Science Service, Albany.)

Figure 10.41   Restoration of the fauna from the Middle Cambrian Burgess Shale in British Columbia. The fauna includes: (a) Archeocyathids, (b) trilobites, (c) sea cucumber, (d) an annelid worm, (e) jellyfish, (f) trilobite-like arthropods. (Courtesy of the Field Museum of Natural History, Chicago; prepared by George Marchand under the direction of I. G. Reimann.)

the Cambrian fauna is probably due to the absence of calcareous shells among Precambrian organisms. However, this raises the question of why so many groups developed calcareous shells in such a relatively short interval of time. A number of theories have been proposed to explain this phenomenon:

1. It has been suggested that the atmosphere may have been relatively low in oxygen during the Precambrian and earliest Paleozoic time. Since ozone is triatomic oxygen, the ozone content of the atmosphere would also have been low. Ozone in the modern atmosphere helps to screen out harmful ultraviolet radiation. Calcareous shells may have provided protection from such radiation, and therefore selection may have operated in favor of organisms with calcareous shells in earliest Paleozoic time.

2. Some paleontologists have suggested that Precambrian seas were too acidic to allow invertebrates to extract carbonate for shell formation. The presence of abundant calcareous stromatolites in Precambrian deposits shows that this was probably not the case. Furthermore, high acidity cannot account for the absence of siliceous skeletons in Precambrian rocks, since silica is readily precipitated in an acid medium. Since siliceous sponge spicules first appear in Cambrian deposits, it is unlikely that the chemistry of the seawater was a significant factor in the appearance of organisms with hard parts.

3. It is possible that shell development is related to the rise of predators. If Precambrian invertebrates were herbivores or scavengers, for example, hard skeletons would not have been advantageous. As predatory forms evolved, shells became essential for the survival of many invertebrates.

4. Finally, the appearance of calcareous shells may have been climatically controlled (56). The ice age at the close of the Proterozoic Era brought about significant changes in the chemistry, temperature, and depth of the oceans. Decreased ocean temperatures during the ice age produced an increase in the solubility of calcium carbonate. Warming of the ocean water following the ice age then resulted in an increase in the calcium carbonate available for shell formation. At the end of the Precambrian, the seas were quite restricted. As sea level rose with the melting of the glaciers, new environments developed into which expanding faunas could migrate. The transgression of the seas into the miogeosynclines during the Early Cambrian coincided approximately with the appearance of fossils with hard parts. With an increase in the area of the seas, there was less competition among marine forms and this, in turn, may have facilitated dispersal and evolution.

Protists

Stromatolites produced by blue-green algae are common in lower Paleozoic deposits (Fig. 10.40). They often formed reefs in the shallow waters of carbonate banks. In British Columbia, green algae and red algae have been found in the Burgess Shale along with a rich invertebrate fauna.

Agglutinated foraminiferans, which built tests by cementing sand grains together, have been reported from strata of Early Cambrian age. Radiolarians have also been reported from Cambrian deposits, but some paleontologists have questioned whether these fossils have been correctly identified. In any event, the remains of foraminiferans and radiolarians are too rare and too poorly preserved in lower Paleozoic sequences to be useful as guide fossils.

Archeocyathids and Sponges

The spongelike archeocyathids occur in Early and Middle Cambrian sequences throughout the world (Fig. 10.41). Since they are restricted to this time range, they are very useful for intercontinental correlations. Archeocyathids are the oldest known sessile organisms and were among the first to secrete a hard, calcareous shell or test.

Sponges first appear in Lower Cambrian sequences and are abundant enough in Silurian deposits to be useful as guide fossils. Complete sponges with interlocking networks of spicules have been found in the Middle Cambrian Burgess Shale.

Coelenterates

Soft-bodied anemones were present in Cambrian seas, and stony corals were common during the Ordovician and Silurian. Ordovician corals were mainly colonial, tabulate corals that formed both the horizontally bedded biostromes and occasionally moundlike bioherms (Fig. 10.42). True coral reefs are common in Silurian limestones and dolostones. These reefs were a mixture of colonial corals, solitary corals, stromatoporoids, and a variety of other invertebrates such as gastropods, brachiopods, and trilobites (Fig. 10.43).

Bryozoans and Brachiopods

Bryozoans are not known with certainty from

Figure 10.42 Reconstruction of the sea bottom as it might have appeared during the Middle Ordovician in Illinois. The fauna includes: (a) Straight-shelled nautiloid cephalopods, (b) trilobites, (c) colonial corals, (d) solitary corals, (e) gastropods, (f) and brachiopods. (Courtesy of the Field Museum of Natural History, Chicago.)

Cambrian strata, but they are abundant in Ordovician and Silurian strata. In shaly limestones or limey shales they frequently comprise a significant percentage of the rock (Fig. 10.44).

Cambrian inarticulate brachiopods were generally only a fraction of an inch in size. They first appear in the fossil record in Early Cambrian deposits. Articulate brachiopods also occur in rocks of Early Cambrian age, but they are less abundant than the inarticulate variety. In some sequences, brachiopods occur stratigraphically below the lowest trilobites. Possibly brachiopods developed

hard shells before the trilobites did. During the Ordovician, articulate brachiopods increased and the inarticulate variety decreased in importance. The abundance of brachiopods in lower Paleozoic deposits makes them useful as guide fossils (Fig. 10.44).

Mollusks

Pelecypods first appear in rocks of Ordovician age and are fairly common in Upper Ordovician sequences. However, they are not useful as guide fossils. Gastropods are found in Early Cambrian

Figure 10.43   Reconstruction of a coral reef as it might have appeared during the Middle Silurian in Illinois. The fauna includes: (a) Colonial tabulate coral *Favosites,* (b) colonial tabulate coral *Halysites,* (c) solitary corals, (d) trilobites, (e) brachiopods, (f) cystoids, and (g) a nautiloid cephalopod. (Courtesy of the Field Museum of Natural History, Chicago; prepared by George Marchland under the direction of I. G. Reimann.)

strata and are relatively abundant in Ordovician and Silurian sequences. Since some species of gastropods have relatively short time ranges, they may be useful as guide fossils.

The oldest fossils that have been classed as cephalopods are small conical shells in rocks of Early and Middle Cambrian age in Europe (57). However, it is not certain whether these are true cephalopods. Fossils unquestionably identified as cephalopods have been found in Late Cambrian deposits in North America and Asia. The shells of these cephalopods are between 0.5 in. (1 cm) and

Figure 10.44   Middle Silurian (Niagaran) fossils from Western New York: (a) *Eucalyptocrinites caelatus,* crinoid calyx (X 3.5), (b) *Rhynchotreta americana,* brachiopod (X 1.5), (c) *Fardenia subplana,* brachiopod (X 1.5), (d) *Heliolites,* sp., tabulate coral (X 5), (e) *Trematopora tuberculosa,* (X 5), (f) *Dalmanites limulurus,* trilobite (X 2), (g) *Fenestella* (lower right), bryozoan (X 3), (h) *Dictyonema retiforme,* a dendroid graptolite (X 2). (Photograph courtesy of Carlton Brett.)

(a)

(b)

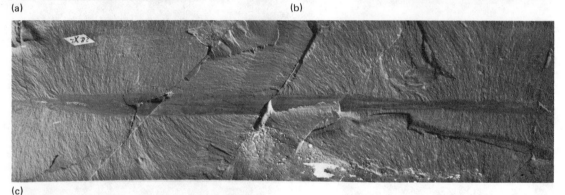

(c)

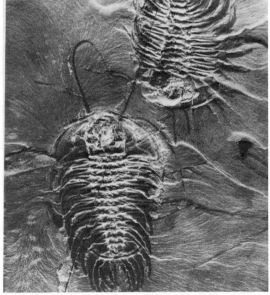

(d)

Figure 10.45   Fossils from the Middle Cambrian Burgess shale from British Columbia: (a)-(c) Annelids, (d) trilobites belonging to the species *Olenoides serratus.* (Courtesy of Smithsonian Institution.)

several inches long. During the Early Ordovician, many new groups appeared and some developed shells as long as 17 ft (5 m) (Fig. 10.42a). The first coiled shells appeared during the Ordovician, but the straight-shelled forms dominated for the remainder of the early Paleozoic.

Annelids and Similar Organisms

Carbon residues of segmented worms have been found in the Middle Cambrian Burgess Shale (Fig. 10.45a-10.45c). Conodonts became abundant during the Silurian and are useful in intercontinental correlations because of their widespread occurrence and distinctive characteristics.

Arthropods

Trilobites are the most abundant fossils in Cambrian sequences (Fig. 10.45d). Since they evolved rapidly and ranged widely, they are good index fossils for Cambrian and Ordovician sequences. They reached their climax in the Late Cambrian and declined with the rise of predators such as cephalopods and fish. Trilobites remained abundant throughout the Ordovician and Silurian, but their numbers were proportionally less than during the Cambrian. Some trilobites were blind burrowers; others crawled about on the sea bottom or were adapted to swimming or floating. They did not have a biting mouth, which restricted their diet to relatively soft material. They may have been scavengers feeding on decaying matter, or they may have fed on microorganisms.

Ostracodes appeared first in the Cambrian and became very abundant during the Ordovician. These microscopic, bivalved crustaceans have proven very useful in subsurface correlations.

Eurypterids have been found in beds of Ordovician age in association with normal marine assemblages. During the Silurian, they appear to have migrated into brackish-water environments. Specimens from Silurian deposits in the vicinity of Buffalo, New York, reach a maximum of 9 ft (3 m) long, but most species were just under 1 ft (25 cm) long (Figs. 10.46 and 10.47).

It has been suggested that the earliest land animals were scorpionlike arthropods which first appeared during the Silurian. However, the ecology of these organisms has not been established definitely. Undoubted land-dwelling animals do not appear in the fossil record until the Devonian.

Echinoderms

The oldest known echinoderms are found in Early Cambrian strata in southern California. These include a primitive cystoid and a spindle-shaped form quite unlike later echinoderms. Crinoids first appeared during the Ordovician, and from then until the end of the Paleozoic, they were abundant in shallow epicontinental seas (Fig. 10.44). Starfish, which first arose during the Ordovician, have

Figure 10.46 *Eurypterus remipes lacustris,* most common eurypterid in vicinity of Buffalo, New York, during Late Silurian. (Courtesy Buffalo Museum of Science.)

never been an important component of marine faunas.

Graptolites

Graptolites first occur in rocks of Cambrian age, and they became abundant during the Ordovician. They are common in many black shales where they are preserved as carbon residues (Fig. 10.48). Since graptolites evolved rapidly and were quite widely dispersed in Ordovician and Silurian seas, they are useful in making correlations both within and between the continents. Early Ordovician graptolites include *Dictyonema,* which was a colonial form having a large number of branches attached to a floating structure. *Tetragraptus,* a four-branched graptolite, appeared somewhat later in the Early Ordovician, and *Didymograptus,* a two-branched form, is found in Middle Ordovician deposits. The evolutionary trend toward simple

Figure 10.47　Reconstruction of eurypterids that lived during the Late Silurian in the Buffalo area. (Courtesy of Buffalo Museum of Science; diorama prepared by George and Paul Marchand, under the direction of I. G. Reimann.)

forms continued into the Silurian with the appearance of *Monograptus,* which had but a single row of thecae.

Vertebrates

The oldest fossil remains of vetebrates are fragments of fish from the Lower Ordovician deposits both in the Baltic region of northern Europe and in Missouri. Fragments of fish are relatively abundant in the Middle Ordovician Harding Sandstone in the central Rockies. Like the modern lamprey, these fish had no jaws and belong to the class Agnatha. These bony-plated forms, which are known as ostracoderms, were seldom more than a few inches long (Fig. 10.49). They were probably filter feeders, since the head contained a set of gill pouches that could have been used for straining out small particles of food. Unlike many of the more primitive filter feeders, the ostracoderms had the advantage of being mobile and therefore were able to search out sources of food. Attempts to trace the evolution of the vertebrates from the invertebrates have led to the suggestion that the invertebrate ancestor may have been a filter feeder closely related to the echinoderms (Fig. 10.50).

Plants

Seaweedlike fossils, such as *Newlandia,* are found in many Cambrian and Ordovician age limestones, and there have been reports of land plants of Cambrian age. However, nearly all reports of pre-Silurian land plants are probably in error, either because the plants were not vascular or because the age of the enclosing strata was not correctly determined (59). The earliest authenticated fossils of vascular plants have come from Middle Ordovician deposits of central Europe (60). These remains included spores and plant tissues. Spores that may have been derived from vascular land or semiaquatic plants have also been found in Early Silurian deposits from western New York (60). Well-preserved fossil land plants have been found in Late Silurian strata.

EARLY PALEOZOIC CLIMATES

The widespread occurrence of thick limestones and coral reefs of early Paleozoic age has led some geologists to conclude that the entire earth had a warm and equable climate during this interval. However, this evaluation was made by comparing climatic indicators with modern latitudes, not paleolatitudes. Furthermore, the recent discovery of widespread glacial deposits of Late Ordovician age in the Sahara Desert has shown that there were severe climatic fluctuations during part of the early Paleozoic.

Cambrian

Since the glaciation that occurred at the end of the Late Proterozoic was very extensive, it is probable that the climate at the beginning of the Cambrian was rather cool throughout much of the world. With widespread transgressions of the seas during the Cambrian, the climate warmed considerably.

In the Canadian Rockies, southern California, and southwestern Nevada, Cambrian deposits are dominantly carbonates of considerable thickness. These areas were very close to the Cambrian equator (Fig. 10.3a). The Cambrian and Ordovician rocks of the Appalachian Miogeosyncline are also very thick and are dominantly carbonates. This region was between 20 and 35° south during depo-

sition of the carbonates. Relatively thick limestones were deposited at 75° south in Morocco and between 30 and 50° south in China. Reefs or reeflike bodies formed by algae and archeocyathids generally occur within 30° of the Cambrian equator, but are found at 75° south in Morocco. Important salt deposits occur between 20 and 30° north in Iran and at about 15° north in Siberia. Gypsum and anhydrite were deposited at 30° south in the Andes, between 5 and 25° north in Australia and the Soviet Union, and at 75° south in Morocco. The thick limestones, reefs, and evaporites at relatively high latitudes in China and Morocco indicate that the climate may have been warmer than average in the middle to high latitudes during the Cambrian.

Ordovician

Glacial deposits of Ordovician age occur within 50° of the Ordovician South Pole in Africa and South America. The deposits in the Saharan region are so extensive that they must have been de-

Figure 10.48 Graptolite (*Nemagraptus*) from the Lower Paperville Shale of Ordovician age. These carbonized remains on a bedding plane represent a colony that contained many individual organisms on each stipe. (Courtesy of Smithsonian Institution.)

Figure 10.49   A restoration of the ostracoderm *Hemicyclaspis.* (Courtesy of the Field Museum of Natural History, Chicago.)

posited by continental rather than alpine glaciers (Fig. 10.3b). The Ordovician red beds of the northeastern United States and Siberia were depos-

ited between 10 and 30° of the paleoequator, and rather thick deposits of salt and gypsum accumulated during the Ordovician in the Arctic Islands of northern Canada at a paleolatitude of approximately 10° south. Ordovician reefs generally fall within 30° of the paleoequator. Thus the climate for most of the Ordovician was not abnormal with respect to temperature and humidity. However, during the Late Ordovician glaciation, the climate at middle and high latitudes in the southern hemisphere was colder than average, perhaps being similar to that during the Pleistocene. Presumably the climate in the northern hemisphere would also have cooled somewhat at the time of the glaciation.

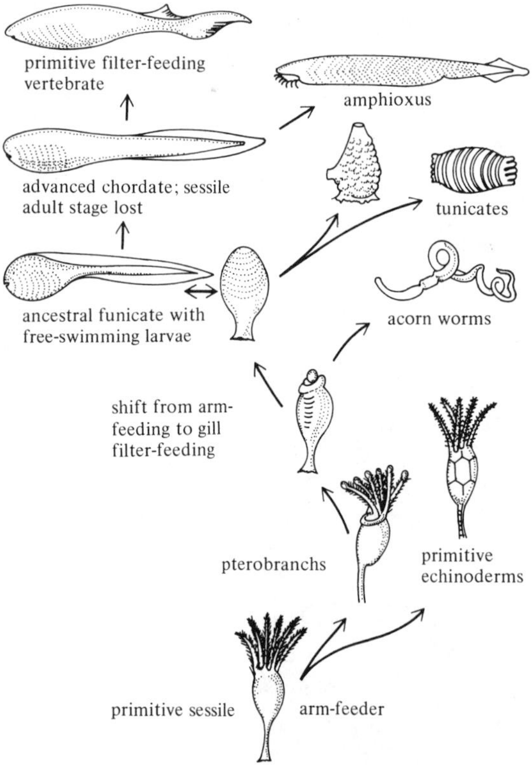

Silurian

Coral reefs, evaporites, and red beds generally fall within 40° of the Silurian equator, whereas glacial deposits occur only within 25° of the Silurian South Pole (Fig. 10.3c). Thus climates during the Silurian probably resembled modern climates.

SUMMARY

At the beginning of the Cambrian Era, the five ancestral continents were separated from each other by deep oceans. These oceans gradually became smaller throughout the early Paleozoic as the ancestral continents moved closer together. The epicontinental seas were largely confined to the outermost margins of the continents at the begin-

primitive filter-feeding vertebrate

amphioxus

advanced chordate; sessile adult stage lost

tunicates

ancestral funicate with free-swimming larvae

acorn worms

shift from arm-feeding to gill filter-feeding

pterobranchs

primitive echinoderms

primitive sessile   arm-feeder

Figure 10.50   Diagrammatic family tree suggesting the possible mode of evolution of the vertebrates. (Copyright © 1967 by American Association for the Advancement of Science (From Romer, Ref. 58.)

ning of the Cambrian, but they transgressed toward the centers of the continents during the Cambrian. By the Late Cambrian, seas covered more than half of the area of the continents.

Sedimentation was most rapid in the geosynclines bordering the ancestral continents. The eugeosynclines received thick deposits of clastic sediments and volcanics, while thick limestones were deposited in most miogeosynclines. Sedimentation on the platforms was rather slow and there was deposition of limestone, shale, or sandstone.

Extensive deformation at the end of the Cambrian resulted in the formation of mountains in much of the southern hemisphere and parts of the northern hemisphere as well. Deformation during the Middle and Late Ordovician in the Appalachian, Caledonian, and Uralian regions resulted in the uplift of elongated ridges within the geosynclines and deposition of clastic wedges both in the miogeosyncline and on the platform. Near the close of the Silurian, the deposits of geosynclines were deformed and metamorphosed. The early Paleozoic deformations may have been caused by the progressive joining of Ancestral North America, Europe, Siberia, and China.

The Cambrian marks the first appearance of fossils with calcareous and siliceous skeletons. The trilobites were the most important invertebrates in the Cambrian, whereas the Ordovician saw the rise of the brachiopods, graptolites, bryozoans, crinoids, and mollusks. The first fish appeared during the Ordovician, but they did not become an important part of the marine fauna until the late Paleozoic. The first land plants and the first land animals are found in Ordovician or Silurian rocks and in Silurian or Devonian rocks, respectively.

The climate at the beginning of the Cambrian was rather cool, but it warmed up rapidly during the period. The Early and Middle Ordovician were warm, but during the Late Ordovician, the climate cooled, especially in the middle and high latitudes of the southern hemisphere. At that time, extensive glaciation occurred in Africa and possibly in South America as well. The climate remained cool in the middle and high latitudes of the southern hemisphere during the Silurian.

REFERENCES CITED

1. R. W. Fairbridge, 1970, Ice age in the Sahara: *Geotimes,* v. 15, no. 6, p. 18.

2. L. L. Sloss, E. C. Dapples, and W. C. Krumbein, 1960, *Lithofacies Maps:* Wiley, New York.

3. J. Rodgers, 1968, The eastern edge of the North American Continent during the Cambrian and Early Ordovician, *in* E-an Zen et al., eds., *Studies of Appalachian Geology: Northern and Maritime:* Wiley-Interscience, New York, p. 141.

   J. Rodgers, 1970, *The Tectonics of the Appalachians:* Wiley-Interscience, New York.

4. E-an Zen, 1967, *Time and space relationships of the Taconic Allochthon and Autochthon:* Geological Society of America Special Paper 97.

5. C. K. Seyfert and D. J. Leveson, 1969, *Speculations on the relation between the Hutchinson River Group and the New York City Group: Geological Bulletin,* v. 3, Queens College Press, p. 33.

6. J. M. Bird, 1949, Middle Ordovician gravity sliding — Taconic region, *in* M. Kay, ed., *North Atlantic — Geology and Continental Drift,* American Association of Petroleum Geologists Mem. 12, p. 630.

7. Y. W. Isachsen, 1964, Extent and configuration of the Precambrian in northeastern United States: *Trans. N.Y. Acad. Sci.,* v. 26, p. 812.

8. G. M. Kay, 1951, *North American Geosynclines:* Geological Society of America Mem. 48.

9. L. D. Meckel, 1970, Paleozoic alluvial deposition in the central Appalachians: A summary, *in* G. W. Fisher et al., eds., *Studies in Appalachian Geology: Central and Southern:* Wiley-Interscience, New York, p. 49.

10. R. B. Neuman, 1968, Paleogeographic implications of Ordovician shelly fossils in the Magog Belt of the Northern Appalachian Region, *in* E-An Zen et al., eds., *Studies of Appalachian Geology: Northern and Maritime:* Wiley-Interscience, New York, p. 35.

11. J. L. Rosenfield, 1960, Rotated garnets and the diastrophic-metamorphic sequence in south-eastern Vermont: *Geol. Soc. Amer. Bull.,* v. 71, p. 1960.

12. J. B. Lyons and H. Faul, 1970, Isotope geochronology of the Northern Appalachians, *in*

E-an Zen et al. eds., *Studies in Appalachian Geology: Northern and Maritime:* Wiley-Interscience, New York, p. 305.

13. B. A. Hall, 1966, Stratigraphy and structure of the Chamberlain Lake region, Maine, *in* D. W. Caldwell, ed., *New England Intercollegiate Geological Conference Guidebook:* p. 42.

   R. S. Naylor and A. J. Boucot, 1965, Origin and distribution of rocks of Ludlow age (Late Silurian) in the northern Appalachians: *Amer. Jr. Sci.,* v. 263, p. 153.

14. P. B. King, 1950, Tectonic framework of the southeastern states: *Amer. Assoc. Petrol. Geol. Bull.,* v. 34, p. 635.

15. R. V. Dietrich, P. D. Fullager, and M. L. Bottino, 1969, K/Ar and Rb/Sr dating of tectonic events in the Appalachians of southwestern Virginia: *Geol. Soc. Amer. Bull.,* v. 80, p. 307.

16. J. Gilluly, 1965, *Volcanism, Tectonism, and Plutonism in the Western United States:* Geological Society of America Special Paper 80.

17. J. R. Patterson and T. P. Storey, 1960, Caledonian earth movements in western Canada: Caledonian Orogeny, *21st International Geological Congress:* Copenhagen, p. 150.

18. W. P. Irwin, 1966, Geology of the Klamath Mountains Province, *in* E. H. Bailey, ed., *Geology of Northern California:* California Division of Mines and Geology, San Francisco, p. 19.

19. P. C. Bateman and J. P. Eaton, 1967, Sierra Nevada Batholith: *Science,* v. 158, p. 1407.

20. G. M. Kay, 1964, Paleozoic facies from the miogeosynclinal to the eugeosynclinal belt in thrust slices, central Nevada: *Geol. Soc. Amer. Bull.,* v. 75, p. 425.

21. R. J. W. Douglas et al., 1963, *Geology and Petroleum Potentialities of Northern Canada:* Geological Survey of Canada Paper 63-31.

22. A. Hallam, 1958, A Cambro-Ordovician fauna from the Hecla Hoek Succession of Ny Freisland, Spitsbergen: *Geol. Mag.,* v. 95, p. 71.

   C. Poulsen, 1951, The position of the East Greenland Cambro-Ordovician in the paleogeography of the North Atlantic region: *Dansk Geol. Foren. Medd.,* no. 12, p. 61.

23. T. N. George, 1965, The geological growth of Scotland, *in* G. Y. Craig, ed., *The Geology of Scotland:* Archon, Hamden, Conn., p. 1.

24. M. R. W. Johnson, 1965, Dalradian, *in* G. Y. Craig, ed., *The Geology of Scotland:* Archon, Hamden, Conn., p. 115.

25. K. Bell, 1968, Age relations and provenance of the Dalradian Series of Scotland: *Geol. Soc. Amer. Bull.,* v. 79, p. 1167.

26. E. K. Walton, 1965, Lower Paleozoic rocks — Stratigraphy, *in* G. Y. Craig, ed., *The Geology of Scotland:* Archon, Hamden, Conn., p. 161.

27. L. L. Sloss, 1963, Sequences in the cratonic interior of North America: *Geol. Soc. Amer. Bull.,* v. 74, p. 93.

28. J. W. Skehan, 1969, Tectonic framework of southern New England and eastern New York, *in* M. Kay, ed., *North Atlantic — Geology and Continental Drift:* American Association of Petroleum Geologists Mem. 12, p. 793.

29. W. D. McCarthy, 1969, Geology of Avalon Peninsula, Southeast Newfoundland, *in* M. Kay, ed., *North Atlantic — Geology and Continental Drift:* American Association of Petroleum Geologists Mem. 12, p. 115.

30. O. Gates, 1969, Lower Silurian-Lower Devonian volcanic rocks of New England Coast and Southern New Brunswick, *in* M. Kay, ed., *North Atlantic — Geology and Continental Drift:* American Association of Petroleum Geologists Mem. 12, p. 484.

   A. R. Palmer, 1969, Cambrian trilobite distribution in North America and their bearing on Cambrian paleogeography of Newfoundland, *in* M. Kay, ed., *North Atlantic — Geolology and Continental Drift:* American Association of Petroleum Geologists Mem. 12, p. 139.

31. W. B. N. Berry, 1968, Ordovician paleogeography of New England and adjacent areas based on graptolites, *in* E-An Zen et al., eds., *Appalachian Geology: Northern and Maritime:* Wiley-Interscience, New York, p. 23.

   A. J. Boucot, 1969, Silurian-Devonian of northern Appalachians — Newfoundland, *in* M. Kay, ed., *North Atlantic — Geology and Continental Drift:* American Association of Petroleum Geologists Mem. 12, p. 477.

32. H. Williams, 1969, Pre-Carboniferous Development of Newfoundland Appalachians, *in* M. Kay, ed., *North Atlantic — Geology and Continental Drift:* American Association of Petroleum Geologists Mem. 12, p. 32.

33. L. Pavlides, A. J. Boucot, and W. B. Skidmore,

1968, Stratigraphic evidence for the Taconic Orogeny in the northern Appalachians, *in* E-an Zen et al., eds., *Studies in Appalachian Geology: Northern and Maritime:* Wiley-Interscience, New York, p. 61.

34. B. Kummel, 1961, *History of the Earth:* Freeman, San Francisco.

35. M. G. Rutten, 1969, *The Geology of Western Europe:* Elsevier, New York.

36. A. J. Boucot, 1969, The Soviet Silurian: Recent impressions: *Geol. Soc. Amer. Bull.,* v. 80, p. 1155.

37. D. A. Brown, K. S. W. Campbell, and K. A. W. Crook, 1968, *The Geological Evolution of Australia and New Zealand:* Pergamon Press, New York.

   A. J. Wright, 1970, Silurian fossils from New Zealand: *Nature,* v. 228, p. 153.

38. A. Gansser, 1964, *Geology of the Himalayas:* Wiley-Interscience, New York.

39. J. Sougy, 1962, West African fold belt: *Geol. Soc. Amer. Bull.,* v. 73, p. 871.

40. H. W. Sundelius, 1970, The Carolina Slate Belt, *in* G. W. Fisher et al., eds., *Studies in Appalachian Geology: Central and Southern:* Wiley-Interscience, New York, p. 351.

41. F. A. Hills, personal communication, 1970.

42. H. J. Harrington, 1962, Paleogeographic development of South America: *Bull. Amer. Assoc. Petrol. Geol. Bull.,* v. 46, p. 1173.

43. H. Burgl, 1967, The orogenesis in the Andean System of Colombia: *Tectonophysics,* v. 4, p. 429.

44. H. L. Allsopp and P. Kolbe, 1965, Isotopic age determinations on the Cape Granite and intruded Malmesbury sediments, Cape Peninsula, South Africa: *Geochim. Cosmochim. Acta,* v. 29, p. 1115.

45. A. L. Du Toit, 1954, *Geology of South Africa:* Oliver and Boyd, London.

46. E. Picciotto and A. Coppez, 1963, Bibliography of absolute age determinations in Antarctica (addendum), *in* R. J. Adie, ed., *Antarctic Geology:* North-Holland Publishing Co., Amsterdam, p. 563.

47. C. Craddock, J. J. Anderson, and G. F. Webers, 1963, Geologic outline of the Ellesworth Mountains, *in* R. J. Adie, ed., *Antarctic Geology:* North-Holland Publishing Co., Amsterdam, p. 155.

48. S. Beuf et al., 1966, Ampleur des glaciations "Siluriennes" au Sahara: Leurs influences et leurs conséquences sur la sédimentation: *Rev. Inst. Franc. Pétrole,* v. 21, p. 363.

49. F. Arbey, 1968, Structures et depots glaciaires dans l'Ordovician terminal des chaines d'Ougarta (Sahara algerien): *C.R. Acad. Sci. Paris,* v. 266, Ser. D, p. 76.

50. P. L. Applin, 1951, *Preliminary report on buried pre-Mesozoic rocks in Florida and adjacent states:* U.S. Geol. Survey Circular 91.

51. C. Milton and R. Grasty, 1969, "Basement" rocks of Florida and Georgia: *Amer. Assoc. Petrol. Geol. Bull.,* v. 53, p. 2483.

52. E. Lopez-Ramos, 1969, Marine Paleozoic rocks of Mexico: *Amer. Assoc. Petrol. Geol. Bull.,* v. 53, p. 2399.

53. J. Pantoja-Alor and R. A. Robinson, 1967, Paleozoic sedimentary rocks in Oaxaca, Mexico: *Science,* v. 157, p. 1033.

54. S. Matushita, 1963, General remarks, *in* F. Takai et al, eds., *Geology of Japan:* University of California Press, Berkeley.

55. Ta Ch'ang, 1963, *The Geology of China:* Translation published by U.S. Department Commerce, Office of Technical Services, Joint Publ. Research Service, Washington, D.C.

56. M. J. S. Rudwick, 1964, The Infra-Cambrian glaciation and the origin of the Cambrian fauna, *in* A. E. M. Nairn, ed., *Problems in Paleoclimatology:* Wiley-Interscience, New York, p. 150.

57. C. Teichert, 1967, Major features of cephalopod evolution, *in* C. Teichert and E. L. Yochelson, eds., *Essays in Paleontology and Stratigraphy:* Special Publ. 2, Department of Geology, University of Kansas, p. 162.

58. A. S. Romer, 1967, Major steps in vertebrate evolution: *Science,* v. 158, p. 1629, Dec. 29.

59. H. P. Banks, 1968, The history of land plants, *in* E. T. Drake, ed., *Evolution and Environment:* Yale University Press, New Haven, Conn., p. 73.

60. J. Gray and A. J. Boucot, 1971, Early Silurian spore tetrads from New York: Earliest New World evidence for vascular plants: *Science,* v. 173, p. 918.

11
The Late Paleozoic Era

The late Paleozoic was a time of great tectonic activity. During this time, many of the world's geosynclines, including the Appalachian, Ouachita, Uralian, and Angara Geosynclines, were gradually transformed into lofty mountain ranges. Uplift of the continents associated with this activity caused a general but irregular withdrawal of the seas.

The emergence of the amphibians and subsequent development of the reptiles during the late Paleozoic was accompanied by widespread colonization of the land by plants such as horsetails, tree ferns, seed ferns, and true conifers. During the Carboniferous, extensive coal-forming swamps developed under tropical conditions along the paleoequator. At the same time the area within 45° of the Carboniferous South Pole was extensively glaciated. This glaciation may also have been related to late Paleozoic orogenic activity.

LATE PALEOZOIC GEOGRAPHY

The relative positions of the continents during the late Paleozoic may be inferred by comparing polar wandering curves, determining radiometric ages of granitic and metamorphic rocks in orogenic belts, and correlating folded sedimentary rocks. Comparison of the recalculated polar wandering curves suggests that all the continents were in close proximity during the Carboniferous and Permian (Fig. 7.41).

The last major orogeny in the Caledonian and Northern Appalachian Geosynclines occurred during the Middle Devonian, at which time the joining

of Ancestral North America and Ancestral Europe was probably completed. Since deformation within the Uralian Geosyncline did not cease until the Late Permian, the joining of Ancestral Europe and Ancestral Siberia was probably not completed until the end of the Paleozoic. Radiometric dating of orogenic activity in the Angara Geosyncline indicates that the last major metamorphism and intrusion occurred during the Late Permian. Thus Ancestral China and Ancestral Siberia were finally joined by the end of the Paleozoic. Deformation in the Appalachian and Ouachita geosynclines did not cease until the Middle Permian, when Ancestral North America and Gondwanaland were welded together. The proposed sequence of joining of the continents appears in Fig. 11.1.

ANCESTRAL NORTH AMERICA

Throughout most of the late Paleozoic, Ancestral North America was almost completely encircled by geosynclines (Fig. 11.2). Limestones and shales were deposited in the geosynclines and on the adjacent platform during times of little tectonic activity, and coarse clastics were deposited in these areas during orogenic episodes.

Northern Appalachian Geosyncline

At the beginning of the late Paleozoic, the Northern Appalachian Geosyncline consisted of a miogeosyncline and a eugeosyncline separated by a geanticline. Deposition ended in the eugeosyncline at the beginning of the Middle Devonian and in the miogeosyncline during the Early Permian.

(a)

Figure 11.1 Proposed reconstruction of the continents showing paleoclimatic indicators: (a) Devonian, (b) Carboniferous—the alpine glacial deposits in western South America were laid down during the Early Carboniferous, (c) Permian—the alpine glacial deposits in India, Australia, and South America were laid down during the Late Permian.

(b)

(c)

D	desert sandstone	
C	coal	
Rb	redbeds	
Rf	reef	
Rc	reef coral	
E	evaporites	

⌒ 30°N paleolatitude

•→ direction of prevailing winds

▲→ continental glacial deposit with direction of ice movement

△ alpine glacial deposit

Figure 11-2

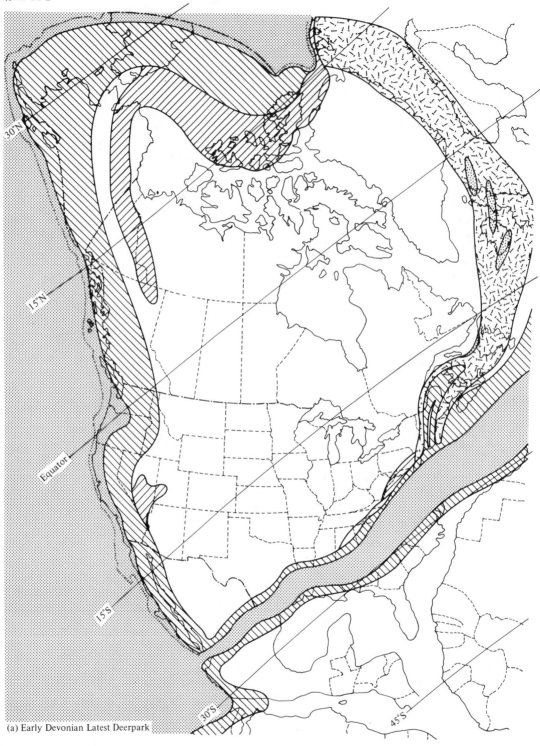

(a) Early Devonian Latest Deerpark

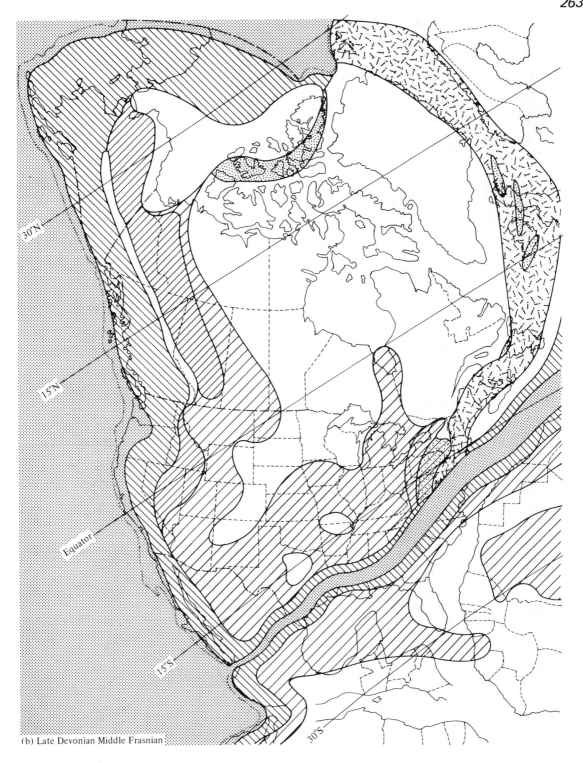

(b) Late Devonian Middle Frasnian

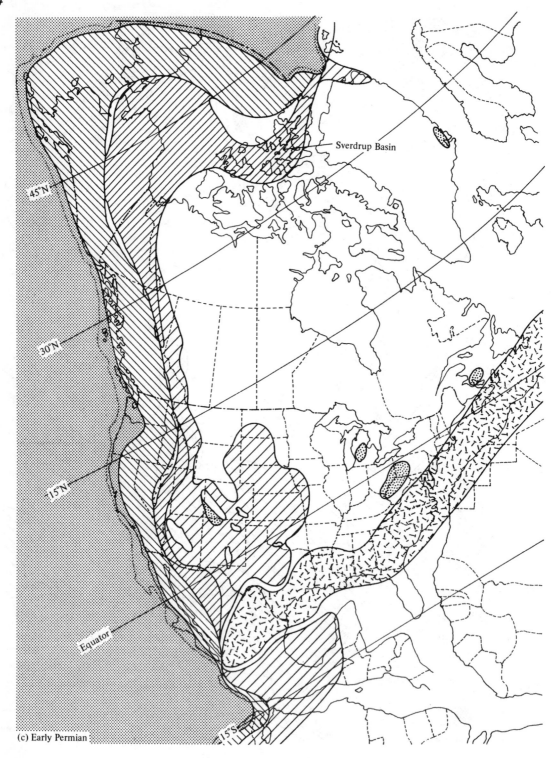

Sverdrup Basin

45°N

30°N

15°N

Equator

15°S

(c) Early Permian

marine miogeosynclinal deposits	generally shallow water	geanticline or island arc	mountains
marine eugeosynclinal deposits	generally moderate depth water	land which is undergoing erosion	
marine platform deposits	shallow water	deep ocean	
terrestrial deposits	land	★ volcanics	

Figure 11.2 Paleogeographic maps for Ancestral North America: (a) Early Devonian, latest Deerpark; (b) Late Devonian, middle Frasnian; (c) Early Permian.

Miogeosyncline. Late Paleozoic miogeosynclinal deposits are almost completely restricted to the southern part of the geosyncline. In southeastern New York, eastern Pennsylvania, and Maryland, Devonian beds overlie the Silurian with apparent conformity. The Lower Devonian sequence is thin and consists mainly of carbonates with minor shales. Evidently, the geanticline to the east was rather low at this time.

The Middle Devonian, Acadian Orogeny produced a mountain range that extended from eastern Pennsylvania northward to Newfoundland and from the Hudson Valley eastward to the Atlantic Ocean (Fig. 11.2b). Sediments derived from the mountains were deposited in a large, subaerial alluvial plain (the Catskill Delta) which was built out into the miogeosyncline (Fig. 11.3). Continental red beds were deposited on the flood plains of rivers that crossed the alluvial plain. These arkosic sandstones, shales, and conglomerates were probably deposited in an arid or semiarid climate. According to paleomagnetic data, this region was located at approximately 20° south during the Devonian. The Middle and Late Devonian red beds are part of an eastward thickening clastic wedge, which reaches a maximum thickness of more than 10,000 ft (3000 m) (Figs. 11.4 and 11.5). The contact between marine and continental facies migrated westward as sediments spread outward from the base of the mountains. By the end of the

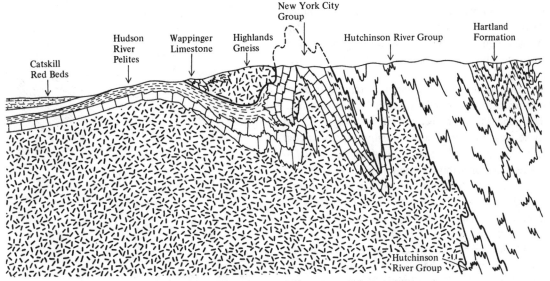

Figure 11.3 Restored cross section across southeastern New York and southwestern Connecticut during the Late Devonian, following the Acadian Orogeny. (After Seyfert and Leveson, Ref. 1.)

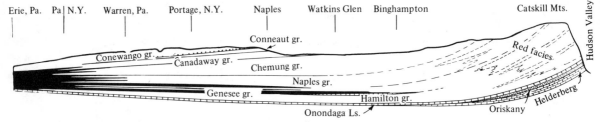

Erie, Pa.   Pa│N.Y.   Warren, Pa.   Portage, N.Y.   Naples   Watkins Glen   Binghampton   Catskill Mts.

Hudson Valley

Conneaut gr.

Conewango gr.
Canadaway gr.
Chemung gr.

Red facies

Naples gr.

Genesee gr.

Hamilton gr.

Onondaga Ls.

Oriskany

Helderberg

Figure 11.4 Cross section from the Catskill Mountains to Erie, Pennsylvania, showing the clastic wedge of sediments deposited in and adjacent to the Catskill delta during the Middle and Late Devonian. (From A. J. Eardley, Ref. 2.)

Devonian, continental red beds were deposited on the platform adjacent to the geosyncline.

Deposition of great thicknesses of sandstone and conglomerate continued during the Carboniferous and Early Permian. Some limestone and coal are intercalated with these clastic sediments. The direction of dip of cross-beds in sandstones and the change in pebble size in conglomerates indicate that the clastics were derived from a mountain range to the east (Fig. 11.6). The mountain range extended from central Mexico through the southeastern United States and into southern Europe (Fig. 11.2). The deposition of great thicknesses of coarse clastics throughout the late Paleozoic required repeated uplift of the mountain range. There is no evidence that the Carboniferous uplifts were accompanied by folding, but radiometric dating of crystalline rocks from the core of the Appalachian Fold Belt indicates that granites were intruded during the Carboniferous (Fig. 10.13 and 11.7).

During the Pennsylvanian, extensive swamps and marshes developed within the miogeosyncline. The deposition and decay of the lush vegetation in these areas produced economically important coal deposits. Although these coal-forming environments were probably restricted to a relatively narrow belt along the coastline, coal deposits cover a considerable area because the seas repeatedly inundated low-lying coastal regions.

By the beginning of the Permian, the seas had withdrawn completely from the geosyncline, and the continental red beds of the Dunkard Series were deposited in West Virginia, Pennsylvania, and Ohio (Fig. 11.2c). The Alleghenian Orogeny, an episode of intensive folding, uplift, and intrusion which probably occurred during the Middle Permian, terminated deposition in the geosyncline and converted it into a mountain range.

Eugeosyncline. At the beginning of the late Paleozoic, a eugeosyncline extended from southern New England through the Maritime Provinces of Canada to central Newfoundland. A very thick sequence of clastic sediments and volcanic rocks was de-

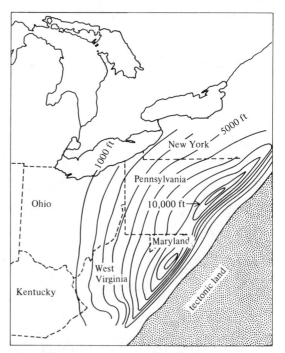

Figure 11.5 Isopach map of Middle and Upper Devonian strata in Pennsylvania and adjacent states. (After Kay, Ref. 3.)

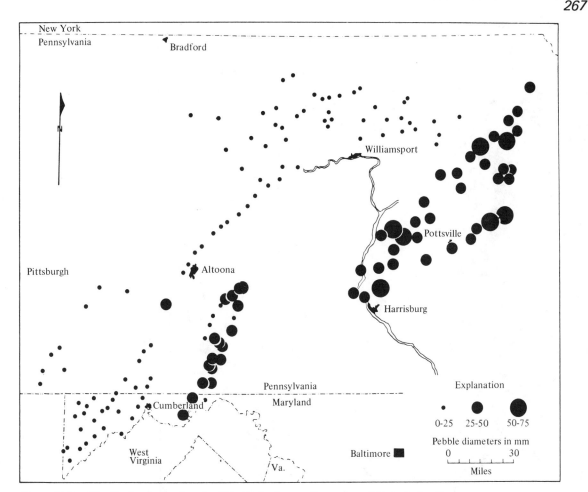

Figure 11.6 Distribution in the size of pebbles in the Pocono Formation of Early Carboniferous age in Pennsylvania. The westward decrease in the size of the pebbles indicates that they were derived from an eastern source. The large size of some of the pebbles indicates that the source had a considerable relief. (After Pelletier, Ref. 4.)

posited in the eugeosyncline during the Early Devonian. The sediments were derived mainly from geanticlines within and adjacent to the eugeosyncline (Fig. 11.2a). The rhyolitic composition of many of the volcanic rocks suggests that they were deposited on continental rather than oceanic crust.

The eugeosynclinal deposits were folded, metamorphosed, and intruded during the Acadian Orogeny. In central Maine, slightly deformed sandstones and conglomerates of probable Middle Devonian age rest with angular unconformity on

rocks as young as Early Devonian (5). In central New England, sediments and volcanics were metamorphosed into schists, quarzites, and amphibolites. Early students of New England geology believed that these rocks were of Precambrian age. However, some of these units have been traced northward into less-metamorphosed, sedimentary rocks containing Paleozoic fossils.

Laboratory studies indicate that the minerals within the intensely metamorphosed Paleozoic rocks of the eugeosyncline were formed under very high temperatures and pressures. Sillimanite,

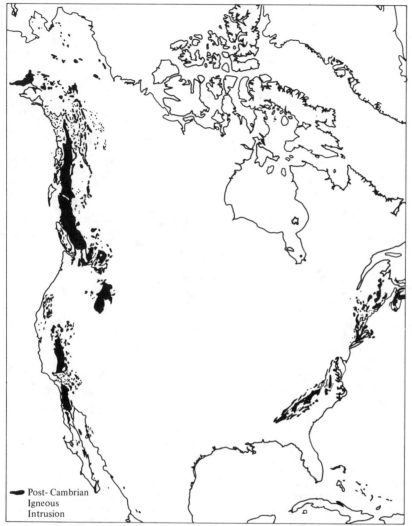

Figure 11.7   Granitic rocks (in black) in the Appalachian and Cordilleran regions. Most of the granites of the Appalachian region were emplaced during the late Paleozoic; most of those of the Cordilleran region were emplaced during the Mesozoic. (Data from the Geologic Map of North America, published by the U.S. Geological Survey.)

a common mineral in many of these rocks, forms under conditions of regional metamorphism at a minimum temperature of about 900°F (500°C) and a pressure equivalent to 50,000 ft (15,000 m) of rock (6). Such conditions might reasonably be expected in the core of a mountain range. Since rocks containing sillimanite are now exposed at the earth's surface, a considerable amount of rock must have been eroded from the mountains. The

weathering of this rock produced the clastics that were deposited in the miogeosyncline during the Middle and Late Devonian.

Although the Acadian Orogeny ended deposition in the eugeosyncline, both marine and non-marine sedimentary and volcanic rocks accumulated in local basins along the trend of the eugeosyncline. For example, Carboniferous and Permian sediments were deposited in basins in Massachusetts, Rhode Island, the Maritime provinces, and Newfoundland. Angular unconformities in these sequences indicate that the deformation continued during the Carboniferous and Permian. In Massachusetts and Rhode Island, sediments of Pennsylvanian age were folded and metamorphosed prior to the deposition of nearby Upper Triassic sediments. Potassium-argon dating of granitic and metamorphic rocks in western Connecticut indicates that an episode of intrusion and metamorphism occurred in Middle Permian time, approximately 250 million years ago (7). The metamorphism of the Carboniferous rocks in Massachusetts and Rhode Island and the folding of the deposits of the miogeosyncline (Alleghenian Orogeny) may have occurred at this time.

Southern Appalachian Geosyncline

Late Paleozoic clastics of the Southern Appalachian Geosyncline interfinger with carbonates to the west. Pebbles in conglomerate include granitic and metamorphic rocks derived from a crystalline highland to the east. Marine and continental clastics of Devonian age, which are very thick in the northern part of the geosyncline, thin rapidly southward (Fig. 11.5). These clastics were derived from highlands uplifted during the Acadian Orogeny (8). Miogeosynclinal deposits in eastern Tennessee were tilted and eroded during the Middle or Late Devonian. The overlying Chattanooga Shale is black, has few fossils, and was probably deposited in a reducing environment. The shale contains small percentages of uranium and constitutes a major reserve of low-grade uranium ore. It is approximately 500 ft (150 m) thick in the miogeosyncline, but thins to a few tens of feet on the platform to the west.

Uplift to the east of the geosyncline resulted in the deposition of clastic wedges during the Mississippian, Pennsylvanian, and Permian. The Pennsylvanian clastic wedge is especially widespread and reaches a maximum of 10,000 ft (3000 m) in northern Alabama. The culminating deformation in the geosyncline was the Allegheny Orogeny (Figs. 11.8 and 11.9). Deformation was

Figure 11.8 Limestones in West Virginia which were folded during the Allegheny Orogeny. (Photo by Ralph Yalkovsky.)

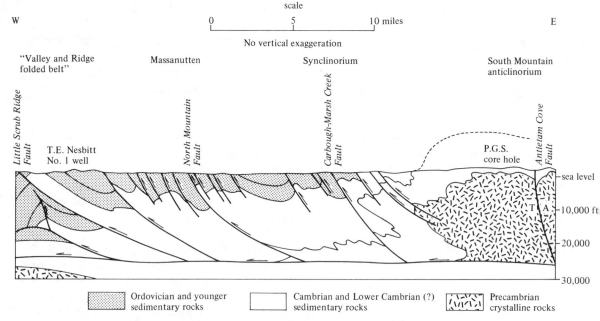

W

scale

0 ———— 5 ———— 10 miles

No vertical exaggeration

E

"Valley and Ridge folded belt"

Massanutten

Synclinorium

South Mountain anticlinorium

Little Scrub Ridge Fault

T.E. Nesbitt No. 1 well

North Mountain Fault

Carbough-Marsh Creek Fault

P.G.S. core hole

Antietam Cove Fault

sea level

10,000 ft

20,000

30,000

☷ Ordovician and younger sedimentary rocks   ▢ Cambrian and Lower Cambrian (?) sedimentary rocks   ⟨⟩ Precambrian crystalline rocks

Figure 11.9 Generalized cross section from the Valley and Ridge Province to the Blue Ridge Province in Pennsylvania. Note that the angle of dip of these thrust faults decreases at depth. (From Root, Ref. 9.)

so intense that the geosyncline was converted into a great mountain range. The Appalachian Mountains are the deeply eroded remains of this range.

Ouachita Geosyncline

From Early Devonian until Late Mississippian time, the rate of sedimentation was very slow in the area that was to become the Ouachita Geosyncline. However, toward the end of the Late Mississippian, a great flood of clastic sediments was deposited. These deposits, which include the Stanley Shale and the Jackfork Sandstone, are part of a 25,000 ft (8000 m) thick sequence whose age ranges from Late Mississippian to Early Pennsylvanian. Small fragments of granitic and metamorphic rocks in these deposits and a southward thickening of the sequence indicate that the source was a crystalline highland that lay to the south. Radiometric dating of basement rocks from the Gulf coastal plain shows that these rocks were metamorphosed during the late Paleozoic and may

therefore have been part of the crystalline terrain from which the clastics were derived.

During the Middle Pennsylvanian, folding and faulting of the Ouachita Geosyncline produced a mountain range that extended along the southern border of Ancestral North America (Fig. 11.10). Folding also occurred during the late Pennsylvanian, but since Permian strata are relatively undisturbed, deformation had ended by that time. Subsequent erosion has largely leveled these mountains, leaving only the Ouachita Mountains in southwestern Arkansas and southeastern Oklahoma, and the Marathon Mountains of southwestern Texas. All deposition in the Ouachita Geosyncline had ceased by the end of the Permian.

Cause of Deformation in the Appalachian and Ouachita Geosynclines

Deformations of Late Devonian, Carboniferous, and Permian age in the Appalachian and Ouachita geosynclines were probably caused by a collision of continental plates. Paleomagnetic evidence indicates that Ancestral North America and Gondwanaland were separated during the early Paleozoic but were joined from the Carboniferous to the Triassic. Thus the beginning of intense deformation in the Appalachian and Ouachita geosynclines coincides approximately with the initial contact of the continents. This contact may have occurred at about the same time as deposition of the Chattanooga Shale (uppermost Devonian and lowermost Mississippian). During the Middle Carboniferous, the zone of contact may have extended southward to the Ouachita Geosyncline.

Cordilleran Geosyncline

The late Paleozoic deposits of the eastern Cordilleran Geosyncline consist largely of carbonates and clastics, but deposits to the west were dominantly clastics and volcanics (Fig. 11.11).

Miogeosyncline. The complete absence of earliest Devonian marine deposits from the miogeosyncline suggests that the seas had withdrawn from the miogeosyncline at that time. However, the seas returned before the end of the Early Devonian. Devonian miogeosynclinal deposits are principally carbonates. Large coral reefs within this sequence indicate deposition in a warm climate. Paleomagnetic data place the miogeosyncline close to the equator throughout the Devonian (Fig. 11.2).

At the end of the Devonian, geosynclinal deposits were deformed along a linear zone in central Nevada known as the Antler Orogenic Belt. Eugeo-

Figure 11.10   Oblique aerial view of the Peña Blanca Hills, near Marathon in southwestern Texas. The white beds are cherts of Silurian and/or Devonian age. The outcrop pattern is a result of the erosion of plunging folds. A faint line of vegetation on the second hill in the foreground marks a thrust fault. (Photo by Earle F. McBride.)

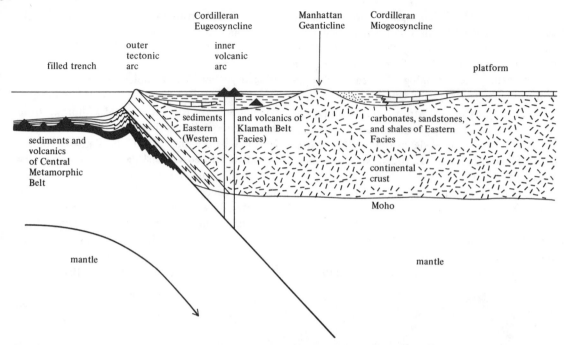

Figure 11.11   Restored cross section across the Cordilleran Geosyncline during the Early Carboniferous.

synclinal deposits were folded and thrust over miogeosynclinal deposits in a manner similar to that of the Taconic Allochthon in the Appalachians. The displacement along the Roberts Mountain Thrust, which separates the two sequences, has been estimated at more than 50 miles (80 km) (10). Since the faulting does not involve basement rocks, it may have resulted from the sliding of giant slices from uplifted highlands to the west of the miogeosyncline. An angular unconformity at the base of Carboniferous limestones in northern Alaska may have been formed during the same orogenic episode (Fig. 11.12a).

Throughout most of the Carboniferous and Permian, the Manhattan Geanticline was an island that shed sediments eastward into the miogeosyncline and westward into the eugeosyncline (Fig. 11.11). The Malay Peninsula in the Indonesian Island Arc may be a modern analog of the Manhattan Geanticline. This peninsula is bordered by volcanic islands on the west and south, and the deposits in the vicinity of the volcanic islands are eugeosynclinal in character, whereas those on the opposite side of the peninsula are miogeosynclinal.

In central Nevada, the Mississippian Tonka Conglomerate is 2500 ft (800 m) thick and grades eastward into thinner shale beds. Deposits in this clastic wedge were derived from the Manhattan Geanticline to the west. The presence of conglomerates throughout the Mississippian section implies continued uplift of the geanticline. Intermittent uplift continued during the Pennsylvanian, and occasionally deformation extended into the miogeosyncline (Fig. 11.12b). By Early Permian time, the geanticline had been eroded almost to sea level, and limestone was deposited in the western part of the miogeosyncline. In contrast to the abundant clastics in the miogeosyncline in Nevada, the deposits in the Canadian segment of the miogeosyncline are dominantly limestones and dolostones. These carbonates are exposed in towering cliffs in Banff and Jasper National Parks in the Canadian Rockies (Fig. 11.13).

(a)

(b)

Figure 11.12   Late Paleozoic deposits of the Cordilleran Geosyncline: (a) Gently dipping carbonates of Carboniferous age unconformably overlying steeply dipping beds of early Paleozoic or Late Proterozoic age. (Photo by J. T. Dutro, Jr., U.S. Geological Survey.) (b) a tilted angular unconformity exposed along U.S. 40 west of Elko, Nevada. The more gently dipping beds are limestones and siltstones of latest Pennsylvanian and Permian age; the underlying steeply dipping beds are Late Mississippian conglomerates. This area is close to the Manhattan Geanticline and shows a record of three phases of deformation. The first is recorded in the thick conglomerates, which require a nearby uplift. The angular unconformity was produced and the unconformity was tilted during the second and third phases of deformation, respectively.

Figure 11.13 Mount Rundle in Banff National Park in the Canadian Rockies, showing late Paleozoic sedimentary rocks. The resistant unit capping the mountain is the limestone-dolostone Rundle Formation of Early Carboniferous age. The underlying unit is the less resistant Banff formation, of limestone, argillaceous limestone, and calcareous shale also of Early Carboniferous age. The lowest cliffs are composed of massive limestone and dolostone of the Palliser Formation, which is Devonian. (Photo by T. L. Tanton, courtesy of the Geological Survey of Canada, Ottawa.)

Toward the end of the Paleozoic, large faults displaced eugeosynclinal deposits over miogeosynclinal deposits. These faults, which include the Golconda Thrust in Nevada, do not involve basement rocks, and they may be gravity thrusts that slid from uplifted areas to the west of the miogeosyncline.

Eugeosyncline. A very thick sequence of clastic sediments and volcanic rocks was deposited in the eugeosyncline to the west of the Manhattan Geanticline. Volcanic and tectonic islands in the eugeosyncline also contributed sediments. Episodes of deformation, uplift, and erosion occurred within the eugeosyncline during the Late Devonian

or Early Mississippian, during the Pennsylvanian, and at least twice in the Permian. Radiometric dating indicates that these deformations were accompanied by emplacement of granitic rocks.

In the western part of the eugeosyncline, an intensely deformed sequence of rhythmically bedded cherts, clastics, and volcanics contains late Paleozoic fossils (Fig. 11.14). These deposits include the Western Paleozoic and Triassic Belt in the Klamath Mountains of northern California and the western part of the Calaveras Formation in the Sierra Nevada Mountains of eastern California. Radiometric dates on rocks from the Central Metamorphic Belt in the Klamath Mountains (Fig. 10.17) indicate that metamorphism and deformation occurred during the Devonian and Carboniferous (11). The late Paleozoic volcanic and orogenic activity may have been due to the movement of an oceanic plate under the continent during the late Paleozoic.

Franklin Geosyncline

From the Early to the Middle Devonian, carbonates and minor shales were deposited in the Franklin Geosyncline. Coral reefs are abundant in Middle Devonian deposits, even though this area is now located between 75 and 80° North latitude. Paleomagnetic measurements show that this area was at approximately 15° North latitude during Devonian time (Fig. 11.2).

Marine sedimentation ended in the Middle Devonian when a geanticline was uplifted on the northern border of the geosyncline (12). Continental clastics spread southward from this uplift into the miogeosyncline. These deposits reach a thickness of 10,000 ft (3000 m) and include coal-bearing sandstones and shales. In the latest Devonian or earliest Mississippian, the deposits of the geosyncline were strongly folded. This deformation, the Ellesmerian Orogeny, ended sedimentation in the Franklin Geosyncline. Subsequently, the Sverdrup Basin developed approximately along the former boundary between the miogeosyncline and the eugeosyncline.

Figure 11.14 Aerial view of the Klamath Mountains in northern California. The rocks in the foreground are metasedimentary and metavolcanic and belong to the Western Paleozoic and Triassic Belt. The white cliff in the middle is composed of marble. Mount Shasta, a relatively young volcano, appears on the skyline. (Photo courtesy of W. P. Irwin, U.S. Geological Survey.)

Sverdrup Basin

The seas transgressed over the folded rocks of the Franklin Geosyncline during the Carboniferous and deposited a thick sequence of sediments in the Sverdrup Basin (Fig. 11.2c). Thick deposits of carbonates and evaporites interbedded with clastics indicate that this region had an arid climate during the late Paleozoic. Paleomagnetic data show that the Sverdrup Basin probably lay between 25 and 50° North latitude during the Carboniferous (Fig. 11.2). The Permian deposits of the Sverdrup Basin appear to be conformable with those of Triassic age, so that, at least in this region, the Paleozoic Era was not terminated by orogeny.

Continental Interior

The continental interior of Ancestral North America was periodically inundated by shallow seas during the late Paleozoic. Following withdrawal in the Late Silurian, the seas returned to cover much of the continent and a sequence of limestone and shale was deposited (Fig. 11.15).

An extensive evaporite basin developed east of the Cordilleran Geosyncline in western Canada during the Middle Devonian. The basin, which was partially enclosed by reefs and geanticlines, contains halite, gypsum, anhydrite, and some of the largest deposits of sylvite in the world (14, 15). Sylvite (potassium chloride) requires almost complete evaporation of seawater before it precipitates. Thus this region must have been very arid during the Middle Devonian. However, according to paleomagnetic data, it was located very near the equator during the Middle Devonian.

During the Late Devonian and Early Mississippian, a very extensive black shale was deposited over much of the interior of the continent. This

Figure 11.15 Middle and Upper Devonian shale and limestone at 18 Mile Creek in western New York. There is a disconformity between Middle and Upper Devonian strata just below the highest overhang. (Courtesy of the Buffalo Museum of Science; from E. J. Buehler & I. H. Tesmer, Ref. 13.)

shale is continuous with the Chattanooga Shale and was probably derived from highlands to the east. The seas became shallower during the late Early Mississippian, and limestones were deposited over most of the southern and western continental interior. These include the Madison Limestone of Wyoming and Montana and the equivalent Redwall Limestone of the Grand Canyon region.

Throughout most of the Paleozoic, there was very little deformation in the interior of the continent. However, folding and uplift during the late Mississippian and Early Pennsylvanian formed the Arbuckle, Wichita, and Amarillo mountains in northern Texas and southern Oklahoma and the Ancestral Rockies in Colorado, New Mexico, and eastern Utah (Fig. 11.16). Great thicknesses of clastic sediments were eroded from these uplifts and deposited in the adjacent basins. The Central Paradox Basin, located west of the Uncompahgre Uplift, received up to 13,000 ft (4000 m) of red and gray sandstones, conglomerates, siltstones, salt, and gypsum during the Pennsylvanian and Permian (Figs. 11.17 and 11.18).

Highlands to the east and south increased in width and height during the Pennsylvanian, and clastic sediments spread to the eastern, central, and southern part of the continental interior. Alternating transgressions and regressions of the seas produced a series of cyclothems in these regions. A typical cyclothem consists of nine different rock types each representing a successive depositional environment at any one locale during sea-level changes (Fig. 11.19). Such changes in sea level may have been the result of: *(a)* glaciation in the southern continents, *(b)* variations in the rate of sediment deposition, *(c)* variations in the rate of subsidence, or *(d)* changes in the location of major rivers.

The seas withdrew from the eastern part of the continental interior in the Permian. During the Middle Permian, thick accumulations of salt were laid down in coastal lagoons in Kansas and Oklahoma. The uplift of the Marathon Mountains in southern Texas created the Delaware Basin in western Texas and southeastern New Mexico. The deeper parts of this basin were bordered by reefs

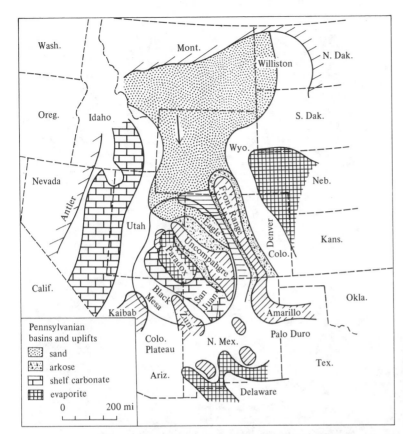

Pennsylvanian
basins and uplifts

- ▨ sand
- ▨ arkose
- ⊞ shelf carbonate
- ⊞ evaporite

0   200 mi

Figure 11.16 Late Paleozoic
uplifts and basins in Colorado,
Oklahoma, and Texas. (From
Peterson and Hite, Ref. 16.
Used with the permission of
American Association of
Petroleum Geologists.)

Figure 11.17 Red shales overlain by red sandstones, both of Permian age, in the Monument Valley,
southeastern Utah. (Santa Fe Railway photo.)

277

Figure 11.18 Well-bedded Late Carboniferous limestones, shales, and sandstones in southeastern Utah. These clastic sediments and those in Figure 17 were eroded from the Ancestral Rockies. (Courtesy of the U.S. Department of the Interior, Bureau of Reclamation.)

during the Middle and Late Permian (Fig. 11.20). One of these, El Capitan Reef, is comprised mainly of the remains of calcareous algae, sponges, bryozoans, pelecypods, brachiopods, and foraminiferans (Fig. 11.21). Broken fragments of fossils and limestone are found as much as 1800 ft (600 m) below the top of the reef. Limestone and dolostone were deposited in shallow lagoons behind the reefs, and gypsum, anhydrite, and shales were deposited in the deeper parts of the basin. Groundwater percolating through the limestones has carved vast networks of caves, including Carlsbad Caverns, one of the largest and most beautiful

caves in the United States. Toward the end of the Middle Permian, the waters of the Delaware Basin became highly saline and thick layers of halite and sylvite were deposited in what was presumably a very arid climate. Paleomagnetic studies indicate that the Delaware Basin was at a latitude of about 15° North during the Middle and Late Permian.

ANCESTRAL EUROPE

Ancestral Europe was bordered by the Hercynian Geosyncline on the south and the Uralian Geosyncline on the east during most of the late

Paleozoic. By the end of the Paleozoic, orogenic movements had converted these geosynclines into mountains. This orogenic activity was probably associated with plate collisions. Deformation in the Uralian Geosyncline seems to have been associated with the joining of Ancestral Europe and Ancestral Siberia, and deformation in the Hercynian Geosyncline may have been caused by the joining of Ancestral Europe and Gondwanaland.

Hercynian Geosyncline

Marine eugeosynclinal deposits of the Hercynian Geosyncline grade northward into marine and continental platform deposits with no intervening miogeosynclinal deposits. The Devonian and Car-

boniferous deposits are largely shale and sandstone, with minor limestone and volcanic rocks. Early Devonian sediments were derived mainly from the Caledonian Mountains to the northwest, but as these mountains were worn down, an increasing amount of sediment was derived from highlands to the south.

A few local episodes of folding occurred during the Devonian, but the geosyncline was not subjected to strong folding until the end of the Devonian. Deformation was largely restricted to the southern portion of the geosyncline, but toward the end of the Early Carboniferous, the entire geosyncline was intensely folded, metamorphosed, and intruded by granites (Fig. 11.22).

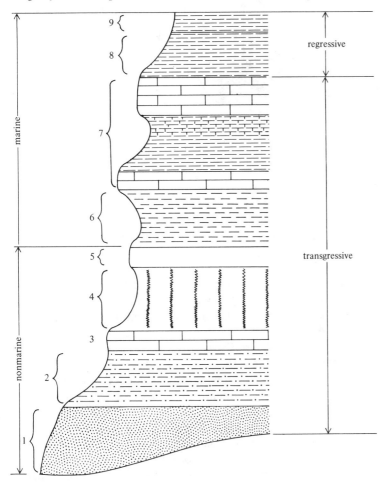

Figure 11.19 Idealized stratigraphic section showing a completely developed Late Pennsylvanian cyclothem. It is rare for all members to be present at any one locality. (After Weller, Ref. 17.)

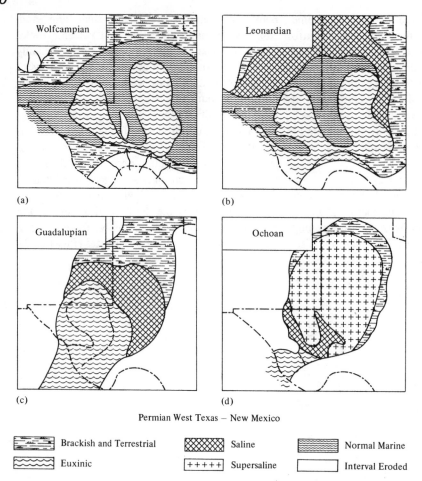

(a) Wolfcampian

(b) Leonardian

(c) Guadalupian

(d) Ochoan

Permian West Texas — New Mexico

≣≣ Brackish and Terrestrial	▨ Saline	≋ Normal Marine
≋ Euxinic	+++++ Supersaline	☐ Interval Eroded

Figure 11.20   Changes in paleogeography in the vicinity of the Delaware Basin during the four stages of the Permian Period. (After Sloss, Ref. 18.)

This deformation, which was the main phase of the Hercynian Orogeny, ended marine sedimentation in the geosyncline and produced a mountain range extending across the southern part of Ancestral Europe. The Ore Mountains, on the border between Germany and Czechoslovakia, are remnants of this ancient range. During the Late Carboniferous, coal swamps developed in intermontane basins north of the new mountain range. Subsequent episodes of folding occurred toward the end of the Late Carboniferous and between the Early and Middle Permian. All episodes of deformation correlate with deformations in the Southern Appalachian and Ouachita geosynclines and may have been caused by the "collision" of Ancestral Europe with Gondwanaland.

Uralian Geosyncline

Late Paleozoic sedimentary and volcanic rocks attain geosynclinal thickness in the Ural Mountains in easternmost Ancestral Europe. A thick sequence of limestone, coal, and clastic sedimentary rocks in the western part of the geosyncline grades eastward into an even thicker sequence of clastics, lavas, and tuffs of the same age. Accordingly, the deposits in the west are miogeosynclinal in charac-

Figure 11.21   Oblique aerial view of El Capitan and Guadalupe Peak at the southern end of the Guadalupe Mountains in western Texas. (Photo by P. B. King, U.S. Geological Survey.)

Figure 11.22   Anticline in sandstone and shale, Cornwall County, southwestern Britain. These rocks were deformed during the Hercynian Orogeny during the Carboniferous. (Crown Copyright Geological Survey photo. Reproduced by permission of the Controller, Her Britannic Majesty's Stationery Office.)

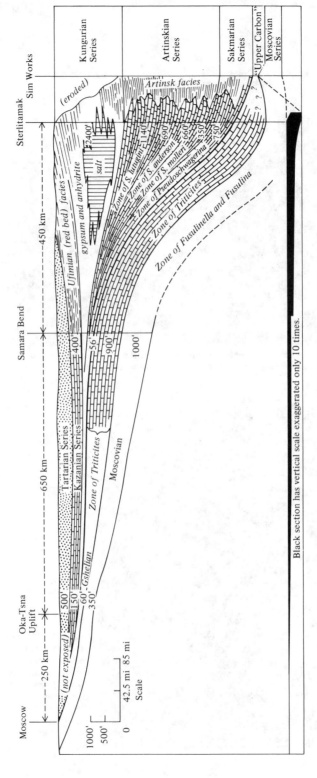

Figure 11.23 Restored cross section across the western part of the Uralian Geosyncline during the Late Permian. (After Dunbar, Ref. 19. Used with the permission of the American Association of Petroleum Geologists.)

ter, whereas those in the east are eugeosynclinal. The climate of this region was probably warm and moist. According to paleomagnetic data, the Uralian Geosyncline was near the equator during the Carboniferous (Fig. 11.1b).

The uplift of a geanticline along the trend of the eugeosyncline late in the Early Carboniferous ended sedimentation and resulted in deposition of clastic wedges in the eastern part of the miogeosyncline. In the Permian, an increasing amount of arkosic sediment was deposited in the miogeosyncline. These clastics grade westward into limestones and evaporites (Fig. 11.23). The salt basin of the Upper Kama region of the Soviet Union is said to be the largest in the world (20). Large quantities of sylvite in these evaporites were probably deposited under conditions of extreme aridity. Paleomagnetic studies indicate that the basins were located at 25° North latitude, the same latitude occupied by the Sahara Desert today. The entire geosyncline was transformed into a mountain range during intense deformation in Middle

Permian time. The deformation was presumably the result of compression between the plate containing Ancestral Siberia and the plate containing Ancestral Europe.

Caledonian Mountains

Following the final phase of the Caledonian Orogeny, continental sediments accumulated between and adjacent to the new mountain ranges. These deposits, which are collectively known as the Old Red Sandstone, are largely deltaic and lacustrine in origin. Although similar in character, only the Middle and Upper Old Red Sandstones are the time-stratigraphic equivalents of the Devonian red beds of New York and Pennsylvania. The Lower Old Red Sandstone reaches a maximum thickness of 40,000 ft (13,000 m) and rests unconformably on folded early Paleozoic strata (Fig. 11.24). In Scotland, the Upper Old Red Sandstone of Late Devonian age rests unconformably on mildly deformed Lower Old Red Sandstone and on Middle Devonian granitic rocks (Fig. 11.25). The

Figure 11.24 Angular unconformity between vertical Silurian beds and gently dipping deposits of the Old Red Sandstone. Near Cockburnspath on the coast of southeastern Scotland. (Crown Copyright Geological Survey photo. Reproduced by permission of the Controller, Her Britannic Majesty's Stationery Office.)

Figure 11.25   Angular unconformity between the moderately dipping beds of the Lower Old Red
Sandstone and the gently dipping beds of the Upper Old Red Sandstone, in the central part of eastern
Scotland. (Crown Copyright Geological Survey photo. Reproduced by permission of the Controller,
Her Britannic Majesty's Stationery Office.)

deformation indicated by this unconformity oc-
curred at approximately the same time as the
Acadian Orogeny in the northern Appalachian
Geosyncline. However, the deformation was much
less intense in Scotland than it was in New
England.

The Carboniferous deposits of the Caledonian
region are rather thin and include extensive coal-
bearing clastics interbedded with lavas. Movement
along large-scale strike-slip faults was contemporaneous with deposition of these sequences.
One such fault, the Great Glen Fault in northern
Scotland, has a left-lateral displacement of approximately 65 miles (100 km).

Continental Interior
Sediments derived from the erosion of the Cale-

donian Mountains were deposited over much of
the interior of Ancestral Europe in Devonian time.
Continental sequences grade eastward and southward into marine deposits. As the mountains were
leveled, finer sediments were deposited and the sea
transgressed onto the platform. By the early Carboniferous, limestone covered much of the continental interior.

Uplift of the Hercynian Mountains on the
southern border of the platform during the late
Early Carboniferous resulted in the deposition of
clastic wedges on the adjacent platform. By mid-
Late Carboniferous time, the seas had withdrawn
from all but the eastern continental interior. The
seas became even more restricted in the Permian,
and a large basin of evaporation formed in central
Europe. The evaporites, which include thick de-

posits of halite and sylvite, formed in an arid region that lay, according to paleomagnetic data, $10°$ North of the equator during the Permian.

GONDWANALAND

Thick upper Paleozoic sequences were deposited in the geosynclines which encircled Gondwanaland (Fig. 10.28). The eugeosynclinal sequences contain andesitic volcanic rocks and were deformed several times during the late Paleozoic. It is likely that the continent was bordered by island arcs and trenches, under which oceanic plates were moving. Upper Paleozoic sequences of the continental interiors of South America, Africa, Antarctica, India, and Australia are remarkably similar. A typical sequence includes Devonian shales and sandstones overlain unconformably by tillites of Carboniferous and/or Permian age, which in turn are followed by coal-bearing shales of Permian age. The similarity of these units provides one of the strongest arguments that the southern continents were part of a single landmass during the late Paleozoic.

Tasman Geosyncline

Situated on the east coast of Australia, the Tasman Geosyncline consists of a western miogeosyncline and an eastern eugeosyncline. Miogeosynclinal sandstones, limestones, and shales contain fossils indicating deposition in shallow water. Coral reefs and evaporites of Devonian age were deposited at a latitude of approximately $35°$ South according to paleomagnetic data (Fig. 11.1a). It is interesting to note that at the present time this region is at approximately the same latitude and has extensive living coral reefs nearby.

Eugeosynclinal deposits include graywacke, shale, and felsic to mafic volcanics, with only minor limestone. Early to Middle Devonian terrestrial volcanics are surrounded by marine deposits of the same age (21), suggesting deposition in a volcanic island arc. Toward the end of the Middle Devonian, an episode of intense deformation affected the entire geosyncline, with granites intruding the eastern part of the geosyncline. This deformation resulted in an eastward shift of the axis of the geosyncline.

From the Late Devonian until the end of the Permian, the eugeosyncline was restricted to the easternmost border of Australia. During the Late Devonian and Early Carboniferous, it received both marine and continental sediments. At the end of the Early Carboniferous, a very intense deformation converted much of the geosyncline into a mountain range. This orogeny was contemporaneous with the main phase of the Hercynian Orogeny in Europe and was accompanied by numerous granitic intrusions. Tillites and fluvioglacial sediments were deposited in the eugeosyncline during the Late Carboniferous and Permian. Some of these deposits may have resulted from alpine glaciers which originated in nearby highlands. A final deformation at the end of the Permian ended sedimentation in the Tasman Geosyncline.

Buller and New Zealand Geosynclines

During the Early and Middle Devonian, quartz sandstone, shale, limestone, and volcanic rocks were deposited in the Buller Geosyncline of western New Zealand. The limestones contain a rich fauna of brachiopods, corals, and trilobites, which indicates deposition in relatively shallow water (21). The deposits of the geosyncline were uplifted and intruded during the Late Devonian and Carboniferous. Sedimentation did not resume until the Permian when clastics, limestone, and volcanics were deposited. At the same time, thick deposits of graywacke, siltstone, and volcanics accumulated in the New Zealand Geosyncline east of the Buller Geosyncline. The volcanics here are principally mafic in composition. Ultramafic intrusives are also common. It has been suggested that the Buller Geosyncline formed on a sialic (continental) basement, and the New Zealand Geosyncline formed on a simatic (oceanic) basement (22).

Timor

A relatively thin sequence of Permian sedimentary

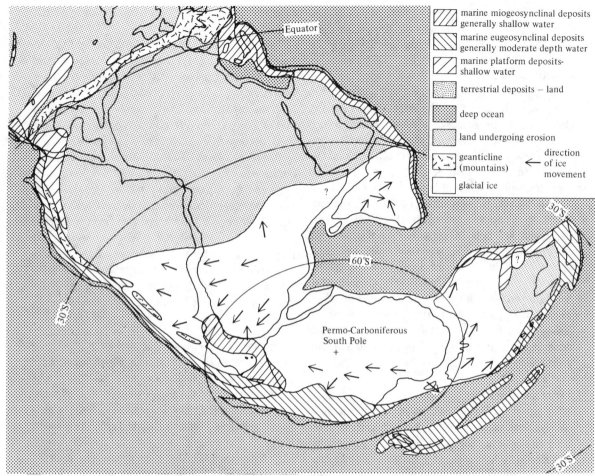

Figure 11.26 Paleogeographic map of Gondwanaland during the earliest Permian, showing the inferred distribution of ice sheets.

The legend for the figure reads:

- marine miogeosynclinal deposits generally shallow water
- marine eugeosynclinal deposits generally moderate depth water
- marine platform deposits- shallow water
- terrestrial deposits – land
- deep ocean
- land undergoing erosion
- geanticline (mountains)
- direction of ice movement
- glacial ice

and volcanic rocks occurs on the island of Timor. The volcanics include basalts and related rocks having an oceanic affinity. These deposits may have accumulated at bathyal to abyssal depths on oceanic crust. Possibly they were deposited seaward of the margin of Australia and were subsequently carried into the Java Trench.

Southern Tethyan Geosyncline

Upper Paleozoic deposits in the Himalayas are miogeosynclinal in character. Although Devonian fossils have not been found in the Himalayas, an unfossiliferous quartzite, which occurs between Silurian and Lower Carboniferous beds, may be of Devonian age. Thick limestones of Lower Carboniferous age grade upward into marine clastics, which in turn are disconformably overlain by continental deposits, including tillites and basaltic lavas of Carboniferous and Permian age (23). Since the volcanics do not include andesites and are associated with thick limestones, the sequence is unlike that of a typical eugeosyncline. Permian continental beds containing a *Glossopteris* flora are interbedded with fossiliferous marine strata.

Thick upper Paleozoic deposits are also found in the Southern Alps of northern Italy. Paleomagnetic measurements indicate that the location of the Permian pole as determined from the Southern Alps is quite different from the Permian pole determined from northern and central Europe (23). However, the Permian pole as determined from the Southern Alps is very similar to the Permian pole as determined from Africa. Therefore, it is likely that the Southern Alps and most of Italy were part of Gondwanaland during the Permian.

West African Geosyncline

In Mauritania, Devonian rocks of the West African Geosyncline are intensely folded. Since Mesozoic rocks in this region are relatively undisturbed, the Devonian sequence was presumably deformed during the late Paleozoic (24).

Southern Appalachian Geosyncline

In the late Paleozoic, the western half of the Southern Appalachian Geosyncline bordered Gondwanaland and linked the Andean and West African geosynclines (Fig. 10.28). Either the late Paleozoic sequences of the geosyncline are covered by coastal plain deposits of Mesozoic and Cenozoic age, or they are so strongly metamorphosed that fossils have not been preserved. Radiometric dates in the Piedmont province of Virginia, North Carolina, South Carolina, and Georgia generally fall into the range of 350 to 250 million years, that is, latest Devonian to Middle Permian. Thus some of the parent sediments may be of late Paleozoic age. Deposition in the southern Appalachian Geosyncline had ended by the close of the Paleozoic.

Andean Geosyncline

Deposits in the Andean Geosyncline record a long history of glaciation during the late Paleozoic. In western Argentina, Early Devonian conglomerates may represent outwash deposits from local alpine glaciation (25). Tillites and fluvioglacial deposits of Early Carboniferous age which occur in southern Bolivia and western Argentina also may have been deposited by alpine glaciers (26). However, Late Carboniferous glacial deposits in the eastern

part of the Andean Geosyncline were almost certainly deposited by a very large continental ice sheet (Fig. 11.26). Tillites containing faceted pebbles grade westward into marine deposits, which in Bolivia reach a thickness of 6500 ft (2000 m). Glacial deposits of Early Permian age, which occur in southern Bolivia and western Argentina, may have been deposited by alpine glaciers. Somewhat younger limestones in Bolivia indicate that warmer climates succeeded the glacial episode in the Andean Geosyncline (Fig. 11.27).

Repeated deformation during the late Paleozoic produced highlands that became the source of many alpine glaciers. An episode of folding and faulting affected the southern part of the geosyncline during the Late Devonian (25). Before separation of the continents, this fold belt was continuous with a Late Devonian fold belt on the

Figure 11.27 Tight folding in thin bedded limestones of the Copacabana Formation of Permian age in western Peru. (Photo by C. R. Peterson, La Oroya, Peru.)

southern tip of Africa. Near the end of the Early Carboniferous, the deposits of the geosyncline were again folded. Potassium-argon dates on granitic rocks indicate that an intrusive episode occurred during the Late Carboniferous (27). Deformation and uplift during the Middle and Late Permian resulted in the widespread deposition of continental red beds in the geosyncline.

Cape Geosyncline

In the Cape Geosyncline marine clastics of Early Devonian age are succeeded by quartzites and shales of Middle and perhaps Late Devonian age. These units were folded prior to the deposition of the overlying Dwyka Series. A thick sequence of tillites and fluvioglacial deposits in the lower Dwyka Series is overlain by the Ecca Series of lowermost Permian age. Apparently the glaciation occurred during Carboniferous and possibly earliest Permian time.

The Ecca Series is comprised of marine and coal-bearing continental deposits containing a *Glossopteris* flora. Since the fossil wood within the continental deposits exhibits seasonal growth rings, the coals were probably deposited in a moist, temperate climate (Fig. 11.28). Paleomagnetic data indicate that this area was at approximately 60° South latitude during the Permian. The source of the Ecca Series was south of the Cape region. Since separation of the continents occurred after the Permian, the source may have been in Antarctica. Folding during the Permian in the southern part of the Cape Geosyncline ended sedimentation in the geosyncline.

Antarctic Geosyncline

Thick sequences of late Paleozoic rocks have been found in the Ellsworth Mountains, the Pensacola Mountains, and the Horlick Mountains of Antarctica. The deposits in these ranges were once part of a major geosyncline that extended from the Wedell Sea to the Ross Sea (28). In the Horlick Mountains, Early Devonian marine sandstones and shales rest unconformably on granitic basement and are overlain by the Buckeye Tillite (29). The tillite is 900 ft (300 m) thick, and shales near the top of

Figure 11.28 Fossil wood showing growth rings from Permian deposits in Antarctica. (Photo by W. E. Long.)

the tillite contain spores that are probably of Permian age. However, the lower part of the unit may be Carboniferous (30). Elsewhere, correlative tillites rest unconformably on grooved and striated basement. Shales, sandstones, and coals that overlie the tillites have yielded a *Glossopteris* flora of Late Permian age (Fig. 11.28).

No evidence of late Paleozoic deformation has been found in the Antarctic Geosyncline, although radiometric ages of granitic rocks indicate that intrusive episodes occurred 350 million years ago (Late Devonian) and 280 million years ago (earliest Permian) (28).

Continental Interior

Thick accumulations of sediment were laid down during the late Paleozoic in basins adjacent to geosynclinal belts. For example, nearly 10,000 ft (3000 m) of Devonian strata were deposited in eastern Brazil, and 10,000 ft of Permian sediments were deposited in northwestern Australia. The rate of sedimentation of these deposits is equal to that of many geosynclines, but the basins are not significantly elongated.

Evidence of glaciation is found in late Paleozoic deposits in all the southern continents and in India as well (Fig. 11.29). During the Middle Devonian, glaciers formed on the northern edge of the Central Brazilian Shield and moved into both the Amazon and Parnaiba basins, where they deposited tillites containing striated and faceted pebbles (25).

The most extensive of the late Paleozoic glaciations occurred during the latest Carboniferous and earliest Permian. Most of the features associated with Pleistocene glaciers have also been found in these glacial deposits:

1. Poorly sorted tillites contain striated and faceted pebbles and rest on grooved, striated, and polished basement (Fig. 11.29).

2. In some areas, the separation of several tillites by marine deposits suggests alternating glacial and interglacial episodes (34).

3. Banded shales associated with tillites may represent varves that were deposited in proglacial lakes (35).

4. Linear bodies of sand interbedded with tillites may be buried eskers.

The glacial deposits are preserved in a series of unconnected basins. Although it has been suggested that glaciation was confined to these basins (36), studies of the orientation of glacial striations and *roches moutonnées* indicate that a single large ice sheet originated in two principal centers, one in southwestern Africa and one in eastern Antarctica (Fig. 11.26). Both centers are located near the Permo-Carboniferous South Pole as determined from paleomagnetic studies. The wide distribution of the glacial deposits and the radial pattern of the ice movement indicate continental rather than alpine glaciation. Although local alpine glaciers probably formed in upland areas at this time, most of the drift may be attributed to continental glaciers. This glaciation extended to within 30° of the Permo-Carboniferous equator in northern India (Fig. 11.26). By comparison, Pleistocene ice sheets reached to within 40° of the equator. Thus the extent of glaciation in the Permian and Carboniferous time is comparable to that of the Pleistocene.

Glacial deposits are often overlain by continental and marine deposits of Early Permian age. The continental deposits commonly contain coal with a *Glossopteris* flora. In Antarctica some of these coal deposits are located within 5° of the Permian South Pole (Fig. 11.30). Presumably these deposits formed under a cool climate in swamps that developed on the uneven, glaciated terrain.

ANCESTRAL SIBERIA

During the Devonian and Early Carboniferous, geosynclines bordered Ancestral Siberia in essentially the same locations as those of the early Paleozoic (Fig. 10.38). The Uralian and Angara geosynclines were converted into mountain ranges toward the end of the late Paleozoic. Deformation

(a)

(b)

Figure 11.29   Permo-Carboniferous glacial features: (a) pebbly tillite from Natal, southern Africa; (b) striated facet on cobble in bouldery tillite, Wynyard, Tasmania, Australia; (c) striated pavement at Iriani, in central India (scale given by hand lens); (d) glacially polished basalt, Kimberly, South Africa (Bushmen etching on the left); (e) exhumed *roche moutonnée* of Precambrian basalt, Kimberly, South Africa (ice movement was toward upper left); (f) bouldery tillite, Antarctica. [(a), (b), (d), (e), and (f) courtesy of Warren Hamilton, U.S. Geological Survey, from Hamilton and Krinsley, Ref. 31; (c) courtesy of A. J. Smith, University College, London, Ref. 32.]

(c)

(d)

(e)

(f)

Figure 11.30 Late Paleozoic sedimentary rocks on the face of Mount Weaver, Queen Maud Mountains, Antarctica. Coals containing a *Glossopteris* flora are present near the top of the mountain. (Photo by George A. Doumani, *in* Ref. 33.)

in the Angara Geosyncline may have been caused by plate collisions involving the joining of Ancestral Siberia and Ancestral China. By the end of the late Paleozoic, Ancestral Siberia was firmly connected to the other continents that comprise Pangaea.

Angara Geosyncline

A sequence of continental red beds and minor volcanic rocks was deposited in the northern part of the Angara Geosyncline after an intense episode of deformation at the close of the Silurian. This sequence, which ranges in age from Devonian to Early Carboniferous, grades southward into marine limestones and shales. It was folded and uplifted near the end of the Early Carboniferous, and sub-

sequent deposits are almost entirely continental. Deformation during the Permian ended deposition in the Angara Geosyncline.

Northern Pacific Ocean Geosyncline

The upper Paleozoic deposits of the Northern Pacific Ocean Geosyncline are almost entirely marine and consist of limestone, shale, sandstone, and minor volcanics. The Verkhoyansk Complex, in the northern part of the geosyncline, ranges from Permian to Middle Triassic and reaches a thickness of more than 30,000 ft (10,000 m).

Taimyr Geosyncline

A sequence of limestones and shales ranging from Middle Devonian to Early Carboniferous rests unconformably on Silurian deposits in the Taimyr Geosyncline. Folding near the close of the Early Carboniferous may have been caused by movement of an oceanic plate under the geosyncline. According to paleomagnetic data, Late Carboniferous coal-bearing beds were deposited at approximately 45° North. Permian strata have not been reported from the Taimyr Geosyncline, and Mesozoic deposits are not of geosynclinal thickness. Evidently, deposition in the geosyncline had ceased by the end of the Carboniferous.

Continental Interior
Devonian red beds crop out in the western part of the Siberian Platform. Since they grade eastward into limestones, it is probable that they were derived from highlands in or adjacent to the Uralian Geosyncline. Carboniferous and Lower Permian deposits are mainly limestone with some interbedded clastics and coals. The Angara flora found associated with Permian coals is similar to the *Glossopteris* flora. However, paleomagnetic studies indicate that the Angara flora grew at approximately 35° North latitude, whereas the *Glossopteris* flora grew at higher paleolatitudes. Very thick volcanic rocks and minor continental sediments were deposited over most of the western part of the Siberian platform during the Late Permian.

ANCESTRAL CHINA
The arrangement of geosynclines during the late Paleozoic was essentially the same as that of the early Paleozoic (Fig. 10.39). However, when Ancestral China was joined to Ancestral Siberia as a result of plate collisions, the Central Asia Geosyncline was converted into a mountain range.

Southern Pacific Ocean Geosyncline
Deposits in the Southern Pacific Ocean Geosyncline have been thoroughly studied in Japan, where thick layers of clastics and felsic to mafic volcanics were deposited during the late Paleozoic (Fig. 11.31). The clastics are almost entirely marine, but the presence of several unconformities in late Paleozoic sequences provides evidence of local emergence. Uranium-lead dates on sphene and zircon indicate that an episode of regional metamorphism occurred in northwestern Japan in the Middle Permian, approximately 240 million years ago (38). This metamorphism was accompanied by uplift and erosion of a granitic basement. Granitic pebbles in Permian deposits were probably derived from this terrain.

Japan is now separated from the mainland of Asia by the Sea of Japan, which is more than 6300 ft (2000 m) deep in most parts. Three theories have been proposed for the origin of this sea:

1. If Japan was at one time connected with the Asian mainland, a small block of continental crust may have subsided to form the Sea of Japan. However, the subsidence of such a block would have produced a large negative gravity anomaly, which would be easily detected by seismic refraction profiling. Geophysical surveys in the Sea of Japan have yet to detect a layer with a thickness and density comparable to that of continental crust.

2. Japan may have drifted southeastward away from Asia (39). In this case, Korea would have been located along the trend of structures in Japan prior to separation from the mainland. However, Japan is not geologically comparable to Korea. Whereas Korea was largely above sea level during the Paleozoic, Japan received a great thickness of sediment during that time.

3. The Sea of Japan may have formed behind an island arc in essentially the same position it occupies today. In this case, Japan would be a relatively young continental block composed of volcanics, sediments derived from the weathering of the volcanics, and granites formed by the partial melting of the sediments and volcanics.

Central Asia Geosyncline
The late Paleozoic was a time of great tectonic activity in the Central Asia Geosyncline. Erosion of uplifts within and adjacent to the geosyncline resulted in the deposition of an average of 15,000

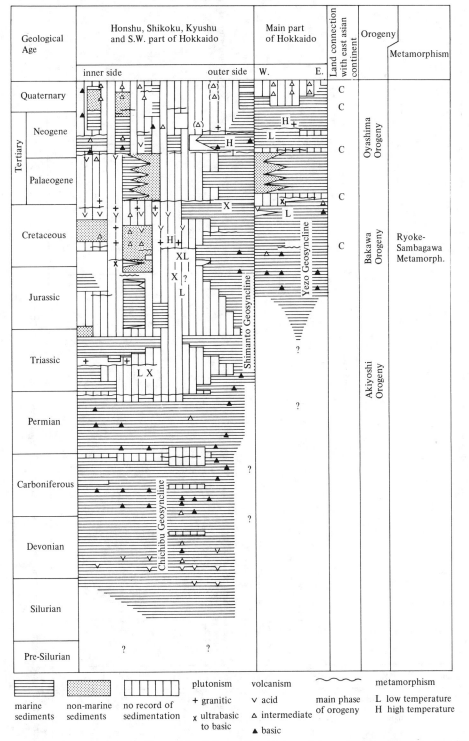

Figure 11.31   A schematic illustration of the geologic history of Japan. (From Matsushita, Ref. 37.)

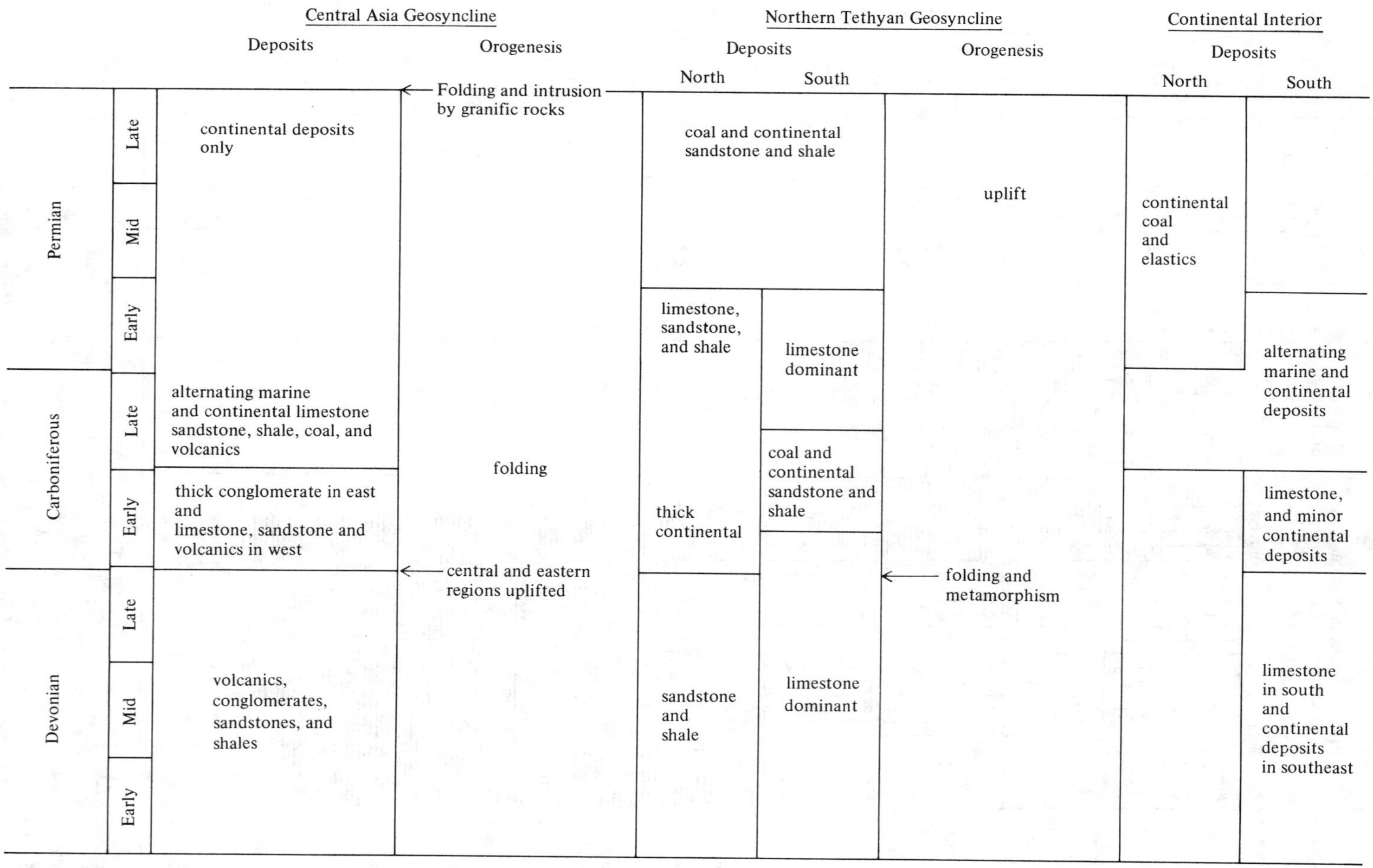

Central Asia Geosyncline

Deposits | Orogenesis

Northern Tethyan Geosyncline

Deposits | Orogenesis

Continental Interior

Deposits

	North	South

Permian

- Late / Mid: continental deposits only
- Early: —

Folding and intrusion by granitic rocks

Deposits — North / South: coal and continental sandstone and shale

uplift

North: continental coal and clastics

South: alternating marine and continental deposits

Carboniferous

- Late: alternating marine and continental limestone sandstone, shale, coal, and volcanics
- Early: thick conglomerate in east and limestone, sandstone and volcanics in west

folding

central and eastern regions uplifted

North: limestone, sandstone, and shale

South: limestone dominant

coal and continental sandstone and shale

folding and metamorphism

North: limestone, and minor continental deposits

Devonian

- Late / Mid / Early: volcanics, conglomerates, sandstones, and shales

thick continental

sandstone and shale

limestone dominant

limestone in south and continental deposits in southeast

Figure 11.32 A schematic illustration of the late Paleozoic geologic history of Ancestral China.

296

ft (5000 m) of late Paleozoic clastic sediments and volcanic rocks (Fig. 11.32).

Northern Tethyan Geosyncline

Geanticlinal uplifts within the Northern Tethyan Geosyncline furnished great thicknesses of clastic sediment during the Devonian and Early Carboniferous. More than 15,000 ft (5000 m) of Devonian clastics were deposited in the central part of the geosyncline (39). After tectonic activity subsided during the Late Carboniferous and Permian, limestone and coal were deposited in the geosyncline (Fig. 11.32).

Indonesian Geosyncline

The oldest known deposits in the Indonesian Geosyncline are Devonian marine clastics, which crop out in western Borneo. Carboniferous deposits in Sumatra contain gneiss and schist pebbles, which indicate that a pre-Carboniferous basement is present in the region. Carboniferous and Permian deposits are dominantly clastics (probably eroded from nearby geanticlines) and volcanics (40).

Continental Interior

Limestones were deposited over much of the southern part of the continental interior during the Devonian and Early Carboniferous, but a fluctuating withdrawal of the seas during the Late Carboniferous and Permian resulted in the deposition of coal-bearing clastics in many areas (Fig. 11.32). Both the Carboniferous and Permian coals contain floras similar to those of Carboniferous and Permian coal deposits in North America and Europe. However, according to paleomagnetic studies, the continental interior of Ancestral China was at approximately 45° North latitude, whereas the Carboniferous coals of North America and Europe were deposited within 10° of the equator.

LATE PALEOZOIC LIFE

Early in the late Paleozoic, rapid evolutionary changes occurred among invertebrates, vertebrates, and plants. All groups, but particularly the invertebrates, were responding to pressures that resulted from the merging of continents. As geosynclines were uplifted to form mountains, many marine environments were eliminated.

The gradual decline of the trilobites, which began in the Ordovician, continued into the late Paleozoic. Corals and stromatoporoids replaced algae as the principal reef-forming organisms, and the graptolites were well on their way to extinction. The spiny, productid brachiopods and coiled, ammonoid cephalopods exhibited unusual morphologic development during the late Paleozoic. By the end of the Permian, most invertebrate groups had diminished considerably in importance and many became extinct.

Protists

Foraminifera with calcareous tests occur in rocks as old as Devonian. One group, the endothyroids, were so abundant in Mississippian seas that some limestones of that age are composed almost entirely of their shells. The endothyroids gave rise to the fusulinids in the Pennsylvanian (Fig. 11.33). Fusulinids are large, spindle-shaped forms which diversified sufficiently to provide important index fossils for Pennsylvanian and Permian sequences.

Sponges

Large colonies of delicate glass sponges flourished during the Devonian, particularly in western New York (Fig. 11.34). Calcareous sponges were important reef builders in the area of present-day Texas and New Mexico. The rise in importance of sponges as reef builders paralleled the decline of the rugose and tabulate corals in the Permian.

Coelenterates

Tabulate corals, such as *Favosites*, and rugose corals, such as *Heliophyllum*, were abundant in the warm seas of the Devonian (Fig. 11.35). However, the rugose corals declined during the Carboniferous and had become extinct by the end of the Permian. Coral reefs contained abundant colo-

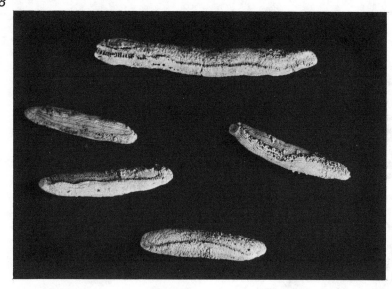

Figure 11.33 *Parafusulina*, fusulinid foraminiferans from Permian deposits near Marathon, Texas. (Courtesy of Smithsonian Institution.)

nial as well as rugose corals during the Devonian. *Pleurodictium*, a small colonial coral, is an excellent index fossil, since certain species are widespread and occupy rather short time spans.

Bryozoans

Delicately branching and lacy bryozoans are very common in late Paleozoic sedimentary rocks. *Fenestella*, whose name means little windows, was

Figure 11.34   Restoration of part of a Devonian coral reef in western New York: (a) Crinoids, (b) glass sponges, (c) cephalopods, (d) corals, (e) seaweed, (f) a gastropod. (Courtesy of the Buffalo Museum of Science.)

Figure 11.35 Restoration of the sea bottom in Michigan during the Middle Devonian. Organisms include: (a) Solitary corals, (b) colonial corals, (c) trilobites, (d) brachiopods, (e) nautiloid cephalopods, (f) crinoids. Some of the genera are *Heliophyllum*, the large solitary corals at the upper left; *Hexagonaria*, the solitary corals at the bottom left; *Atrypa*, the brachiopods at the bottom right; and *Stropheodonta*, the brachiopod at the bottom near the center. (Courtesy of the Field Museum of Natural History, Chicago.)

attached to a screw-shaped support, *Archimedes* (Fig. 11.36). Because of its distinctive shape and widespread occurrence, *Archimedes* is an excellent guide fossil in Mississippian strata. Like the calcareous sponges, bryozoans were important reef-builders during the Permian.

Brachiopods

Devonian marine invertebrates were dominated by the long-ranging, spiriferid brachiopods (Fig. 11.37). These forms characterized the benthonic shelf and deltaic environments. As the sea level fluctuated, the spirifers and their associated faunas migrated within their respective sedimentary facies. With time, the faunas experienced significant evolutionary changes which provided the basis for zonation of the marine Devonian. This zonation, which was developed by H. S. Williams, stands as an early example of the use of facies faunas and population statistics in biostratigraphy.

The productids evolved from the strophomenid brachiopods. They were the dominant brachiopods in the Carboniferous and Permian, but became extinct by the end of the Paleozoic.

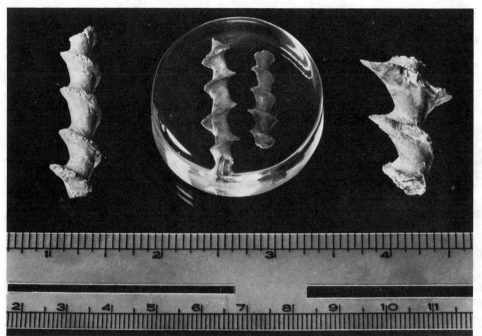

Figure 11.36   Archimedes, the screw-shaped bryozoa. (Courtesy of Wards Natural Science Establishment.)

Productids are characterized by large size and long spines. Some species were extraordinarily large — for example, *Gigantella gigantea*, which grew up to 12 in. (30 cm) or more. Late Permian limestones in the Glass Mountains of western Texas contain large members of productids along with unusual oysterlike, conical, and horn-shaped brachiopods (Fig. 11.38).

Mollusks

Pelecypods were generally far less abundant than brachiopods during most of the late Paleozoic. In the Devonian, some groups of pelecypods moved from marine into freshwater environments, where they are still prevalent. Ammonoids are thought to have evolved from straight-shelled nautiloids. They appeared first in the Early Devonian and progressed rapidly from straight shells to loosely coiled and then tightly coiled shells in the later part of the Early Devonian (41).

The ammonoids became important index fossils in late Paleozoic sequences, mainly owing to the development of distinctive sutures (Fig. 11.37). Devonian and Early Carboniferous ammonoids had relatively simple, goniatite sutures. Ceratite and ammonite sutures developed independently in several families during the Late Carboniferous and Permian (41). Many groups of cephalopods became extinct at the end of the Paleozoic, and only one family with complex sutures survived into the Triassic.

Arthropods

The gradual decline of the trilobites, which began in the Ordovician, continued during the late Paleozoic. In some Devonian sequences, however, trilobites such as *Phacops rana*, are abundant (Fig. 11.39). They became scarce during the Carboniferous and were extinct by the end of the Paleozoic.

Late Paleozoic ostracodes are distinctive and are well represented. Although a number of ostracode families became extinct before the end of the Permian, enough survived to give rise to a great number of new forms in the Mesozoic.

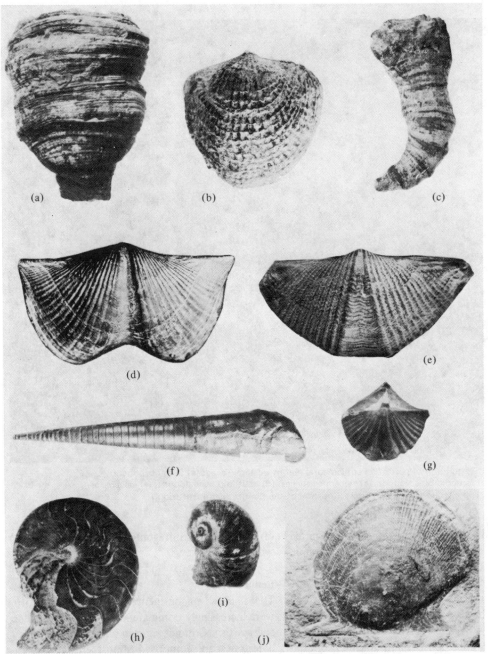

Figure 11.37 Middle Devonian (Hamilton) fossils from western New York: (a) *Heliophyllum halli*, a rugose coral (1X), (b) *Spinatrypa spinosa,* brachial view of a brachiopod (1-1/2X), (c) *Amplexiphyllum hamiltoniae,* a rugose coral (2X), (d) *Spinocytria granulosa,* pedicle view of a brachiopod (1X), (e) *Mucrospirifer mucronatus,* pedicle view of a brachiopod (1-1/2X), (f) *Michelinoceras aldenense*, a nautiloid cephalopod (1X), (g) *Cyrtina hamiltonensis,* brachial view of a brachiopod (2X), (h) *Tornoceras uniangulare,* an ammonoid cephalopod (1X), (i) *Naticonema lineata,* a gastropod (1X), (j) *Pseudaviculopecten princeps,* a pelecypod (1X). (Photo courtesy of Carlton Brett.)

Figure 11.38   Restoration of part of a Permian coral reef in western Texas: (a) Spiny productid brachiopods, (b) leptodid brachiopods, (c) siliceous sponges, (d) nautiloid cephalopods, (e) calcareous sponges, (f) gastropods, and (g) rugose corals. (Courtesy of the Smithsonian Institution.)

Arachnids, which are arthropods with jointed legs and pincerlike appendages in the head region, have been discovered in beds of late Early Devonian age in western Germany (42). These spiderlike fossils are the oldest known nonscorpion arachnids and may be the oldest terrestrial animals. The Rhynie Chert in Scotland contains very small (0.3-3.5 mm) nonscorpion arachnids of Middle or possibly late Early Devonian age.

A great variety of insects developed during the Pennsylvanian, and some reached enormous size (Fig. 11.40). Tropical forests and swamps in North America and Europe contained cockroaches 4 in.

(10 cm) long and dragonflies with 30 in. (75 cm) wingspreads.

Echinoderms

The shallow epicontinental seas of the platform were increasingly dominated by crinoids, blastoids, and starfish. Crinoids were so numerous that many limestones are comprised almost entirely of crinoid fragments. Occasionally, large numbers of complete crinoids are preserved in limestones or shales that were deposited in very quiet waters (Figs. 11.41 and 11.42). Blastoids reached their climax during the Mississippian but died out in

Figure 11.39 The trilobite *Phacops rana* from the Middle Devonian of western New York. (Photograph courtesy of Carlton Brett.)

most areas by the end of the Pennsylvanian. In Indonesia, they persisted until the end of the Permian.

Graptolites

By the end of the Silurian, the number of graptolite genera had declined drastically. Only one group lived on into the Devonian and Carboniferous. Therefore, late Paleozoic graptolites are not useful guide fossils.

Vertebrates

Fish. During the Devonian, fish were abundant and well diversified — the most advanced animals on the evolutionary scale. Consequently, the Devonian has been termed the Age of Fishes. Indeed, one of the major events in the history of the vertebrates was the development of jaws in the placoderms during the Late Silurian and Early Devonian. Jaws are thought to have developed

through ossification of the front set of V-shaped gill arches of the placoderms. The gill arches, which are useful to aid in pumping water through the gill openings, developed hinges to allow for opening and closing. Subsequently, teeth were added for biting and chewing. Released from the necessity of filter feeding, the placoderms could then compete with eurypterids and other predators. The placoderms evolved rapidly during the Devonian, and some became predators of an impressive size. *Dinichthys* was 30 ft (10 m) long.

The Osteichthyes (bony fish), the Chondrichthyes (sharks, rays, and skates), and the first freshwater fish are thought to have evolved from the placoderms. The placoderms and the agnaths declined in importance after the Devonian with the rise of more advanced fish. The Chondrichthyes, which have cartilaginous skeletons, have persisted with few changes into recent times (Fig. 11.43). The Osteichthyes stem from Ordovician ancestors, but the oldest fossil remains have come from Late Devonian shales. These are the dominant fish today and include ray-finned and lobe-finned fish. Most of the modern bony fish descended from ray-finned forms.

Amphibians are thought to have evolved from lobe-finned fish. The history of their evolution is interesting because of its importance in the development of the higher animals and because the evolutionary advances can easily be traced to the demands of the environment.

Arkosic red beds of late Paleozoic age frequently contain the remains of lobe-finned fish. As discussed in Chapter 5, it is believed that such deposits were laid down under arid conditions. Only fish adapted to these conditions could survive. For this reason, most late Paleozoic freshwater fish had lungs and most saltwater forms did not. One order of lobe-finned fish could survive prolonged drought by burrowing deep in the mud and slowing their metabolism in a manner similar to hibernation. Lungfish in South America and Africa still burrow in the mud during dry spells.

The crossopterygians, another order of lobe-finned fish, coped with the problem of drought in

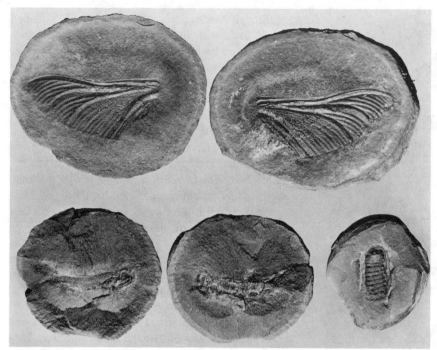

Figure 11.40 Insects in concretions from Late Carboniferous beds at Mazon Creek, Illinois. (Courtesy of the Field Museum of Natural History, Chicago.)

Figure 11.41 Fossil crinoids of Early Carboniferous age from Le Grand, Iowa. (Courtesy of the Buffalo Museum of Science.)

Figure 11.42   Restoration of mid-continent sea-floor during Early Carboniferous. (Courtesy Smithsonian Institution.)

another way. They had strong fins with which they could push themselves from one pool to another. The crossopterygians are anatomically similar to the amphibians. They had a well-ossified skeleton and their skull bones can be matched, bone for bone, with those of amphibians and the higher vertebrates as well. Moreover, the teeth of crossopterygians show a marked similarity to the teeth of the early amphibians. For these reasons, it is almost certain that crossopterygians gave rise to the amphibians.

Amphibians. The earliest known amphibians are found in a Late Devonian red sandstone in eastern Greenland. These amphibians are known as labyrinthodonts because their teeth have intricately infolded enamel which resembles a labyrinth (Fig. 11.44). Some paleontologists suggest that the appearance of the amphibians may have been related to the rise in predatory lobe-finned fish and lungfish. Alternatively, they may have moved into the subaerial environment in search of new sources of food, such as insects and plants. The amphibians flourished during the Carboniferous but then began to decline during the Permian as reptiles became more numerous and more diversified (Fig. 11.45).

Figure 11.43   Upper Devonian fish and associated fauna of western New York. Crinoids in the foreground are attached to a fragment of a Devonian tree. The small fish belong to the genus *Rhadinichthys*. Other genera include *Coccostes* at the bottom left, *Ctenecanthus* the large sharklike form, and the lungfish *Dipterus* at the bottom right. (Courtesy of the Buffalo Museum of Science.)

Reptiles. Before the labyrinthodonts became extinct at the end of the Triassic, they gave rise to the reptiles. The development of the amniote egg was a significant factor in the rise of the reptiles to dominance, since it enabled them to break their dependence on an aquatic existence. The eggs of amphibians are fertilized externally in water, whereas those of reptiles are fertilized internally. Reptile eggs have a tough shell which serves as protection not only from amphibians and fish but against drought as well. The embryo is surrounded by a water-filled membrane, the *amnion*.

The oldest known reptile eggs were found in Early Permian strata, but the oldest known reptile, *Romeriscus*, was discovered in Carboniferous beds

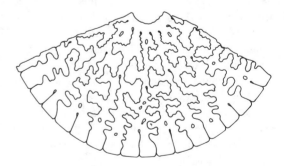

Figure 11.44   Diagrammatic cross-section of part of a labyrinthodont tooth. The sinuous lines are complex infolds of the enamel of the tooth. (After Colbert, Ref. 43.)

Figure 11.45 Skeleton of a large amphibian, *Megacephalus.* (Courtesy of the Field Museum of Natural History, Chicago.)

(44). *Romeriscus* belongs to a group of reptiles known as the cotylosaurs, from which the other reptile groups are thought to have evolved. Some early members of this group may have spent much of their adult lives in ponds and streams, like the amphibians. They probably walked with the sprawling gait that characterized their amphibian ancestors.

The reptiles increased in number in Permian time and broadened their range and habitat as well (Fig. 11.46). One group, the therapsids, are the probable ancestors of the more advanced reptiles and possibly even the mammals. The therapsids appeared during the Carboniferous and by the Permian had greatly improved powers of locomotion. Knees were turned forward and their stride lengthened. The resulting increase in speed was a distinct advantage in escaping predators and running down prey. To judge by their teeth, many of the therapsids were carnivores.

Plants

The Early Devonian plant record is sparse, possibly because the colonization of the land by plants had just begun. Early Devonian plants included primitive varieties or mosses and vascular plants (Fig. 10.53). By the Middle Devonian, the expansion of terrestrial plants was well underway. Forests of tree ferns, such as *Eospermatopteris*, scale trees, such as *Protolepidodendron*, scouring rushes, and seed ferns were abundant at that time.

Recent discoveries indicate a great antiquity of seed-bearing plants. Primitive pinelike conifers arose during the Late Devonian. In the Carboniferous, scouring rushes, such as *Calamites*, achieved tree size, and the scale tree *Lepidodendron* and the tree fern *Sigillaria* grew to heights exceeding 100 ft (30 m) (Figs. 11.47 and 11.48). The more advanced seed-bearing ferns and gymnosperms had achieved dominance over the seedless trees by Late Carboniferous time. Prominent among the gymnosperms were the cordaites, true conifers, ginkgoes, and cycads.

Extinctions

A number of invertebrate groups became extinct near the end of the Paleozoic. These include trilobites, eurypterids, blastoids, and rugose corals. The number of families of foraminiferans and bryozoans declined markedly during the Permian. Among the vertebrates, the agnaths became extinct in the Middle Permian. Amphibians, cotylosaurs, and several groups of plants declined throughout the Permian.

The declines in both fauna and flora are too irregular to be assigned to any single cause. One contributing factor may have been the gradual up-

Figure 11.46 Restoration of a group of late Paleozoic reptiles. The five large finback reptiles are carnivores belonging to the genus *Dimetrodon*. On the left are a group of small pelycosaurs, and the small amphibian *Diplocaulis* can be seen in the water in the bottom right. (Courtesy of the Field Museum of Natural History, Chicago.)

Figure 11.47   Restoration of a Pennsylvanian coal swamp in the mid-continent. Genera of the plants include: (a) *Sigillaria*, (b) *Lepidodendron*, (c) *Cordaites*. Also notice the large cockroach on the tree at the left. (Courtesy of the Field Museum of Natural History, Chicago.)

lift of the land and the resulting withdrawal of the seas, which occurred during the late Paleozoic. The uplifts were associated with deformations such as the Hercynian Orogeny between the Early and Late Carboniferous and the Alleghenian Orogeny in the Permian. The Permian marked the end of marine deposition in the Appalachian, Uralian, Angara, and Hercynian geosynclines. Furthermore, the seas withdrew from the continental interior and portions of the geosynclines bordering the Pacific Ocean and the Tethyan Sea. The uplifts drained many of the coal-forming swamps in both the northern and southern hemisphere. This alone

may have resulted in the decline of swamp-dwelling plants and animals. The uplift of the lands also resulted in a drastic reduction in the area of the shallow seas, which greatly increased competition among their inhabitants. Such competition might account for the decline in the foraminiferans, bryozoans, and brachiopods. For the trilobites it may have been the final blow to their survival.

The principal cause for the decline of the amphibians was probably the rise of the reptiles, which could prey on all but the deepest diving amphibians. Furthermore, the development of the

(a)

(b)

amniote egg insured that a greater percentage of young reptiles survived.

LATE PALEOZOIC CLIMATE

On a late Paleozoic reconstruction of the continents, climatic zones are aligned approximately parallel to paleolatitudes in a pattern similar to that of today (Fig. 11.1). In general, tropical coals, reefs, red beds, evaporites, and dune sands are located within 40° of the paleoequator and glacial deposits occur within 40° of the South Paleopole. The absence of glacial deposits of late Paleozoic age in the northern hemisphere is probably due to the location of the North Paleopole over an open ocean (Fig. 11.1).

Devonian

Glacial deposits are not common in Devonian sequences, but they have been reported from sev-

Figure 11.48   Late Carboniferous fossilized land plants: (a) Part of trunk of the lycopod *Lepidodendron,* showing leaf scars, (b) leaves of the horsetail rush, *Annularia,* from a concretion found at Mazon Creek, Illinois. (Courtesy of the Field Museum of Natural History, Chicago.)

eral areas in South America and one in Africa (Fig. 11.1a). The limited extent of these deposits suggests that they were a result of alpine glaciation. All Devonian glacial deposits occur within 40° of the South Paleopole.

Devonian coals are found on Bear Island north of Norway, in the European part of the Soviet Union, and in the Arctic Islands of northern Canada. All these areas fall within 15° of the paleoequator. Coral and other organic reefs were widespread, and most were formed within 30° of the Devonian equator. However, in northwestern India, well-developed reefs are found at a paleolatitude of approximately 40° South. Red beds and evaporites generally fall within 30° of the paleoequator, but some have been found as far from the equator as 40°. From paleoclimatic indicators, it may be inferred that on the whole the Devonian climate was somewhat like that of today.

Carboniferous

The Carboniferous was a time of gradual cooling in the vicinity of the South Paleopole. There are three distinctly different glacial horizons in the Carboniferous sequences of Bolivia. The first and possibly the second are of Early Carboniferous age (27). Early Carboniferous glacial deposits have also been reported from Argentina (25). The association of these deposits with highlands suggests that they may have been the result of alpine glaciation. The lower sections of the Dwyka Tillite of central and southern Africa may also be of Early Carboniferous age (34).

The climate began to cool rapidly toward the end of the Early Carboniferous, and just prior to the Permian the southern continents were in the grip of a major ice age. Permian and Carboniferous glacial deposits have been found in all the southern continents and India. All these deposits occur within 60° of the South Paleopole (Fig. 11.1b).

Much of North America and Europe experienced tropical, subtropical, or arid conditions during the Carboniferous, since these continents were situated near the paleoequator. Tropical coals formed within 15° of the Carboniferous equator and reefs, evaporites, and red beds generally fall within 40° of this equator (Fig. 11.2b). In Japan, however, small patch reefs are found at a paleolatitude of 65° North. This suggests that the area in the vicinity of the North Paleopole was considerably warmer than that in the vicinity of the South Paleopole during the Carboniferous. The waters near Japan may have been warmed by currents from the south similar to the modern Kuroshio Current.

Permian

The climate was cool at the beginning of the Permian, but it gradually became warmer during the Middle and Late Permian. Continental glaciers formed during the Early Permian in Australia, Antarctica, and possibly Africa. In northeastern Australia, erratic blocks weighing up to several tons occur in Middle and Upper Permian marine deposits and may have been deposited by icebergs originating from alpine glaciers (21). Upper Permian glacial deposits are also present in the Himalayas (23) and in northwestern Argentina (25). These deposits are not extensive and were probably deposited by alpine glaciers.

Very large Permian evaporite deposits occur in the United States and in the Soviet Union between 15 and 30° north of the paleoequator in what was presumably an arid belt. Minor gypsum and salt of Permian age in western Australia indicate local arid conditions. Permian faunas from the Canadian Arctic are thought to have developed in cool water (45), although this area was at a paleolatitude of about 45° North. Possibly the northern hemisphere was cooled as a result of the glaciation in the southern hemisphere. Tropical coals of Permian age are found within 15° of the paleoequator, but coals containing a *Glossopteris* flora occur within 5° of the Permian South Pole in Antarctica. They may have been deposited in an environment similar to that of the modern Arctic muskeg swamps.

SUMMARY

At the beginning of the late Paleozoic there were only two discrete landmasses. These continents were separated by a rather narrow ocean during most of the Devonian, but by the end of the Devonian this ocean was largely closed. Several transgressions and regressions of the seas across the continents took place during the late Paleozoic. At the beginning of the Devonian, the seas were generally restricted to the geosynclines and the adjacent platform. Transgressions occurred during the Middle and Late Devonian and during the Early and Middle Pennsylvanian. The seas began an irregular withdrawal during the latest Pennsylvanian and by the end of the Permian they had largely withdrawn from the continents.

The late Paleozoic was a time of tectonic activity in most geosynclines, especially those located along the junctions between ancestral continents. Uplifts furnished large volumes of clastic sediment to the geosynclines. By the end of the Paleozoic, deformations had ended deposition within the Appalachian, Ouachita, Franklin, Hercynian, Uralian, Angara, Taimyr, Central Asia, and West African geosynclines and had converted them into mountains. These deformations were probably related to collisions of continental plates associated with the completion of joining of the continents into a single landmass, Pangaea. Deformations in the geosynclines on the outer margins of Pangaea were probably associated with movements of oceanic plates under the continent.

The land was largely barren of animal and plant life at the beginning of the late Paleozoic, but insects, vertebrates, and plants rapidly colonized the lands. Dispersal into new and unoccupied environments allowed rapid diversification of certain forms until most of the available habitats were filled. Fish with strong fins gave rise to amphibians, and reptiles evolved from the amphibians during the Carboniferous. Land plants evolved rapidly from a few simple varieties in the earliest Devonian to a rich flora including giant tree ferns, scouring rushes, and conifers.

Among the invertebrates, major trends include the evolution of the ammonoid cephalopods from the nautiloids, the development of the productid brachiopods, and the extinction of many invertebrate groups near the end of the Paleozoic. The groups that became extinct included the rugose corals, two orders of bryozoans, many molluscan families, many groups of stalked echinoderms, several groups of sponges, and the trilobites.

The climate was rather cool during the Devonian, probably similar to that of today. Rapid cooling began in the Late Carboniferous, and glaciation was widespread in the southern hemisphere during the latest Carboniferous and Early Permian. The climate warmed somewhat during the Early Permian, but probably remained cool throughout the period.

REFERENCES CITED

1. C. K. Seyfert and D. J. Leveson, 1969, Speculations on the relation between the Hutchinson River Group and the New York City Group: *Geological Bulletin*, v. 3, Queens College Press.

2. A. J. Eardley, 1962, *Structural Geology of North America:* Harper & Row, New York.

3. M. Kay, 1951, *North American Geosynclines:* Geological Society of America Mem. 48.

4. B. R. Pelletier, 1958, Pocono paleocurrents in Pennsylvania and Maryland: *Geol. Soc. Amer. Bull.*, v. 69, p. 1033.

5. J. Rodgers, 1970, *The Tectonics of the Appalachians:* Wiley-Interscience, New York.

6. A. Hietanen, 1967, On the facies series in various types of metamorphism: *Jr. Geol.*, v. 75, p. 187.

7. G. S. Clark and J. L. Kulp, 1968, Isotopic age study of metamorphism and intrusion in western Connecticut and southeastern New York: *Amer. Jr. Sci.*, v. 266, p. 865.

8. L. E. Long, J. L. Kulp, and F. D. Eckelmann, 1959, Chronology of major metamorphic events in the southeastern United States: *Amer. Jr. Sci.*, v. 257, p. 585.

9. S. I. Root, 1970, Structure of the northern terminus of the Blue Ridge in Pennsylvania: *Geol. Soc. Amer. Bull.*, v. 81, p. 815.

10. R. J. Roberts, 1964, *Stratigraphy and structure of the Antler Peak Quadrangle, Humbolt and Lander Counties, Nevada:* U.S. Geological Survey Prof. Paper 459-A.

11. W. P. Irwin, 1966, Geology of the Klamath Mountain Province, *in* E. H. Bailey, ed., *Geology of Northern California:* California Division of Mines and Geology Bull. 190, p. 19.

12. M. Churkin, Jr., 1969, Paleozoic tectonic history of the Arctic Basin north of Alaska: *Science,* v. 165, p. 549.

13. E. J. Buehler and I. H. Tesmer, 1963, Geology of Erie County, New York: *Buffalo Soc. Nat. Sci. Bull.,* v. 21, no. 3.

14. L. D. Grayston, D. F. Sherwin, and J. F. Allan, 1964, Middle Devonian, *in Geological History of Western Canada:* Calgary, Alberta Society of Petroleum Geologists, p. 49.

15. A. M. Klingspor, 1969, Middle Devonian muskeg evaporites of western Canada: *Amer. Assoc. Petrol. Geol. Bull.,* v. 53, no. 4, p. 927.

16. J. A. Peterson and R. J. Hite, 1969, Pennsylvanian evaporite-carbonate cycles and their relation to petroleum occurrence, southern Rocky Mountains: *Amer. Assoc. Petrol. Geol. Bull.,* v. 53, p. 884.

17. J. M. Weller, 1960, *Stratigraphic Principles and Practice:* Harper & Row, New York.

    J. M. Weller, 1957, Geological Society of America Mem. 67, v. 2.

18. L. L. Sloss, 1953, The significance of evaporites: *Jr. Sediment. Petrol.,* v. 23, p. 156.

19. C. O. Dunbar, 1940, The type Permian: Its classification and correlation: *Amer. Assoc. Petrol. Geol. Bull.,* v. 24, p. 237.

20. D. V. Nalivin, 1960, *The Geology of the U.S.S.R.:* transl., S. I. Tomkeieff, Pergamon Press, New York.

21. D. A. Brown, K. S. W. Campbell, and K.A.W. Crook, 1968, *The Geological Evolution of Australia and New Zealand:* Pergamon Press, New York.

22. C. A. Landis and D. S. Coombs, 1967, Metamorphic belts and orogenesis in southern New Zealand, *in* T. Matsumoto, ed., *Age and Nature of the Circum-Pacific Orogenesis: Tectonophysics,* v. 4, p. 501.

23. J. D. A. Zijderweld et al., 1970, Shear in the Tethys and the Permian paleomagnetism in the Southern Alps, including new results: *Tectonophysics,* v. 10, p. 639.

24. J. Sougy, 1962, West African Fold Belt: *Geol. Soc. Amer. Bull.,* v. 73, p. 871.

25. H. J. Harrington, 1962, Paleogeographic development of South America: *Amer. Assoc. Petrol. Geol. Bull.,* v. 46, p. 1773.

26. L. A. Frakes and J. C. Crowell, 1969, Late Paleozoic glaciation: I, South America: *Geol. Soc. Amer. Bull.,* v. 80, p. 1007.

27. C. Schubert, 1969, Geologic structure of a part of the Barinas Mountain Front, Venezuelan Andes: *Geol. Soc. Amer. Bull.,* v. 80, p. 443.

28. W. Hamilton, 1967, Tectonics of Antarctica, *in* T. Matsumoto, ed., *Age and Nature of the Circum-Pacific Orogenesis: Tectonophysics,* v. 4, p. 555.

29. W. E. Long, 1963, The stratigraphy of the Horlick Mountains, *in* R. J. Adie, ed., *Antarctic Geology:* North Holland Publishing Co., Amsterdam, p. 352.

30. W. E. Long, 1965, Stratigraphy of the Ohio Range, Antarctica, *in* J. B. Hadley., ed., *Geology and Paleontology of the Antarctic:* Antarctic Research Series, v. 6., American Geophysical Union, p. 71.

    L. A. Frakes, J. L. Matthews, and J. C. Crowell, 1971, Late Paleozoic glaciation: Part III, Antarctica: *Geol. Soc. Amer. Bull.,* v. 82, p. 1581.

31. W. Hamilton and D. Krinsley, 1967, Upper Paleozoic glacial deposits of South Africa and southern Australia: *Geol. Soc. Amer. Bull.,* v. 78, p. 783.

32. A. J. Smith, 1963, Evidence for a Talchir (Lower Gondwana) glaciation: Striated pavement and boulder bed at Irai, central India: *Jr. Sediment. Petrol.,* v. 33, p. 739.

33. G. A. Doumani and V. H. Minshew, 1965, General geology of the Mount Weaver area, Queen Maud Mountains, Antarctica, *in* J. B. Hadley, ed., *Geology and Paleontology of the Antarctic:* Antarctic Research Series, v. 6, American Geophysical Union, p. 127.

34. L. A. Frakes and J. C. Crowell, 1970, Late Paleozoic glaciation II, Africa exclusive of the Karoo Basin: *Geol. Soc. Amer. Bull.*, v. 81, p. 2261.

35. L. A. Frakes, P. M. de Figueiredo, and V. Fulfaro, 1968, Possible fossil eskers and associated features from the Parana Basin, Brazil: *Jr. Sediment. Petrol.*, v. 38, p. 5.

36. A. A. Meyerhoff and C. Teichert, 1971, Continental drift III: Late Paleozoic glacial centers, and Devonian — Eocene coal distribution: *Jr. Geol.*, v. 79, p. 285.

37. S. Matsushita, 1963, General remarks, *in* F. Takai et al., eds., *Geology of Japan:* University of California Press, Berkeley.

38. K. Ishizaka and M. Yamaguchi, 1969, U-Th-Pb ages of sphene and zircon from the Hida Metamorphic Terrain, Japan: *Earth Plan. Sci. Lett.*, v. 6, p. 179.

39. S. W. Carey, 1958, *Continental Drift: Symposium:* Geology Department, University of Tasmania, Hobart, p. 177.

40. J. H. F. Umbgrove, 1938, Geologic history of the East Indies: *Amer. Assoc. Petrol. Geol. Bull.*, v. 22, p. 1.

41. C. Teichert, 1967, Major features of cephalopod evolution, *in* C. Teichert and E. L. Yochelson, eds., *Essays in Paleontology and Stratigraphy:* Special Publ. no. 2 Department of Geology, University of Kansas, p. 162.

42. L. Størmer, 1969, Oldest known terrestrial arachnids: *Science*, v. 164, p. 1276.

43. E. H. Colbert, 1961, *Evolution of the Vertebrates:* Science Editions, New York.

44. D. Baird and R. L. Carroll, 1967, *Romeriscus,* the oldest known reptile: *Science*, v. 157, p. 56.

45. J. B. Waterhouse, 1967, Cool-water faunas from the Permian of the Canadian Arctic: *Nature*, v. 216, p. 47.

12
The Mesozoic Era

*T*he supercontinent of Pangaea, created by the joining of the ancestral continents during the Paleozoic, was gradually torn apart during the Mesozoic. The rate of separation of the newly formed continents was very slow, averaging only a few centimeters per year. As the continents moved apart, the rifts between them deepened. At an early stage in the separation, these rifts probably resembled the modern Red Sea. Rifting may have been caused by the development of rising convection currents beneath Pangaea. The Atlantic and Indian oceans widened throughout the Mesozoic, and the Pacific Ocean decreased in area. Movement of the continents away from ridges in the Atlantic and Indian oceans and movement of the sea-floor away from an oceanic ridge in the Pacific Ocean was accompanied by extensive deformation, volcanism, plutonism, and metamorphism in the geosynclines bordering the Pacific Ocean. As a result of this tectonic activity, thick layers of clastic sediment were deposited in these geosynclines.

The widespread extinctions within many plant and animal groups at the end of the Paleozoic were followed by the development of a great variety of new forms during the Mesozoic. Evolution was rapid in a number of groups, but changes were most dramatic among the reptiles. An increase in land area, which resulted from the reduction of the size of the epicontinental seas, may have facilitated the expansion of terrestrial faunas and floras. Reptiles, for example, occupied nearly all the available land, sea, and air habitats. Dinosaurs were so widely dispersed that the Mesozoic is known as the Age of Dinosaurs.

MESOZOIC GEOGRAPHY

During the early Mesozoic, Pangaea was bordered by a more or less continuous system of geosynclines (Fig. 7.13). Paleomagnetic and structural data indicate that the breakup of Pangaea occurred in several stages (1). The initial stage began during the Late Triassic with the separation of North America and Gondwanaland. The breakup of Gondwanaland probably began in the Late Jurassic, when South America and a plate that included Antarctica and Australia separated from a plate that included India and Africa (Fig. 12.1). The separation of North America, Greenland, and Europe probably began during the Jurassic. Since the separation of Antarctica and Australia did not begin until the early Tertiary, the present continental outlines were not established until that time.

NORTH AMERICA

During the Early and Middle Triassic, North America was still part of Pangaea, and the only areas that received thick marine deposits were the Cordilleran and Chukota geosynclines and the Brooks and Sverdrup basins (Fig. 12.2). When North America began to separate from Gondwanaland during the Late Triassic, a series of fault basins developed along the east coast of North

315

Figure 12.1

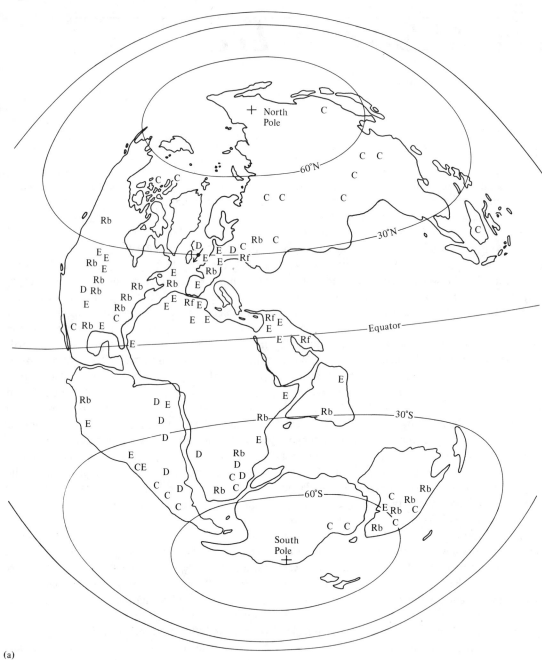

(a)

(b)

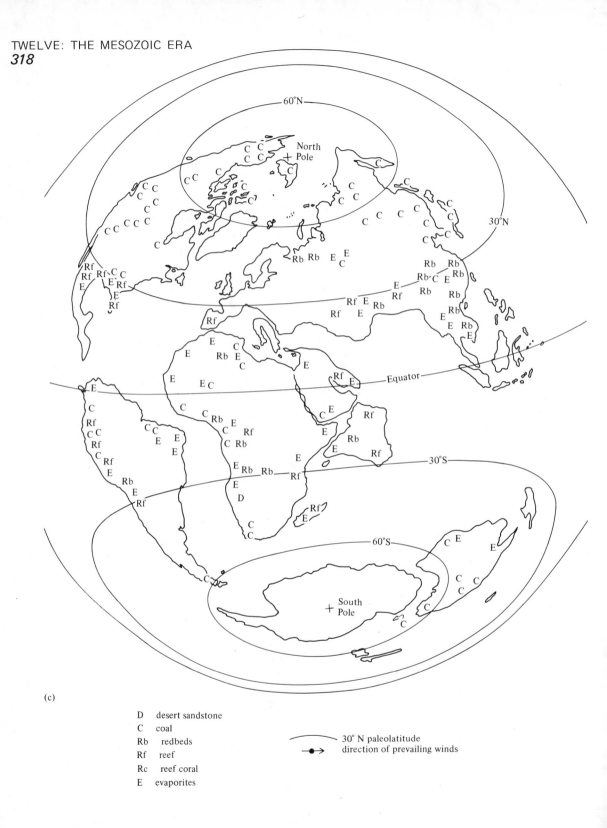

(c)

D   desert sandstone
C   coal
Rb   redbeds
Rf   reef
Rc   reef coral
E   evaporites

30° N paleolatitude
direction of prevailing winds

America. Subsequently, the East Coast and the Gulf Coast geosynclines formed along the Atlantic and Gulf margins of North America (Fig. 12.3). In the Cordilleran Geosyncline an episode of folding during the Late Jurassic resulted in the development of a new depositional trough, the Coast Range Geosyncline.

Triassic Fault Basins

Deposits of Early and Middle Triassic age are not known on the east coast of North America. However, up to 25,000 ft (8000 m) of Late Triassic sedimentary and volcanic rocks were deposited in fault basins paralleling the Atlantic margin of North America (Fig. 12.3). The basins are bounded by normal faults with large vertical displacements.

The Newark Basin extends from southeastern New York through northern New Jersey to southeastern Pennsylvania (Fig. 12.3). The Newark Group, which was deposited in this basin, includes arkosic red beds, black shales, and volcanics. Sandstones and shales grade westward into conglomerates derived from Precambrian crystallines to the west (Fig. 12.4). The black shales of the Newark Group contain fossil branchiopods. These small bivalved crustaceans resemble modern forms found in freshwater lakes. Several varieties of freshwater fish are also present in these deposits. This evidence, along with the lack of marine fossils, indicates that the black shales were probably deposited in a freshwater lake. The red clastics, which contain abundant dinosaur tracks, mud cracks, and raindrop imprints, were presumably deposited on flood plains and subaerial deltas of the streams and rivers that drained into such lakes. Their red color and arkosic composition suggests that they were laid down in an arid region which, according to paleomagnetic data, was located at approximately $20°$ North latitude (Fig. 12.1a).

Three basalt flows in the Newark Group form the prominent ridges of the Watchung Mountains

in central New Jersey. Diabase intrusives in the Newark Group include the Palisades Sill in New Jersey and New York and the Gettysburg Sill of Pennsylvania. The Palisades Sill is up to 1000 ft (300 m) thick and forms a high cliff along the west bank of the Hudson River opposite Manhattan Island (Fig. 12.5). Potassium-argon dates indicate that this sill was intruded during the Late Triassic, between 200 and 190 million years ago (2).

The Connecticut Valley basin contains a sequence quite similar to that of the Newark basin, that is, red clastics and black shales interbedded with three volcanic flows and intruded by a diabase sill. However, the Connecticut Valley basin is bordered on the east by a normal fault, whereas the deposits of the Newark basin are bordered on the west by a normal fault. The deposits of the Connecticut Valley basin dip 5 to $15°$ to the east, and those of the Newark basin dip 5 to $15°$ to the west. Several authors have suggested that the Newark and Connecticut Valley basins were originally connected and later separated as a result of uplift and erosion of the deposits between the basins. However, paleomagnetic data indicate that the lava flows in the Connecticut Valley basin do not correlate with those of the Newark basin. Moreover, the dip of cross-bedding and potassium-argon ages of clastics in the basins suggest that some sediment was transported into the Newark basin from the east and into the Connecticut Valley basin from the west (3).

The tilting of the deposits within the basins may have been caused by differential subsidence during deposition. Since most of the sediments were derived from uplifted blocks bordering the basins, the deposits tend to thicken toward these blocks. Therefore, the maximum subsidence occurred along these margins. The youngest beds have a lower angle of dip than the older beds in the Newark basin, perhaps owing to a smaller amount of subsidence of the younger beds (Fig. 12.6).

Figure 12.1 Proposed reconstruction of the continents showing paleoclimatic indicators: (a) Triassic, (b) Jurassic, (c) Cretaceous. (Considerably modified after Phillips and Forsyth, Ref. 1.)

Figure 12.2   Geosynclines bordering North America during the late Mesozoic.

Other Triassic fault basins are located in North Carolina, Virginia, and Nova Scotia, and subsurface fault basins have been discovered in northern Florida, southeastern Alabama, and southwestern Georgia (Fig. 12.3). Late Triassic deposits near Danville, Virginia, are gray rather than red, which may indicate a different depositional environment. According to paleomagnetic data, this basin was at approximately 10° North during the Late Triassic and therefore may have been within a moist tropical belt.

The faulting and volcanism associated with the Triassic basins probably resulted from stretch-ing of the earth's crust along the entire east coast of North America. It should be noted that while the east coast was being pulled apart, the west coast was being compressed and folded. Both conditions were probably related to the separation of plates. Rising convection currents between North America and Gondwanaland would have resulted in broad arching and consequent normal faulting of the Atlantic margin. As North America moved away from Gondwanaland, the Pacific margin would have been compressed and folded as an oceanic plate moved beneath the continent.

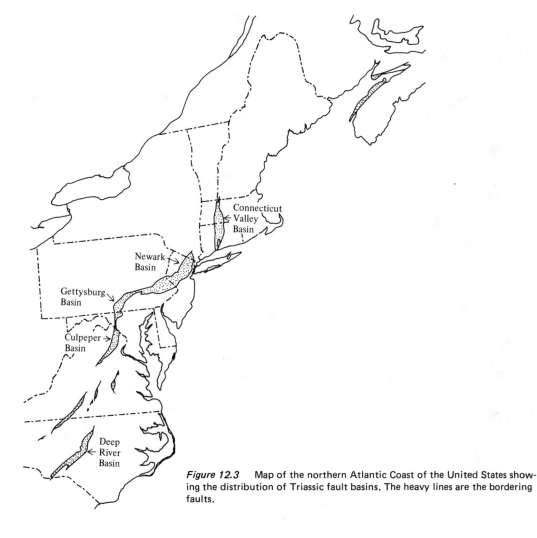

Figure 12.3 Map of the northern Atlantic Coast of the United States showing the distribution of Triassic fault basins. The heavy lines are the bordering faults.

Figure 12.4   Late Triassic conglomerate of the Newark Series in southeastern New York. (Courtesy of Charles Regan.)

Figure 12.5   The Palisades along the Hudson River, west of New York City, is part of a concordant diabase intrusion, the Palisades Sill. The top of the sill was removed by erosion prior to the Cretaceous Period. Notice the prominent columnar jointing. (Photo by Horace Gilmore, courtesy of the Palisades Interstate Park Commission.)

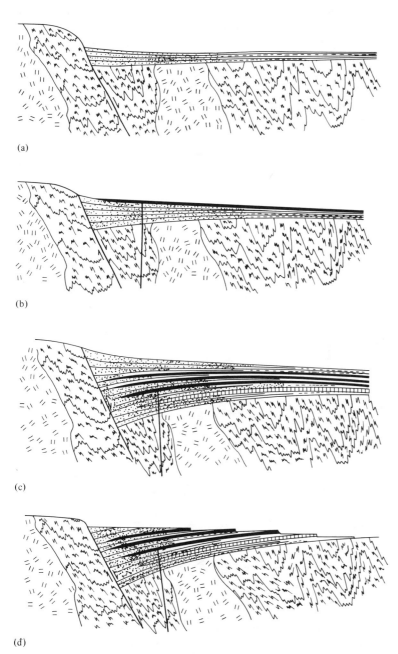

(a)

(b)

(c)

(d)

Figure 12.6 Suggested sequence of events during the formation of the Newark Basin: (a) Normal faulting with deposition of clastic sediments within the basin, (b) continued faulting with extrusion of lavas and deposition of clastic sediments (subsidence is greatest near the fault, which results in a tilting of the beds during deposition), (c) end of movement on the fault, (d) uplift and erosion of the deposits of the basin.

Paleomagnetic evidence supports a Late Triassic age for the beginning of separation of North America and Africa. When the recalculated polar wandering curves of Africa and North America are compared, they agree reasonably well from Carboniferous to Triassic time but diverge thereafter (Fig. 12.7). Thus the separation of North America and Africa probably began during the Late Trias-

sic. Intrusion of diabase in Liberia and Morocco during latest Triassic or earliest Jurassic time may have been associated with the separation of these continents (4). The divergence in the polar wandering curves provides a range of time (from Middle Triassic to Middle Jurassic) during which the initial separation of North America and Gondwanaland took place.

Figure 12.7 Comparison of the recalculated polar wandering curves of North America and Africa from the Cambrian through the Cretaceous. The curves come together near the Carboniferous and diverge after the Triassic. Evidently Africa and North America were separated prior to the Carboniferous, were together from the Carboniferous to the Triassic, and became separated after the mid-Triassic.

Gulf Coast Geosyncline

The Gulf Coast Geosyncline borders the Gulf of Mexico and joins the southern end of the East Coast Geosyncline in Florida (Fig. 12.3). Neither geosyncline contains significant volcanic rocks, nor has either one been affected by compressional deformation. The lack of volcanic and tectonic activity indicates that an oceanic plate has not moved under these geosynclines since their inception. Therefore, both geosynclines would be classed as paraliageosynclines.

A maximum of 60,000 ft (20,000 m) of Mesozoic and Cenozoic sediment has been deposited in the Gulf Coast Geosyncline on a basement of pre-Mesozoic rocks. In addition, as much as 30,000 ft (10,000 m) of sediment has been deposited on oceanic crust in the deeper parts of the Gulf of Mexico. The oldest known sediments in the geosyncline are salt deposits between 2000 and 5000 ft (600 and 1300 m) thick. These deposits have been penetrated in deep wells in the northern part of the geosyncline, and numerous salt domes rise from them in the central and southern parts of the geosyncline. Spores and pollen from a salt dome in southern Louisiana indicate that the salt bed may be as old as Late Triassic (5). Since the Gulf of Mexico probably formed after the separation of South America and Africa from North America, deposition of salt must also have postdated the separation.

In the northern part of the geosyncline, Middle Jurassic salt beds are overlain by a sequence of limestone, sandstone, shale, and salt of Late Jurassic age. According to paleomagnetic data, the Gulf of Mexico was between 20 and 30° North during the Jurassic and therefore was within an arid climatic belt. Cretaceous clastics in this region grade southward into chalks and reef-bearing limestones.

East Coast Geosyncline

The East Coast Geosyncline, which extends along the continental margin from Newfoundland to Florida, received thousands of feet of sediments during the Mesozoic (Fig. 12.8). Seismic refraction profiling of the geosyncline has revealed two structural troughs, one beneath the continental shelf and the other at the base of the continental slope (7) (Fig. 6.). The inner trough contains a maximum of 17,000 ft (5200 m) of Mesozoic and Cenozoic sediments; the outer trough has up to 20,000 ft (6000 m) of sediments.

The oldest sediments in the East Coast Geosyncline are from the Early Cretaceous. However, a deep well at Cape Hatteras, North Carolina, penetrated several hundred feet of red beds below Early Cretaceous deposits. These red beds may be of Jurassic age. A well in the Bahama Islands went through 16,000 ft (5000 m) of limestone, the oldest of which was of Early Cretaceous age (8). The well did not reach the basement, and older sedimentary rocks may lie beneath the limestone.

Cretaceous sediments are exposed in the central parts of the geosyncline, but east of New York City and south of central Georgia, Mesozoic deposits are largely covered by Cenozoic sediments and the shallow waters of the continental shelf. The Cretaceous deposits are generally limestone or chalk in the south, limey shales in the central parts of the geosyncline, and unconsolidated sands and clays in New Jersey and New York. The northward decrease in limestone may have been related to the decrease in seawater temperature, since warmer waters favor precipitation of calcium carbonate.

During the Cretaceous, the seas began a gradual regression from the geosyncline. Spore and pollen zonation of Cretaceous sediments indicates that the coastal plain contains a number of interfingering deltaic wedges, and that each can be related to a specific river system (9). The sediments in the oldest of these deltas were deposited by the Ancestral Potomac drainage during the Early Cretaceous. The sediments deposited by the Ancestral Delaware River are of late Early Cretaceous age, and the sediments in the delta of the Ancestral Schuylkill River in central New Jersey are of early and middle Late Cretaceous age. Finally, the sediments of the Ancestral Hudson River drainage are of Late Cretaceous age. This evidence demonstrates that deltaic facies were de-

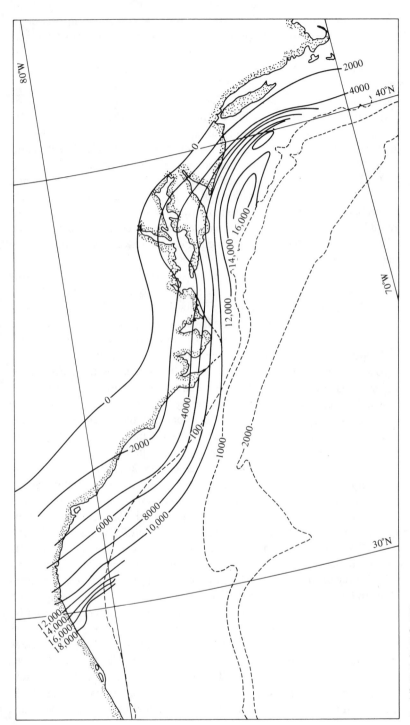

Figure 12.8 Isopach map of Mesozoic and Cenozoic deposits in the East Coast Geosyncline. The heavy lines are structure contours on the top of the crystalline basement. The dashed lines are submarine contours in fathoms. Thickness is given in feet. (After Eardley, Ref. 6.)

posited progressively northward during the Cretaceous.

The Greater Antilles

The islands of the Greater Antilles include Cuba, Puerto Rico, Jamaica, Hispaniola, the Isle of Pines, and the Virgin Islands. Jurassic and Cretaceous graywacke, shale, mudstone, limestone, and volcanics are found on all these islands (10). Most of the clastic sediments were derived from the weathering of volcanic and tectonic islands located near the site of deposition. The Greater Antilles are part of an inactive island arc that is concave toward the south. Active island arcs are concave in the direction of movement of an oceanic plate relative to the island arc. Thus when this island arc was active, the oceanic plate probably moved in a southerly direction beneath South America.

The oldest known rocks in the Greater Antilles are deformed shales, slates, schists, and marbles that crop out in Cuba. These rocks, known as the San Cayetano "Series," range in age from Early Jurassic to Middle Jurassic. Salt domes in northern Cuba contain Jurassic spores, and the salt may be of the same age as the San Cayetano "Series." The Bermeja Complex of Puerto Rico, which contains amphibolites, slates, and serpentinites, is unconformably overlain by deposits of Cretaceous age and it may also be of Jurassic age. Similar metamorphosed sedimentary and volcanic rocks are found on the Isle of Pines, on Hispaniola and on the Virgin Islands. In Cuba, the San Cayetano "Series" is unconformably overlain by less deformed and less metamorphosed limestones of Early Cretaceous (Aptian) age. The absence of Late Jurassic and earliest Cretaceous deposits indicates that folding and metamorphism occurred during latest Jurassic or earliest Cretaceous time.

Cretaceous deposits are widely exposed in the Greater Antilles. In most areas, these deposits include considerable lava and tuff. In Puerto Rico, the lavas are almost entirely andesitic and basaltic (11) and the Cretaceous section contains numerous angular unconformities. Radiometric dates from Cuba point to an Early to Middle Cretaceous

thermal event (12). Near the end of the Cretaceous, an episode of folding, metamorphism, and intrusion affected the Greater Antillean region and may have been caused by the movement of an oceanic plate beneath the islands.

Sverdrup Basin, Brooks Basin, and Chukota Geosyncline

Thick sequences of clastic sediments were deposited during the Mesozoic in the Sverdrup Basin in northern Canada, the Brooks Basin of northern Alaska, and the Chukota Basin of northeastern Siberia (Fig. 12.2). Facies changes in the Sverdrup Basin indicate easterly and southerly sediment sources, possibly the Canadian Shield. More than 10,000 ft (3000 m) of siltstone and sandstone accumulated in the Sverdrup Basin during the Triassic. These deposits grade upward into coal-bearing Jurassic clastics. Explosive volcanism during the Early Cretaceous produced beds of volcanic breccia interbedded with continental sandstones. During the Late Cretaceous, sandstone, shale, red beds, and coals were deposited in the basin.

Cordilleran Geosyncline

The Cordilleran Geosyncline was divided into a miogeosyncline and a eugeosyncline during the Mesozoic. Throughout the era movement of the Pacific Ocean plate under the margin of the continent generated volcanic magmas and resulted in at least two episodes of deformation and metamorphism in the geosyncline. Accretion of sediments and volcanics caused a westward growth of the continent (13).

Miogeosyncline. The seas had withdrawn from the geosyncline at the beginning of the Mesozoic, but they returned early in the Triassic. Marine limestones and sandstones of Early Triassic age grade eastward into a thick sequence of continental clastics. Middle Triassic beds are not preserved in many parts of the miogeosyncline, presumably because of upwarping and erosion. In Late Triassic time the Mesocordilleran Geanticline was uplifted

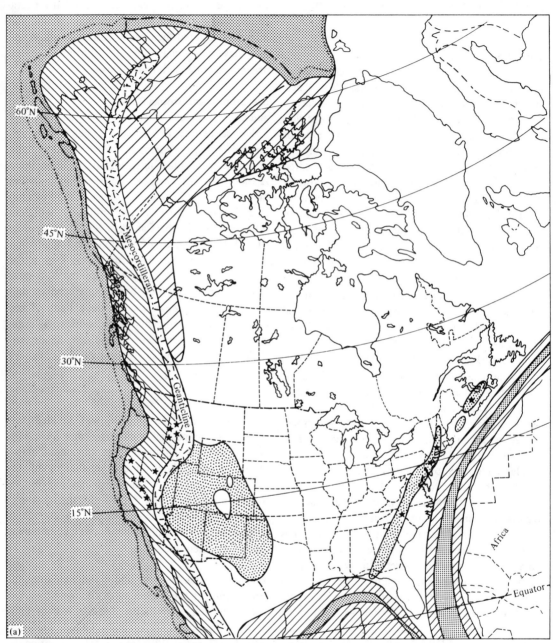

Figure 12.9 Paleogeographic map of North America during; (a) Late Triassic; (b) Late Jurassic, Oxfordian age; (c) Late Jurassic, Portlandian age.

| | marine miogeosynclinal deposits | generally shallow water | | marine eugeosynclinal deposits | generally moderate depth wate |

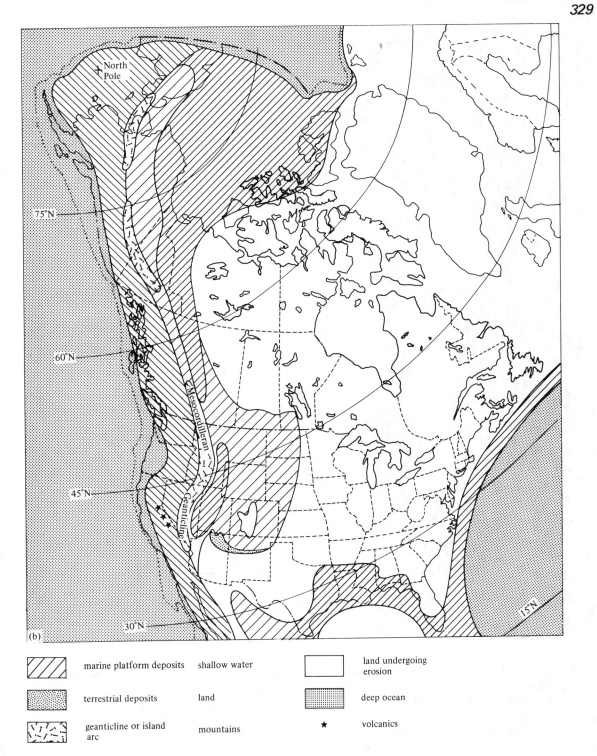

	marine platform deposits	shallow water		land undergoing erosion
	terrestrial deposits	land		deep ocean
	geanticline or island arc	mountains	★	volcanics

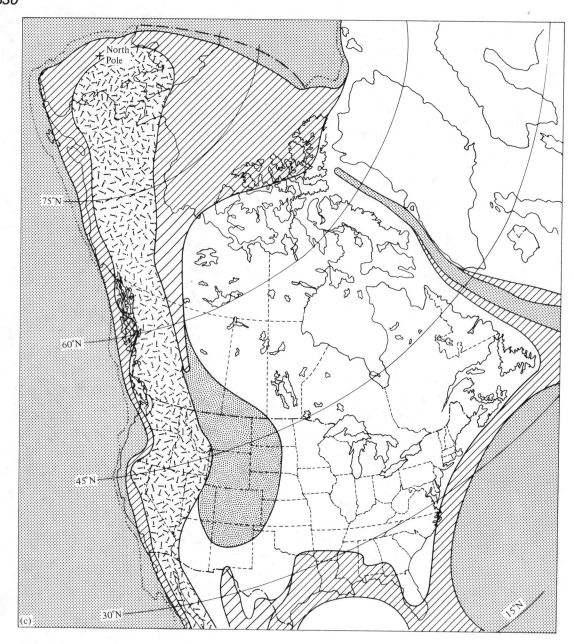

along a zone located approximately at the boundary between the miogeosyncline and the eugeosyncline (Fig. 12.9a). In southwestern Nevada and southeastern California, folding and thrust faulting occurred along the geanticline, and sediments eroded from the resulting uplift spread eastward into the miogeosyncline (14).

The Mesocordilleran Geanticline remained above sea level during the Jurassic and furnished clastic sediments for the adjacent miogeosyncline.

During the Early Jurassic, dune sandstones were deposited near the seacoast in the southeastern part of the miogeosyncline (Fig. 12.10). The seas spread southward during the Middle Jurassic, and by the Late Jurassic they covered much of the miogeosyncline (Fig. 12.9b). The Middle and Late Jurassic sequences include limestone, shale, and sandstone.

Deposition in the Cordilleran Geosyncline ended during the Late Jurassic (Kimmeridgian) as a result of folding and uplift along the entire length of the geosyncline. This orogenic episode broadened the Mesocordilleran Geanticline into a mountain range extending from Central America to Alaska and from California to central Utah. A great thickness of clastic sediment was deposited on both the eastern and western flanks of the mountain range. Numerous unconformities in the eastern sequences record almost continuous uplift of the mountain range during the Cretaceous. Six distinct phases of deformation have been noted, and the entire orogenic episode has been referred to as the Nevadan-Laramide Orogeny (15).

Eugeosyncline. The sedimentary deposits of the eugeosyncline are dominantly shale, sandstone, and conglomerate. Volcanic rocks are locally abundant, but limestones are generally minor. During the Mesozoic, the eugeosyncline may have resembled the modern Indonesian Island Arc.

Figure 12.10 Cross-bedding in dune sandstones of the Navajo Sandstone of Jurassic age in Checkerboard Mesa, Zion National Park, Utah, (Union Pacific Railroad photo.)

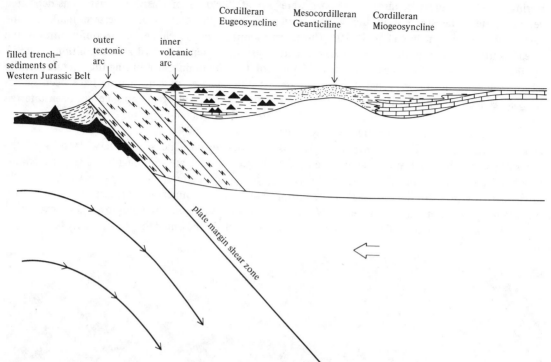

Figure 12.11   Restored cross section across the Cordilleran Geosyncline during the Late Jurassic, Oxfordian Age.

Clastic sediments were eroded from the Meso-cordilleran Geanticline on the eastern margin of the eugeosyncline and from volcanic and tectonic islands within the eugeosyncline. As much as 25,000 ft (8000 m) of Triassic sediments eroded from these highlands was deposited in the eugeosyncline (Fig. 12.11).

Folding, faulting, and uplift occurred during the Late Triassic in central Nevada, the northern Sierras, central Oregon, and northwestern Washington. Radiometric ages of basement rocks indicate that the first widespread intrusion of granites occurred at that time.

In the eastern part of the eugeosyncline, up to 15,000 ft (5000 m) of Jurassic sediments was deposited on the western margin of the continent. Upper Jurassic slates, graywackes, cherts, and volcanics, such as the Galice Formation of southern Oregon, the Western Jurassic Belt of northern California, and the Mariposa Formation in the foothills of the Sierra Nevadas, may have been deposited on ocean crust.

The deposits of the eugeosyncline were intensely deformed during the Nevadan Orogeny. Radiometric dating has shown that numerous granites were emplaced at that time (16). The eugeosynclinal deposits were again folded and intruded during Middle Cretaceous time and were intruded for a third time during the Late Cretaceous (Fig. 11.7). The Sierra Nevada, Idaho, southern California, and Coast Range batholiths are composed of hundreds of separate intrusions that were emplaced during the Mesozoic (Fig. 12.12).

Coast Range Geosyncline

The Coast Range Geosyncline (Fig. 12.2), which extends along the Pacific margin of North America from Central America to Alaska, developed after the Nevadan Orogeny. Its deposits occur in western Mexico, California, western Oregon, western

Figure 12.12 Cretaceous granites are well exposed in Yosemite Valley, Yosemite National Park, California. (Photo by R. Collins Bradley; courtesy of the Santa Fe Railway.)

Washington, western British Columbia, and southern Alaska, and were derived almost entirely from the Mesocordilleran Geanticline to the east.

In California, the Coast Range Geosyncline contains two contrasting sequences, the Great Valley sequence and the Franciscan Formation. The Great Valley sequence is comprised largely of sandstone and shale. Volcanic rocks and chert are found only near the base of the unit (17). The sequence is between 25,000 and 50,000 ft (8000 and 16,000 m) thick and is exposed in two belts, one extending from northern California to the southern end of the Sacramento Valley, and another extending from Point Arena to Santa Barbara (Fig. 12.13). Deposition of the Great Val-

ley sequence occurred between the Late Jurassic (Portlandian) and the Late Cretaceous (Maestrichtian).

Sandstones in the Great Valley sequence are generally poorly sorted, especially in the lower part of the section. An increase in arkosic sandstones in the upper part of the section is thought to be due to a progressive unroofing of the Sierra Nevada Batholith to the east. Conglomerates in the Great Valley sequence contain granitic pebbles derived from this batholith. In central California, the Great Valley sequence was broadly folded during the middle Cretaceous at the time when many of the granitic rocks were emplaced in the Sierras. Folding also occurred at the end of the Cretaceous during the Laramide Orogeny.

The Great Valley sequence rests on granitic basement in the Klamath Mountains and in the northern Sierras. However, along much of the western margin of its outcrop belt, it is underlain by a sequence of basaltic lavas, diabase, gabbro, and serpentinite. It has been suggested that this sequence of mafic and ultramafic rocks was oceanic crust when the part of the Great Valley sequence was deposited (17) (Fig. 12.14).

The Franciscan Formation is a relatively unfossiliferous sequence of sedimentary, volcanic, and metamorphic rocks. It is exposed in two belts just east of the main outcrop belt of the Great Valley sequence (Fig. 12.13). The Franciscan Formation was deposited between Late Jurassic (Portlandian) and early Late Cretaceous (Turonian or Campanian) time (18), nearly contemporaneously with the Great Valley sequence; however, its lithology is quite different (Table 12.1).

Although dominantly graywacke, the Franciscan Formation also contains shale, conglomerate, rhythmically bedded chert, greenstone, and limestone (Fig. 12.15). The graywackes are quite feldspathic and contain fragments of volcanic rocks, shale, and chert. Conglomerates often contain pebbles that can be matched to older Franciscan lithologies. Evidently some of the clastic debris was derived by "cannibalism" of older Franciscan deposits. Furthermore, much of the clastic

Table 12.1   Comparison of the Franciscan assemblage and the Great Valley sequence. (After Bailey and Blake, Ref. 18.)

	Franciscan assemblage, Eugeosynclinal[a]	Great Valley sequence, Miogeosynclinal[a]
Lithology		
Sandstone	Graywacke predominant, both feldspathic and volcanic varieties throughout. Chlorite cement predominates. K-feldspar generally absent, but present in abundance in one coastal unit.	Graywacke predominant in only parts of section, both feldspathic and volcanic in lower $\frac{2}{3}$ of section, arkosic sandstone more abundant in upper $\frac{1}{3}$. Chlorite, clay, or locally calcite cement. K-feldspar: little in Jurassic rock, more than 10% in Late Cretaceous rocks.
Shale	Minor amount, and few sections predominantly shale.	Abundant, and locally forming more than half of thick section.
Conglomerate	Rare, generally in small lenses.	Generally present, and locally in very thick lenses.
Volcanic rocks and chert	Common in most areas.	Absent except in lower part.
Limestone	Associated with volcanic rocks.	Concretions and thin calcareous lenses common in shale sequences.
Metamorphic rocks	Glaucophane schist, jadetized graywacke, and others widespread.	Some zeolites in lowest part.
Serpentinite	Widespread intrusives.	Extensive below lowest part of the sequence.
Sedimentary features		
Bedding	Highly variable; beds ranging in thickness from less than 1 cm to more than a few tens of meters	Thinly bedded, rhythmic alternations of sandstone and shale common, with more massive sandstone lenses interbedded.
Current features	Virtually unknown in most areas, but present locally.	Channeling and cut-and-fill features in upper part of sequence. Ripple marks rare except locally.
Slump features	Unknown.	Slump structures and contorted bedding common.
Turbidity current features	Graded bedding present in only a few areas. Sole markings very rare.	Graded beds and sole markings common in Sacramento Valley.
Fossils	Megafossils very rare; microfossils common in chert and limestone, rare elsewhere. Organic tracks and trails virtually unknown.	Megafossils locally abundant; microfossils common in Cretaceous rocks. Organic tracks and trails common.
Deformation	Highly compacted, broken and sheared; overall structure complex but unknown in detail.	Moderately to slightly compacted, some faults with minor displacement, open folds.
Depositional environment	Marine, probably deep water and dominantly bathyal; deposition by turbidity currents. Shallow-water deposition of some limestone, probably on seamounts.	Marine, upper neritic on east grading to deep water. Deposition of eastern rocks by traction currents and western rocks in part by turbidity currents.

[a]Thickness: 50,000 ft (15,000m).

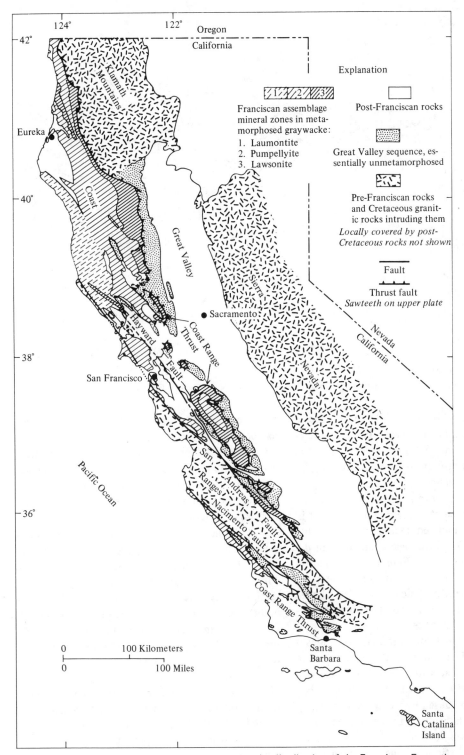

Explanation

| 1 | 2 | 3 |

Franciscan assemblage
mineral zones in meta-
morphosed graywacke:
1. Laumontite
2. Pumpellyite
3. Lawsonite

Post-Franciscan rocks

Great Valley sequence, essentially unmetamorphosed

Pre-Franciscan rocks
and Cretaceous granit-
ic rocks intruding them
*Locally covered by post-
Cretaceous rocks not shown*

Fault

Thrust fault
Sawteeth on upper plate

Figure 12.13   Generalized geological map showing the distribution of the Franciscan Formation and
the Great Valley sequence. (From Bailey et al., Ref. 17.)

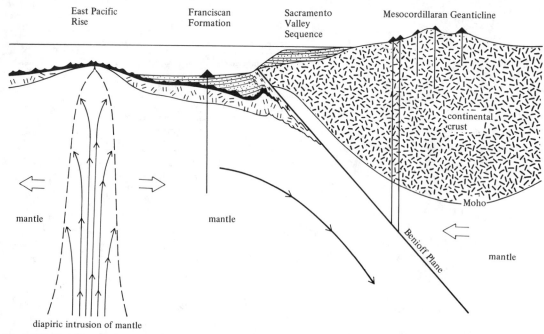

East Pacific Rise · Franciscan Formation · Sacramento Valley Sequence · Mesocordillaran Geanticline · continental crust · Moho · mantle · mantle · mantle · Benioff Plane · diapiric intrusion of mantle

Figure 12.14   Restored cross section across the Coast Range Geosyncline during the Early Cretaceous.

sediment was probably derived from the Meso-cordilleran Geanticline to the east. Potassium feld-spar, which is present in the younger Franciscan beds, may have been derived from the granitic in-trusions of the geanticline.

The Franciscan Formation is intensely de-formed and has been strongly metamorphosed in the vicinity of the fault contact with rocks to the east (Fig. 12.13). The lithologic character and de-gree of deformation of the Franciscan Formation suggest that it was deposited in deep water on oceanic crust. The rhythmically bedded cherts, which often contain fossil Radiolaria, may repre-sent abyssal radiolarian oozes.

The continuing deformation in the Meso-cordilleran Geanticline during the Late Jurassic and Cretaceous indicates that an oceanic plate was moving under North America. It is assumed that this movement was caused by westward drifting of North America relative to Europe and eastward spreading of the sea-floor away from the Pacific Rise. Ages of magnetic anomalies in the Atlantic

and Pacific oceans suggest that sea-floor spreading has been active since at least the end of the Juras-sic. Underflow of the oceanic plate may have caused folding and metamorphism of the Francis-can Formation and may have resulted in accre-tion of the sedimentary and volcanic rocks onto the North American continent.

Radiometric dating has shown that schists just west of the eastern boundary of the Franciscan Formation were metamorphosed during the Late Jurassic; yet they grade into Franciscan rocks con-taining Early Cretaceous fossils (19). Evidently metamorphism of the Franciscan Formation was nearly contemporaneous with its deposition. The metamorphism produced high-pressure, low-temperature minerals such as glaucophane, law-sonite, jadite, and aragonite (18). These high-temperature, low-pressure conditions would have been produced if Franciscan deposits had been carried beneath the continent so rapidly that they failed to reach the temperature of the adjacent rocks.

(a)

(c)

(b)

Figure 12.15 Deposits of the Franciscan Formation: (a) Folded cherts north of San Francisco, (b) pillow lavas north of San Francisco—the shape and orientation indicates that the lava is right-side up and dips steeply toward the right, (c) thick-bedded graywacke near Crescent City, California. [(a) and (b) Photos by Mary Hill; (c) courtesy of Salem Rice.]

Figure 12.16   Fossilized logs at Petrified Forest National Monument. (Photo by N. H. Darton, U.S. Geological Survey.)

Western Continental Interior

Red beds, such as those in the Spearfish Formation of South Dakota and the Moenkopi Formation of Arizona, were deposited over much of the continental interior during the Early Triassic. They were probably deposited on flood plains of slowly moving rivers, which built a vast alluvial plain sloping gently toward the sea. These clastic deposits grade westward into limestones and thus indicate that the source of the clastics must have been to the east.

Uplift of the Mesocordilleran Geanticline during the Late Triassic caused a major change in sedimentation pattern in the western continental interior. For the first time, streams and rivers carried sand, gravel, and silt from the geanticline across the miogeosyncline and onto the adjacent platform. During times of flood, logs were carried onto sand and gravel bars, where they were subsequently buried. Over the years, the wood was replaced by colorful chalcedony and agate. In the Petrified Forest National Monument in Arizona, these petrified logs have been weathered out of the weakly cemented sandstones and conglomerates of the Chinle Formation (Fig. 12.16).

The Wingate Sandstone of uppermost Triassic age in southern Utah and northern Arizona displays large cross-beds of aeolian origin. Paleomagnetic studies indicate that the southwestern United States was between 15 and 20° North latitude during the Triassic, in a climatic zone that was apparently arid or semiarid. Dune sandstones were also widely deposited in the continental interior during the Early Jurassic (Fig. 12.17).

In the Middle Jurassic, an arm of the sea (the Sundance Sea) advanced from the north and covered much of the continental interior. Deposits of this transgression include gypsum, limestone, sandstone, and shale. The Mesocordilleran Geanticline was uplifted during the Late Jurassic, and the seas withdrew from the western continental interior. Streams and rivers deposited thick sequences of clastic sediments in a basin that extended from Montana to northern New Mexico. These deposits include the Morrison Formation, which contains the remains of many large dinosaurs (Fig. 12.18). As the Mesocordilleran Geanticline continued to grow wider and higher during the Cretaceous, a thick layer of shale, sandstone, and conglomerate was deposited east of the geanticline (Fig. 12.19).

The Laramide Orogeny, which began near the end of the Cretaceous, produced a series of uplifts in the western continental interior (Fig. 12.20). In the Medicine Bow Mountains, the earliest locally derived conglomerate is of Late Cretaceous age.

Cretaceous deposits are unconformably overlain by beds of Paleocene age, which are in turn unconformably overlain by undeformed Lower Eocene sediments (20). Evidently the uplift of the Medicine Bow Mountains began during the Late Cretaceous and continued through the Paleocene. Other Laramide uplifts in the western continental interior had a similar structural history.

Southern Continental Interior

During most of the Mesozoic, the seas were confined to the Gulf Coast and East Coast geosynclines. During the Early Cretaceous, however, they spread northward and westward onto the adjacent platform. By the beginning of the Late Cretaceous, this sea was continuous with the sea in the western continental interior. Deposits of this transgression include sandstone, shale, and chalk.

EURASIA

Following the general emergence of the continents at the end of the Paleozoic, the seas again trans-

Figure 12.17 Natural stone arches at Arches National Monument, southeastern Utah, are composed of Jurassic sandstone. (Photo by D. W. Leveson.)

Figure 12.18 Dinosaur bones, exposed by a careful removal of the enclosing rock, at Dinosaur National Monument, northeastern Utah. The enclosing rock is of Late Jurassic age. (Photo by M. W. Williams, courtesy of the U.S. Department of Interior, National Park Service.)

gressed across Eurasia in the early Mesozoic. Thick sequences of sedimentary and volcanic rocks were deposited in the northern Tethyan Geosyncline and in the Indonesian and the Pacific Ocean geosynclines (Fig. 7.13). The seas were restricted to the geosynclines during most of the Triassic, but much of the continent was inundated during the Jurassic and Cretaceous.

Northern Tethyan Geosyncline

The Northern Tethyan Geosyncline developed after the final deformation of the Hercynian Geosyncline. Its axis parallels that of the Hercynian Geosyncline but lies several hundred miles to the south. The Northern Tethyan Geosyncline extends from southern Europe, through the Middle East, and the Himalayas to Indonesia (Fig. 12.1). Inter-

pretation of the geology in this region has been complicated by intense deformation.

The Alpine region, which is the most thoroughly studied section of the geosyncline is considered to have a history and structure typical of the geosyncline as a whole. A thick sequence of Mesozoic carbonates crops out in the northern and central Alps (Fig. 12.21); Mesozoic deposits in the southern Alps consist of a thick sequence of clastics, volcanics, cherts, and minor limestones (21). Thus the Alps fit the familiar pattern of miogeosynclinal deposits grading seaward into eugeosynclinal deposits.

Miogeosyncline. As the seas spread northward into the miogeosyncline during the Early Triassic, continental red beds were covered by marine lime-

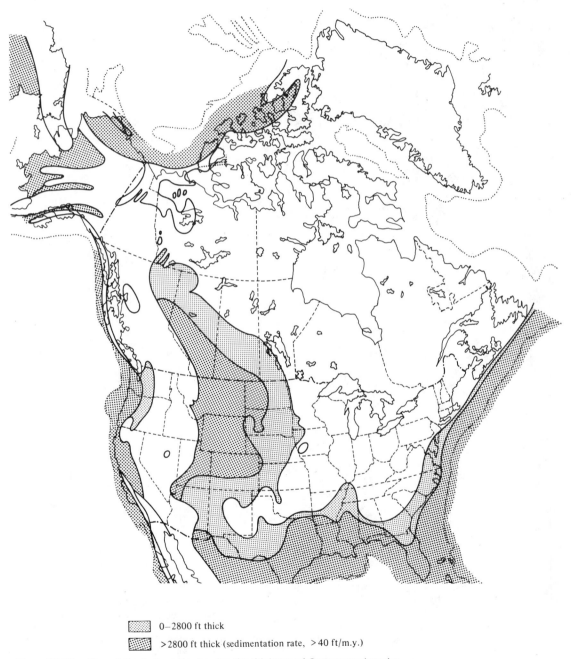

Figure 12.19 Map of North America showing the thickness of Cretaceous deposits.

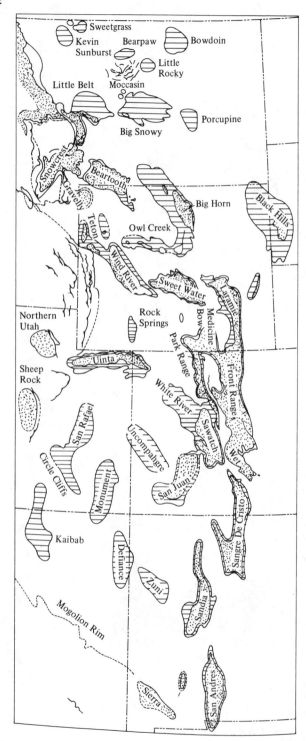

Figure 12.20 Uplifts in the Rocky Mountains and adjacent regions, formed during the Late Cretaceous and early Tertiary. Stippled areas are exposed Precambrian rocks and the horizontal ruling indicates Paleozoic sedimentary rock. (From Eardley, Ref. 6.)

Figure 12.21 Upper Jurassic limestone at Sisteron, France. These beds were folded during the early Tertiary.

stones and dolostones. These carbonates grade southward into sandstones and shales, presumably derived from highlands to the south. Folding occurred in southern Europe between Middle and Late Triassic times.

The seas withdrew from the miogeosyncline at the end of the Triassic, but returned in the early part of the Jurassic. Facies changes within Jurassic deposits indicate that a series of geanticlinal ridges was uplifted at the southern margin of the miogeosyncline during the Jurassic. Although these ridges were generally emergent, they were occasionally covered by shallow seas. Limestone, marls, and cherts of a moderately deep-water facies were deposited in troughs between the geanticlines.

Throughout much of the geosyncline, Cretaceous deposits appear to be conformable with those of Jurassic age. In Turkey, however, folding and uplift during the Late Jurassic or Early Cretaceous produced a widespread unconformity at the base of the Cretaceous sequence (22). Geanticlinal ridges persisted into the Cretaceous, and limestone was laid down between the ridges (23). The deposits of the miogeosyncline were folded once between the Early and Late Cretaceous and again during the

Late Cretaceous. Near the end of the Cretaceous, the core of the western Alps was involved in recumbent folding, and a new depositional basin developed north of this rising landmass. This basin received a thick *flysch* sequence, which consists of thousands of feet of rapidly deposited marine clastics. *Wildflysch,* a deposit containing exotic blocks of sedimentary rocks of various compositions, is commonly associated with the flysch. The exotic blocks may have originated as large fragments that slid from thrust sheets advancing onto the sea floor.

Eugeosyncline. During the Middle Triassic, volcanism was widespread along a geanticline that separated miogeosynclinal deposits from a very-deep-water facies to the south. In this region, reef-bearing carbonates are interbedded with lavas and tuffs of andesitic to basaltic composition. Volcanic and tectonic islands along the geanticline may have been part of an island arc on the southern border of Europe.

While relatively shallow-water deposits accumulated on the southern margin of the European continent, radiolarian cherts, limestones, and

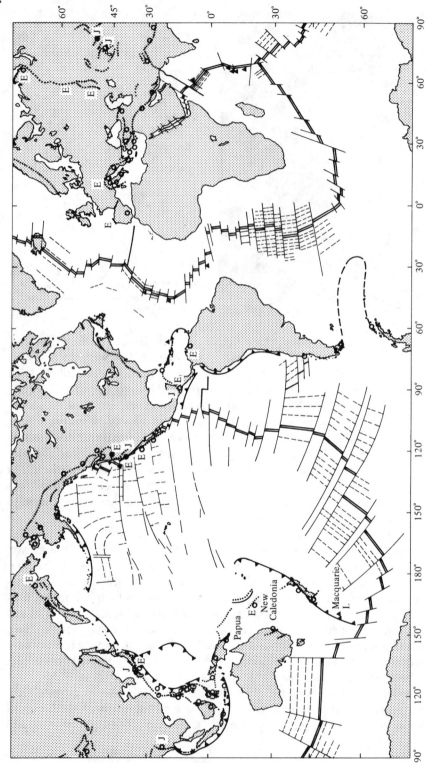

Figure 12.22   Map of the world showing the distribution of ophiolites (serpentinites, greenstones, and related rocks by solid squares, blueschists by open circles, pure jadite by J. and eclogites by E.). (From Coleman, U.S.Gr.S., Meno Park, Ref. 24.)

shales were being deposited in a deep trough to the south. Some geologists have suggested that the cherts and limestones represent lithified oozes that were deposited on oceanic crust at a depth of several kilometers (25). The limestones contain only pelagic microfossils, and little detrital sediment is present in the cherts. Evidently these deposits were laid down far from their continental sources. Many of the fossil radiolarians in the cherts are similar to forms that live today only at depths below 12,000 ft (4000 m). The concentration of silica is high in such areas, and therefore an ample source of silica is available for radiolarian tests. A further indication of a deep-water depositional environment for the cherts is their association with manganese nodules, which are also found in the deeper parts of modern oceans. There is no known granitic basement, but these deposits are associated with ophiolites, a suite of greenish igneous rocks including pillow greenstone, serpentinites, and gabbros. The ophiolites, radiolarian cherts, and associated deposits are found in a belt extending from northern Italy through southwestern Yugoslavia, Albania, southwestern Greece, Cyprus, southeastern Turkey, and the Middle East to the Himalayas (Fig. 12.22). It is possible that the ophiolites are part of a basement of oceanic crust on which the radiolarian cherts and associated deposits were laid down. The oceanic crust may have been located between Eurasia and Gondwanaland when these continents were separated by the Tethyan Sea. Evidence from paleomagnetic and structural studies suggests that Africa and Eurasia were widely separated during the Early and Middle Triassic. The movement of an oceanic plate beneath Eurasia as these continents moved toward each other may have caused folding and volcanism on the southern margin of Eurasia.

Continental Interior of Europe

Relatively thick Mesozoic deposits are found in the Germanic, London, and Paris basins, and in a basin in Spain. These basins developed between broad uplifts that formed during the Hercynian Orogeny. The continental interior was above sea level during most of the Triassic, but short transgressions occurred during the Middle Triassic and near the end of the Late Triassic. Triassic continental deposits are mostly conglomerates and sandstones. Marine deposits include siltstone, evaporites, and limestone, which formed in environments that ranged from shallow lagoons to an open shallow sea.

Shallow seas covered much of Europe during the Jurassic. Limestone and marls were deposited near the centers of basins, and sandstones and shales settled near the shorelines. Sponge reefs were abundant in Jurassic seas, and well-preserved fossils are found in lithographic limestones that were deposited within the reefs. One such limestone near Solenhofen, Germany, contains fossils of the oldest known bird, *Archaeopteryx*, as well as crabs, insects, fish and swimming reptiles.

An unconformity at the base of the Cretaceous sequence in the continental interior indicates that the seas withdrew near the end of the Jurassic. The seas began to transgress across the continental interior during the Early Cretaceous, and by the middle of the Late Cretaceous they covered approximately the same area as did the Jurassic seas. Cretaceous deposits are dominantly chemical and organic precipitates, and chalk is especially abundant in the London and Paris basins (Fig. 12.23).

A very rapid regression of the seas occurred at the end of the Cretaceous. Both this regression and the one near the end of the Jurassic seem to be associated with episodes of mountain building in the Tethyan Geosyncline to the south.

Indonesian Geosyncline

During the Triassic, volcanic eruptions, deformation, and intrusion occurred in a linear belt along

Figure 12.23 Chalk cliffs of Cretaceous age at Beer Cove, Devon, England. (Crown Copyright Geological Survey photo. Reproduced by permission of the Controller, Her Britannic Majesty's Stationery Office.)

the axis of the Indonesian Geosyncline. The zone of volcanic and tectonic activity shifted progressively southward during the Mesozoic (26). Facies relationships indicate that there were islands within the eugeosyncline. Thus Indonesia was probably an active island arc during the Mesozoic, just as it is today.

In southwestern Borneo, folded Permian deposits are unconformably overlain by sedimentary and volcanic rocks of Late Triassic age. Andesitic and basaltic volcanics occur along a belt extending from Cambodia to southwestern Borneo. Near the end of the Triassic, the deposits of the geosyncline were folded, intruded, and uplifted along a belt extending from the Celebes Islands through central Borneo to the Malay Peninsula. The resulting geanticline persisted as a positive feature throughout much of the Mesozoic and was a source of clastic sediment for the adjacent troughs. Geanticlines were also uplifted in the geosyncline during the Jurassic and Cretaceous. At the

end of the Cretaceous, folding in the southern part of the geosyncline created a geanticline that extended from Java and Sumatra to Flores.

Lower Cretaceous deposits in Sumatra include radiolarian cherts, limestones, tuffs, and ophiolites. This sequence may have been deposited in the deep sea far from land (26). At the end of the Cretaceous, the deposits of the geosyncline were folded and uplifted, perhaps also becoming accreted to the continent. The cause of the Mesozoic deformation, volcanism, and intrusion in the Indonesian Geosyncline may have been the movement of an oceanic plate under Asia as Australia moved away from Africa and toward Asia. The Late Triassic deformation and volcanism in the Indonesian Geosyncline may have been associated with the initial separation of the continents.

Pacific Ocean Geosyncline

The Pacific Ocean Geosyncline extends from eastern Siberia through Japan, Taiwan, and the Philip-

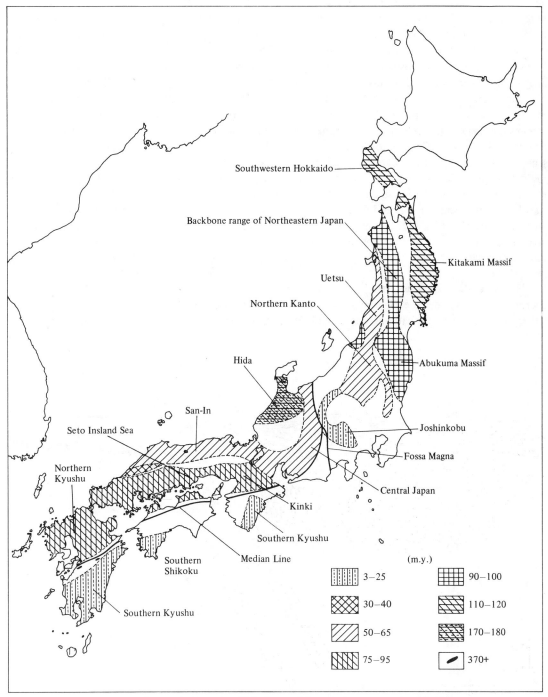

Southwestern Hokkaido

Backbone range of Northeastern Japan

Kitakami Massif

Uetsu

Northern Kanto

Hida

Abukuma Massif

San-In

Seto Insland Sea

Joshinkobu

Fossa Magna

Northern
Kyushu

Central Japan

Kinki

Southern Kyushu

Southern
Shikoku

Median Line

Southern Kyushu

(m.y.)

3–25   90–100

30–40   110–120

50–65   170–180

75–95   370+

Figure 12.24 Arrangement of zones in Japan in which the granitic rocks have approximately the same
potassium-argon dates. (From Kawano and Ueda, Ref. 27.)

pines to the Indonesian Geosyncline (Fig. 7.13). Thick sequences of sedimentary and volcanic rocks were deposited during the Mesozoic in Japan, eastern Siberia, Taiwan, and the Philippines. The Mesozoic geosynclinal deposits of Japan have been divided into near-shore and deep-water facies. The near-shore facies is composed largely of coarse continental and marine clastics and volcanics; the deep-water facies contains cherts, limestones, and fine-grained clastics. These two facies are now separated by a major fault zone, the Median Tectonic Line (Fig. 12.24). Numerous unconformities within the near-shore facies indicate that Japan was the site of repeated uplift and folding throughout the Mesozoic (Fig. 11.31). Potassium-argon ages of granites indicate that the deformations were accompanied by intrusion of granitic rocks (Fig. 12.24).

Mesozoic deposits southeast of the Median Tectonic Line include the Sambosan Group, a rather thin sequence of bathyal cherts and fine clastics with only minor limestone. Fossils are not common in this sequence, but enough have been found to indicate that the sequence was deposited more or less continuously from the Permian through the Jurassic. The lithologic character of the Sambosan Group suggests deposition in a marine environment, possibly in the deep ocean. The Shimatogawa Group ranges from Triassic or Jurassic to Miocene in age. The Mesozoic deposits in this sequence include shale, sandstone, radiolarian chert, pillow greenstone, and limestone. These rocks have been intensely folded and are remarkably similar in lithology and structure to the Franciscan Formation in California. Both the Shimatogawa Group and the Sambosan Group may have been deposited in the deep ocean and incorporated into the continent as an oceanic plate moved under the continent (Fig. 12.25).

GONDWANALAND

Geosynclines bordered Gondwanaland during the

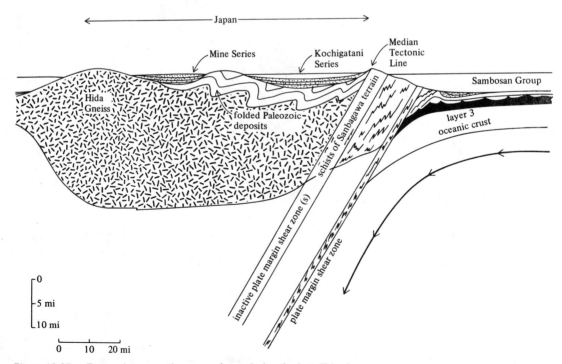

Figure 12.25 Restored cross section across Japan during the Late Triassic.

early part of the Mesozoic (Fig. 12.1). Following the breakup of Gondwanaland, geosynclines formed on the Atlantic and Indian ocean borders of South America, Africa, Australia, Antarctica, and India. There were several episodes of widespread volcanism in the continental interior during the Mesozoic. This volcanic activity may have been the result of stretching of the crust as the continents separated.

New Zealand Geosyncline
Two distinct facies have been recognized in the Mesozoic of the New Zealand Geosyncline, a relatively shallow-water "marginal facies" and a deepwater "axial facies." These facies are separated by a major fault zone which, as in Japan, is called the Median Tectonic Line (Fig. 12.26). The marginal facies contains clastics, volcanic flows, and volcanic tuffs ranging in age from Early Triassic to Late Jurassic. The lack of Late Jurassic (Oxfordian) deposits in the marginal facies indicates that during this interval the geosyncline was uplifted and the seas withdrew. The deformation culminated during the Late Jurassic or Early Cretaceous, when the deposits of the marginal facies were folded. Intrusives into this sequence have been dated as Late Jurassic and Middle Cretaceous.

The axial facies of the New Zealand Geosyncline contains a thick layer of relatively unfossiliferous graywacke with only minor amounts of basalt, chert, and limestone. The deposits of the axial facies are more intensely deformed than those of the marginal facies (Fig. 12.27). The lithologies suggest that these rocks were deposited on oceanic crust (29). Ultramafic intrusions, which are common in this sequence, may be fragments of the upper mantle which underlay this facies during its deposition (Fig. 12.28).

High-pressure, low-temperature minerals such as glaucophane and lawsonite also occur in the axial facies. It is possible that these minerals were produced when deep ocean deposits were carried beneath New Zealand during underflow of an oceanic plate. The final deformation of the axial facies occurred during the late Jurassic or Early

Cretaceous. Subsequently, sedimentation shifted to the New Zealand East Coast Geosyncline east of the New Zealand Geosyncline.

New Zealand East Coast Geosyncline
Sedimentary deposits in the New Zealand East Coast Geosyncline are dominantly marine clastics, such as mudstone, conglomerate, and sandstone. Throughout the Cretaceous, the geosyncline was bordered on the east by a chain of volcanic islands. Evidently then, as now, this region was an island arc. The seas withdrew during the Late Cretaceous, and coal-bearing continental sediments were laid down.

Timor
The island of Timor is located between Australia and Indonesia. Few unconformities occur within a sequence that ranges in age from Permian to Late Cretaceous. The Mesozoic deposits in Timor include limestone, clastics, and cherts that have been intruded by ultramafic rocks such as serpentinites. Some of the sedimentary units are very fossiliferous. Over 900 species of marine invertebrates have been found in Triassic beds. The Mesozoic sequence is very thin, which suggests a slow rate of deposition in deep water away from highlands. However, the abundance of fossils is generally associated with deposition in shallow water. Some of the deposits contain a mixture of species of different ages, which suggests reworking of the sediments. Possibly these deposits were originally laid down in shallow water and were redeposited in deep water by submarine currents. The limestones in the sequence may have been derived from erosion of reefs on the outermost margin of Australia. At the end of the Cretaceous, the deposits of Timor were folded and then accreted to the Asian continent, probably as a result of compression between Asia and Australia.

Southern Tethyan Geosyncline
The Southern Tethyan Geosyncline extends along the outer margin of Gondwanaland from northern India through the Middle East and southern

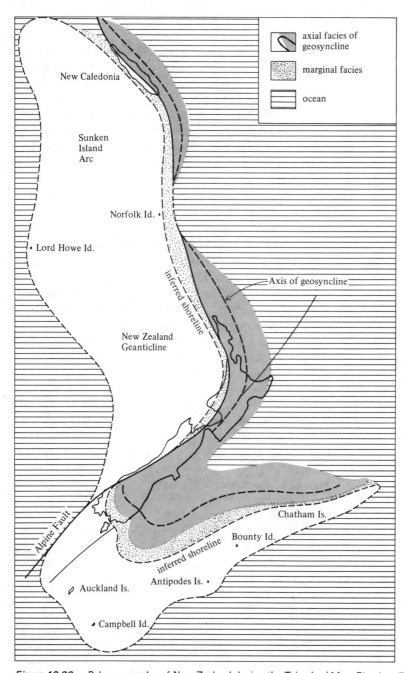

Legend:
- axial facies of geosyncline
- marginal facies
- ocean

New Caledonia

Sunken
Island
Arc

Norfolk Id.

Lord Howe Id.

inferred shoreline

New Zealand
Geanticline

Axis of geosyncline

Alpine Fault

Chatham Is.

Bounty Id.

inferred shoreline

Auckland Is.

Antipodes Is.

Campbell Id.

Figure 12.26   Paleogeography of New Zealand during the Triassic. (After Fleming, Ref. 28.)

Figure 12.27 Folded graywackes of the axial facies in the Southern Alps of New Zealand (Photo by D. L. Homer, courtesy of the New Zealand Geological Survey.)

Europe to northern Africa (Fig. 7.13). In the early Mesozoic, a wide, deep ocean separated this trough from the Northern Tethyan Geosyncline on the Eurasian border. As Gondwanaland separated from North America, the Tethys Sea narrowed, and deformation occurred along the margin of Gondwanaland.

During the Triassic and Jurassic, carbonates were deposited in the Southern Tethyan Geosyncline (Fig. 12.29). However, thick shales were also deposited in the Himalayan region in Late Triassic and Late Jurassic time.

The Cretaceous was a time of increasing deformation. Thick sequences of Cretaceous clastics are found in the Himalayas, the Middle East, and southern Europe (Fig. 12.30). In Iran and Iraq, Late Cretaceous conglomerates contain pebbles of radiolarian cherts and igneous rocks. Since these deposits grade southward into limestones, the source must have been to the north, possibly an island arc. In Italy, Albania, western Yugoslavia, and western Greece, unconformities in Cretaceous sequences provide evidence of deformation and uplift during that interval (25).

Andean Geosyncline

During the Mesozoic, carbonates and clastics were deposited in the Andean Miogeosyncline, and clastics and volcanics were laid down in the eugeosyncline to the west (Fig. 12.31). The boundary between these two troughs lies more or less along the Central Ranges of the Andes Mountains. In some areas a geanticline separated the miogeosyncline and eugeosyncline.

Miogeosyncline. Seas occupied only a small part of

Figure 12.28 Dun Mountain ultramafic mass, New Zealand. Notice the almost complete absence of forest cover on the ultramafic works (Photo by S. N. Beatus, courtesy of the New Zealand Geological Survey.)

Figure 12.29 The Rock of Gibraltar is composed of limestone of Jurassic age. Gibraltar is located at the junction between the Northern and Southern Tethyan geosynclines. (Photograph by *The Times,* London.)

Figure 12.30   Mesozoic deposits in Zardeh Kuh, southwestern Iran. The whole Jurassic System crops out as a pale band at the bottom of the cliff. The Cretaceous deposits are dominantly clastics. (Photo from Aerofilms Limited, by courtesy of British Petroleum Co.)

the Andean Miogeosyncline during the Triassic and Jurassic. Continental red beds were deposited in some areas, but elsewhere uplift and erosion prevailed (Fig. 12.32). In Cretaceous time, the seas covered much of the northern and central regions of the miogeosyncline. Cretaceous deposits are largely carbonates and fine clastics, but they grade into sandstones to the east.

Eugeosyncline. Volcanic activity was especially intense in the eugeosyncline during the Late Triassic, Late Jurassic, and Late Cretaceous. Graywackes interbedded with the volcanics may have been deposited by turbidity currents, and deposits in the western part of the eugeosyncline may have been laid down on oceanic crust (Fig. 12.31).

Deformation. Widespread faulting and plutonism occurred in the Andean Geosyncline during the Late Triassic (30). It is quite possible that the deformation was caused by the movement of an oceanic plate under the geosyncline during the initial stages of separation of Gondwanaland and North America. A Late Jurassic deformation may have been related to an increase in the rate of movement of the plate under the continent as South America began to separate from Africa. The principal metamorphism and deformation in the geosyncline began during the late Early Cretaceous and culminated in the Late Cretaceous and early Tertiary (31). Radiometric dating has shown that numerous granite bodies were emplaced during this interval (30) (Fig. 12.33). Deformation near

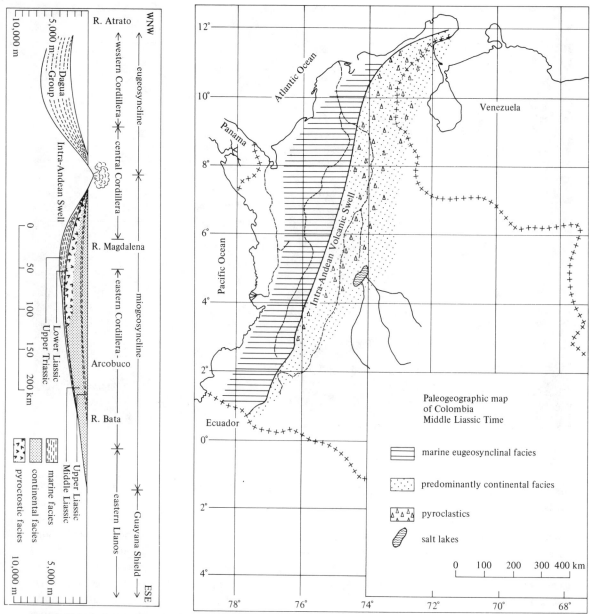

Figure 12.31   Paleogeographic map and cross section of Colombia during the Early Jurassic, Pliensbachian Age. (After Burgl, Ref. 30.)

the end of the Cretaceous resulted in a withdrawal of the seas from much of the geosyncline. The seas did not return to the geosyncline during the remainder of the Mesozoic.

Antarctic Geosyncline

Mesozoic sedimentary and volcanic rocks of geosynclinal thickness are found in peaks rising above the continental ice sheet on the Pacific margin of

Antarctica. These sequences were folded, uplifted, and eroded several times during the Mesozoic. An especially intense deformation occurred during the middle Cretaceous. Radiometric dating of granitic intrusions shows that granites were emplaced at that time (32) (Fig. 19.28).

Continental Interior

The continental interior of Gondwanaland was subjected to widespread volcanic activity during the Jurassic and Cretaceous (Fig. 12.34). This volcanism was apparently related to the breakup of the southern continents. The most widespread activity occurred during the Late Jurassic (Table 12.2). This is the approximate time of the initial separation of South America and Africa, based on the age of the oldest sediment in the South Atlantic (Fig. 7.35c) and on comparison of recalculated polar wandering curves of these two continents (36). The separation of Australia and Antarctica from Africa and India may also have begun during the Late Jurassic.

Ocean Basins

An intensive effort has been made to recover pre-Tertiary sediments from the deeper parts of the ocean basins. Cretaceous sediments are widespread, but no samples older than Middle Jurassic have been obtained, even in areas where holes have been drilled to the basement. It appears, then, that the ocean floors are relatively young — a conclusion that is consistent with the concept of sea-floor spreading.

Atlantic Ocean. Sediments of Mesozoic age have been recovered by the Glomar Challenger from 16

Figure 12.32 Banded sediments of the Tonel Formation of Jurassic age in the Andes Mountains of northern Chile. (Photo by R. J. Dingman, U.S. Geological Survey.)

Table 12.2   Radiometric dates of basalts and dolerites from the southern continents

Continent	Formation	Radiometric age (m.y.)	Geologic age	Ref.
South America	Serra Geral	161 - 111	Late Jurassic - Early Cretaceous	32
Antarctica	Ferrar dolerites	170 - 147	Early Jurassic - Late Jurassic	33
Australia	Tasmanian dolerites	167 - 143	Middle Jurassic - Late Jurassic	34
Africa	Karroo	190 - 154	Early Jurassic - Late Jurassic	35

sites in the North Atlantic Ocean. The oldest of these is of Late Jurassic (Callovian) age, taken just east of the Bahama Islands (37). The absolute age of these sediments is approximately 162 million years. They rest on a basaltic basement and provide a *minimum* age for the opening of the Atlantic Ocean. As discussed earlier, the opening of the Atlantic Ocean probably began about 190 million years ago with the separation of Africa from North America. The rate of separation of Africa and North America probably averaged about 4.0 cm per year throughout most of the Mesozoic (37).

Mesozoic deposits from the deeper parts of the Atlantic Ocean basin include clays, foraminiferal and radiolarian oozes, turbidites, and cherts. Cores taken in water depths greater than 12,000 ft (4000 m) generally have no calcareous sediments in their upper levels, since the solubility of calcium carbonate is greatly increased at these depths. However, calcium carbonate is generally present near the bottom of Mesozoic sequences in regions which are now below 12,000 feet deep. It may be that these areas were originally less than 12,000 ft deep and have subsided as the sea-floor moved away from the Mid-Atlantic Ridge.

Caribbean Sea. Extensive drilling in the Caribbean Sea by the *Glomar Challenger* has indicated that the oldest sediments are of Cretaceous (Coniacian) age (38). These deposits include chalks, siliceous clay, volcanic ash, and volcanic sands. Some of the chalks have been lithified near the bottom of the section where they pass into limestone interbedded with cherts. This horizon forms a prominent seismic reflector.

Labrador Sea. A partly buried oceanic ridge, the Mid-Labrador Sea Ridge, separates Greenland and North America (39). This ridge is not seismically active at the present time, but it probably was at one time a zone of active sea-floor spreading. The oldest sediment recovered from the Labrador Sea is of Middle Jurassic (Bajocian) age, taken from northeast of Newfoundland. If this sediment was deposited in place, the separation of Greenland and Europe from North America began during or before the Middle Jurassic. However, the major phase of drifting may have occurred from the Late Cretaceous to the Late Eocene (37).

Bay of Biscay. Paleomagnetic evidence (40) indicates that the Bay of Biscay was formed by the rotation of Spain relative to Europe some time after the Triassic (Fig. 12.35). The oldest sediments cored from the Bay of Biscay are of Late Cretaceous (Maestrichtian) age, so that the opening of the bay must have begun before that time (42). The middle Cretaceous folding in the Pyrenees Mountains may have been associated with the rotation of Spain relative to Europe. Before the beginning of separation of the continents, the fracture which was to become the Labrador Sea was approximately aligned with the fracture which was to become the Bay of Biscay (Fig. 7.3). It is possible that both opened at the same time.

Gulf of Mexico. The Glomar Challenger penetrated typical cap-rock sediments of a salt dome in the Gulf of Mexico in water almost 12,000 feet deep (43) (Fig. 12.36). Pollen from the cap rock indi-

cates a Late Jurassic age for the salt bed. Since these rocks come from near the middle of the Gulf of Mexico, the sea-floor closer to the margins of the continents is probably somewhat older.

Pacific Ocean. Although the lithology of the Mesozoic sediments from the Pacific Ocean is essentially the same as that of the sediments from the Atlantic Ocean, the age of the oldest sediments in Pacific Ocean cores increases away from the East Pacific Rise (Fig. 12.37). The oldest known sediments are of Late Jurassic age and were recovered by the *Glomar Challenger* approximately 4800 miles (8000 km) from the crest of the East Pacific Rise. Cretaceous sediments and volcanic rocks have been dredged and cored from the western and central Pacific, but none have been found within about 3000 miles (5000 km) of the rise.

MESOZOIC LIFE

Many invertebrate groups prospered during the Mesozoic and then declined or became extinct at the end of the Cretaceous. For example, the ammonoids diversified and became very abundant during the Jurassic and Cretaceous. A few forms became rather bizarrely coiled, and others developed a nearly straight shell, but by the end of the Cretaceous all were extinct.

The reptiles evolved rapidly during the early Mesozoic and adapted to most of the available environmental niches. Some developed the ability to fly, others became adapted to an aquatic environment. Some terrestrial reptiles ran on two legs, and others lumbered about on four. Among them were vicious predators, armored herbivores, and scavengers. In spite of their success, however, almost all the reptile groups had become extinct by the end of the Cretaceous.

The plants also underwent dramatic changes during the Mesozoic. Of greatest importance was the evolution and rise to dominance of the flowering plants, the angiosperms. These advanced forms appeared during the Jurassic, and by the Late Cretaceous had replaced the gymnosperms in abundance in most areas.

Figure 12.33 Mesozoic and Mesozoic? granitic rocks in South America. (Data from *Carte Geologique de l'Amérique du Sud* prepared by the Commission de la Carte Geologique du Monde, published by the Geological Society of America.)

Figure 12.34   Sill of Ferrar Dolerite (dark band) intruding sedimentary rocks of the Beacon Group in South Victoria Land, Antarctica. (U.S. Navy photo, courtesy of the U.S. Geological Survey.)

Protists

Foraminifera proliferated in the Jurassic and Cretaceous seas. They were dominantly microscopic varieties, and their tests make up a significant portion of the Cretaceous chalk deposits of northern Europe and the Gulf Coastal region. Foraminiferal oozes were deposited in many parts of the ocean during the Cretaceous.

Two types of one-celled, planktonic plants, the diatoms and the coccoliths, appeared for the first time in the Mesozoic fossil record. They were present in such quantities in Cretaceous seas that their tests were important rock forming sediment. Diatoms, which have ornate, siliceous tests, are abundant in diatomaceous earth and in many cherts. Coccoliths secrete rings and groups of plates, each of which is a single calcite crystal (Fig. 12.38). Very fine-grained coccolith debris is a major component of many Cretaceous chalks.

Sponges and Coelenterates

Siliceous sponges occur in Mesozoic deposits that were laid down in moderately deep water. The chert nodules in the Cretaceous chalks of northern

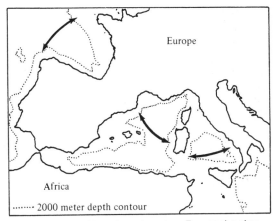

Europe

Africa

········ 2000 meter depth contour

Figure 12.35 Map of southwestern Europe showing the rotations that were suggested by. S. W. Carey. (Redrawn from S. W. Carey, Ref. 41.)

Europe may have been formed by the replacement of chalk by silica from siliceous sponge spicules. Calcareous sponges grew in muddy waters in shallow, near-shore environments during the Mesozoic.

Mesozoic coelenterates include both jellyfish and hexacorals. One species of Cretaceous jellyfish, *Kirklandia texana*, resembles the problematical *Brooksella* from Precambrian deposits in the Grand Canyon (Fig. 12.39). Hexacorals, which succeeded tabulate and rugose corals when these forms became extinct, have a sixfold symmetry — their septa are arranged in multiples of six (Fig. 12.40). During the Late Triassic, hexacorals were widely distributed in a narrow equatorial belt. In the Late Jurassic, however, coral reefs had spread throughout Europe and Africa in a wide latitudinal

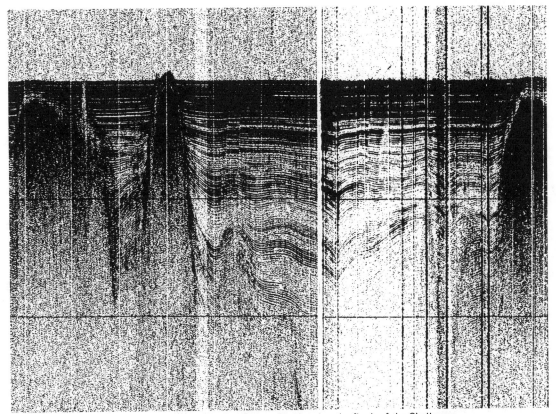

Figure 12.36 Southwest-northeast seismic reflection profile showing the flank of the Challenger Knoll and adjacent dome intruding the sediments of the Sigsbee Abyssal Plain. This profile is 6 miles (10 km) long. (From Burk et al., Ref. 43. Reproduced with the permission of The American Association of Petroleum Geologists.)

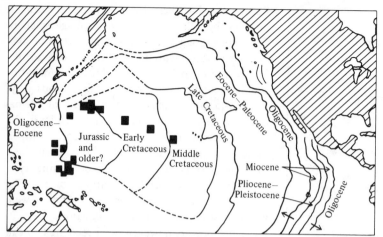

Figure 12.37 Ages of basement rocks in the north Pacific as determined by drilling on *Glomar Challenger's* Leg 6. (From Fischer, Ref. 44. Copyright © 1970 by the American Association for the Advancement of Science.)

belt. Corals were also abundant in the shallow, epicontinental seas during the Cretaceous.

Bryozoans and Brachiopods

The two dominant orders of Paleozoic bryozoans are not found in Mesozoic deposits. Instead, Mesozoic bryozoans are represented by two new orders. Bryozoans were abundant in Cretaceous seas and have remained so since that time.

The number of brachiopod species in early Mesozoic deposits was greatly reduced from that in late Paleozoic sequences, and it continued to decline throughout the Mesozoic. The simple, unspecialized forms have been the most persistent. *Lingula*, for example, has changed very little between the Paleozoic and the present. The terebratulids, which comprise most of the brachiopods living today, became numerous for the first time during the Mesozoic. They are important in correlating marine sequences, especially in the Tethyan region.

Mollusks

The pelecypods, which expanded and diversified during the late Paleozoic, replaced the brachiopods as the dominant group in the near-shore, benthonic environment. They had already become well established in brackish and fresh waters as well.

Important shallow-water marine pelecypods included oysterlike forms, such as *Exogyra* and *Gryphaea*, and forms closely related to the modern scallop. The pelecypod *Inoceramus*, which was abundant in Cretaceous seas (Fig. 12.41), grew to be as large as 3 ft (1 m) long — about the same size as the modern giant clam. Many Cretaceous chalks contain tiny prisms of calcite derived from the shells of this form. Some pelecypods had an external morphology similar to that of horn corals, and like horn corals, they attached themselves permanently to the sea bottom.

The ammonoids were close to extinction at the end of the Paleozoic. Only two families of ceratites persisted into the Triassic, and one of these did not survive beyond the Early Triassic (45). The remaining ammonoids dispersed and diversified so rapidly during the Triassic that this group has provided excellent index fossils. The ammonoids declined significantly in the Late Triassic and by the beginning of the Jurassic, only one family remained. However, a number of new ammonoid groups evolved from this stock during the Jurassic, and these spread throughout the oceans of the world. As in the Triassic, the rapid evolution and widespread occurrence of the Jurassic ammonoids makes them ideal guide fossils for intercontinental correlations. The 58 ammonoid

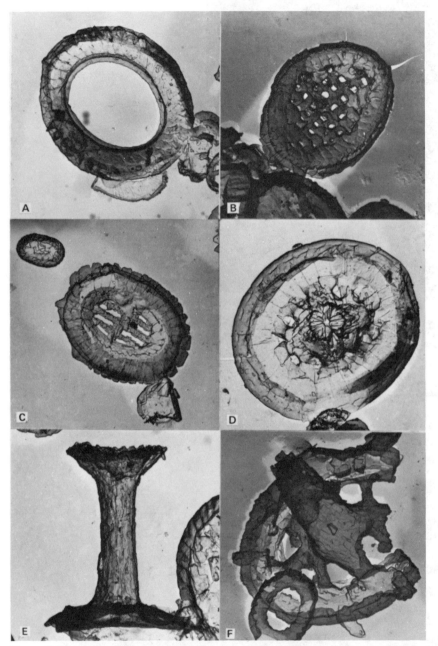

Figure 12.38   Replica electronmicrographs of Early Cretaceous, late Albian, coccoliths recovered by the *Glomar Challenger* from site number 5 in the Atlantic Ocean, east of the Bahama Islands. Magnification approximately 10,000 X. (a) *Apertapetra gronosa,* (b) *Arkhangelskiella erratica,* (c) *Costacentrum horticum,* (d) *Cretarhabdus crenulatus,* (e) *Cretarhabdus decorus,* (f) *Cretarhabdus descussatus.* (Photo by David Burkry, U.S. Geological Survey, courtesy of the National Science Foundation.)

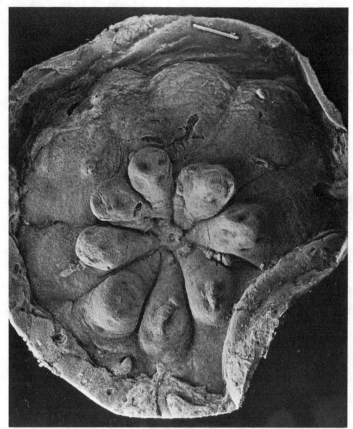

Figure 12.39   Latex mold of fossil jelly-fish *Kirklandia texana* from the Lower Cretaceous of Texas (X ½). It is possible that these forms are not a true jellyfish, but large hydrozoan jellies similar to forms living today in the cold waters around Antarctica. (Photo by K. E. Castor.)

Figure 12.40   *Thamnasteria*, a Cretaceous hexacoral from Texas, magnification 3X. (Photo courtesy of J. W. Wells.)

Figure 12.41 Upper Cretaceous fossils from the Ripley Formation in Tennessee: (a) *Nostoceras* sp., an aberrant ammonoid cephalopod (X 1.1); (b) *Trigonia eufaulensis*, interiors of pelecypod valves showing original nacreous luster (X 1); (c) *Trigonia eufaulensis*, exterior of a right valve (X 1); (d) *Inoceramus sagensis*, typical large pelecypod (X 1); (e) *Anchura*, gastropod (X 2); (f) *Turritella verte-broides*, high, spired gastropod *Baculites* sp., straight shelled ammonoid, and other mollusks (X 1.3). (Photo courtesy of Carlton Brett.)

Figure 12.42 Restoration of the sea bottom during the Late Cretaceous showing ammonoids that are tightly coiled, loosely coiled (in weeds), and have straight shells. (Courtesy of the Field Museum of Natural History, Chicago.)

zones that have been established for Jurassic marine deposits permit very accurate age determinations.

Most ammonoids are bilaterally symmetrical (Fig. 12.42), but quite a few of the Mesozoic forms diverged from this pattern (Fig. 12.41). As early as the Triassic, some ammonoids developed a spirally coiled shell, much like that of gastropods. Other forms developed straight shells or assumed a variety of shapes (Fig. 12.41). A few species grew to very large sizes. One specimen found near Graybull, Wyoming, measures more than 5 ft (2 m) across. Belemnoids, a group of squidlike cephalopods, were abundant in Cretaceous seas, and their cigar-shaped internal skeletons are common marine fossils (Figs. 12.43).

Arthropods

With the extinction of the trilobites at the end of the Paleozoic, the only important arthropods in the Mesozoic were crustaceans and insects. Mesozoic crustaceans include ostracodes, branchiopods,

barnacles, and lobsters; Mesozoic insects include flies, moths, lice, bees, beetles, and ants. The insects achieved a proficiency which has enabled them to persist relatively unchanged to modern times.

Echinoderms

The Mesozoic echinoderms are represented mainly by crinoids and echinoids, and both are important guide fossils. Only one family of crinoids survived the mass extinctions at the end of the Paleozoic. This family gave rise to free-swimming, rootless forms which probably lived in communities much like modern crinoids. One species of rooted crinoid had a calyx more than 2 ft (0.6 m) across and a stem 50 ft (16 m) long.

Fish and Amphibians

Mesozoic fish are dominated by the ray-finned fish, the actinopterygians. The early Mesozoic forms had heavy scales and skeletons composed mostly of cartilage; later forms developed small scales and bony skeletons, which are characteristic of modern fish. The lungfish and lobe-finned fish were of little importance during the Mesozoic.

The amphibians continued to decline during the Mesozoic and the labyrinthodonts and stegocephalians became extinct by the end of the Triassic. Frogs appeared in the Jurassic and salamanders in the Cretaceous. These small amphibians are useful in ecological studies but are of little stratigraphic value.

Terrestrial Reptiles

The Mesozoic terrestrial reptiles included cotylosaurs, therapsids, thecodonts, dinosaurs, snakes, lizards, and crocodiles. The cotylosaurs are thought to be the group from which most Mesozoic reptiles stemmed. The therapsids, which evolved from the cotylosaurs during the Permian, were the dominant reptiles during the Early Triassic. However, as the period progressed, they dwindled in numbers and variety. The therapsids finally became extinct during the Jurassic, but before they died out, they had given rise to primitive mammals. The thecodonts, which were the first archosaurs, also evolved from the cotylosaurs. Thecodonts walked on their hind feet, using their tails for balance. As they developed longer and stronger hind legs, the forelegs were used mainly for holding. Walking on two

Figure 12.43 Restoration of the Late Jurassic sea bottom in South Dakota with the squidlike belemnoid *Pachyteuthis* swimming above a bank of oysterlike pelecypods. (Courtesy of the Field Museum of Natural History, Chicago.)

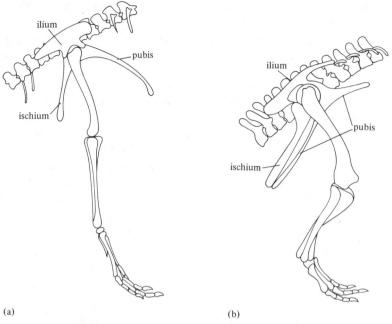

ilium

pubis

ischium

ilium

pubis

ischium

(a)   (b)

Figure 12.44   Differences in the pelvis of the two types of dinosaur: (a) Saurischian, (b) ornithis-chian. (From Romer, Ref. 46.)

legs gave the carnivorous thecodonts added speed and agility. The decline and extinction of the therapsids can probably be related directly to competition from the more efficient thecodonts.

The more specialized and advanced forms that evolved from the thecodonts were far more efficient than the thecodonts which could not survive the competition with their "offspring." This group became extinct by the end of the Triassic. However, they gave rise to the other four orders of archosaurs, as well as Crocodillia (crocodiles), Pterosauria (flying reptiles), and two orders of dinosaurs, Saurischia and Ornithischia.

The word dinosaur comes from the Greek words *deinos* (terrible) and *sauros* (lizard). The saurischians and ornithischians are separated on the basis of differences in the structure of their hip bones. The saurischians have a reptilelike pelvis and the ornithischians have a birdlike pelvis (Fig. 12.44). The saurischians include both carnivorous and herbivorous forms, but almost all the ornithischians were herbivores. Like the the-

codonts, the early dinosaurs walked on their hind legs, but as they evolved, some became quadrupedal.

The saurischians first appear in Late Triassic deposits, and they evolved rapidly during the Jurassic and Cretaceous. Notable carnivorous saurischian dinosaurs were the bipedal – *Allosaurus*, a Jurassic form, and the largest of them all, *Tyrannosaurus*, which lived during the Cretaceous. *Tyrannosaurus rex* reached a length of 50 ft (16 m), stood 20 ft (6 m) tall, and had teeth the size of railroad spikes (Fig. 12.45). Herbivorous saurischians include the well-known *Apatosaurus* (*Brontosaurus*) (Fig. 12.46) and *Diplodocus* which lived during the Jurassic. *Diplodocus* grew to a length of 80 ft (26 m) and weighed more than 50 tons. *Branchiosaurus*, a Cretaceous form, was 80 ft long and stood 40 ft (13 m) high. It is quite likely that the large herbivorous saurischians lived in swamps where plants were abundant and where they were relatively safe from predators. *Branchiosaurus* had nostrils on the top of its head and

Figure 12.45 Reconstruction of a confrontation between the largest of the carnivorous dinosaurs, *Tyrannosaurus rex* (right) and the horned dinosaur *Triceratops* (left). (Courtesy of the Field Museum of Natural History, Chicago.)

367

Figure 12.46 Reconstruction of the skeleton of *Apatosaurus (Brontosaurus) excelsus.* (Courtesy of the Field Museum of Natural History, Chicago.)

apparently remained submerged much of the time.

The ornithischian dinosaurs reached their peak during the Cretaceous, and they occupied a wide variety of environments. Early bipeds included *Camptosaurus*, which had a birdlike beak, and *Iguanodon*, a similar but larger form. Another unusual biped was the Upper Cretaceous "bone-head," *Pachycephalosaurus*, which had a nearly solid skull; *Trachodon* was a duck-billed form with birdlike feet (Fig. 12.47). Fossil tracks of bipedal ornithischians are locally abundant in the sandstones and shales that formed from the sands and muds through which they walked (Fig. 12.48).

Quadrupedal ornithischians included *Stegosaurus*, which had a spiked tail and a double row of bony plates along its back (Fig. 12.49). *Ankylosaurus*, another quadruped, had a bony knob at the end of its tail which could be used as a bludgeon. The spiked-tail of *Stegosaurus* and the bludgeon of *Ankylosaurus* were probably used for defense against carnivorous dinosaurs.

The horned dinosaurs, the ceratopsians, comprise a separate order. The most famous of the ceratopsians was *Triceratops* which had three horns on its massive head (Figs. 12.45 and 12.50).

Reptile eggs are not commonly preserved as fossils, but several nests of fossilized eggs of *Protoceratops*, a Middle Cretaceous ceratopsian, have been found in Mongolia. Some of the eggs even contain the bones of the embryos.

Marine Reptiles

When fierce competition for food developed among the terrestrial reptiles, some adapted to a marine environment. Their bodies became streamlined and their legs were modified into flippers. Some retained a reptilelike appearance, as did the mosasaurs; others developed a fishlike morphology, as did the ichthyosaurs (Fig. 12.51). The plesiosaurs are characterized by the presence of a small head, a long neck, and a rather fat body (Fig. 12.52). They first appeared during the Triassic, and some forms reached 20 ft (6 m) during the Jurassic. Marine turtles had a morphology quite similar to that of modern turtles. One species which lived during the Cretaceous was 11 ft (3.5 m) long.

Flying Reptiles

The vertebrates dominated the land and both fresh

Figure 12.47 Restoration of a Late Cretaceous landscape with several kinds of dinosaurs. On the right are the duck-billed dinosaurs, *Trachodon*, in the center foreground is the armored *Ankylosaurus*, at the left in the water is *Corythosaurus*, and in the background at the left are the crested duckbills, *Parasaurolophus*. (Courtesy of the Field Museum of Natural History, Chicago.)

Figure 12.48 Part of the exposed footprint horizon at Dinosaur State Park in central Connecticut. White crosses on the photo are 10-ft grid marks. The fine chalk lines represent an attempt to trace the path of individual animals. (Courtesy of the Connecticut Geologic and Natural History Survey.)

and salt waters at the beginning of the Mesozoic. However, they had not yet invaded the last major environment available to them, the air. This realm had been the exclusive domain of the insects since the Carboniferous. The reptiles were the first to fly, but before they were able to do so, they had to develop the ability to glide for short distances. One form had short winglike structures as early as the Late Triassic, but the first flying vertebrates did not appear until the Jurassic.

The flying reptiles or pterosaurs had a very long fourth finger from which a wide membranous wing extended. Their rather thin bones were sometimes hollow, like those of modern birds. The ability to fly allowed the pterosaurs to exploit new supplies of food, such as the insects. Furthermore, it provided some measure of protection against carnivorous dinosaurs. The teeth of one small

Figure 12.49 Reconstruction of a Jurassic landscape in the western part of the United States with two stegosaurs. (From a painting by Charles R. Knight, courtesy of the Field Museum of Natural History, Chicago.)

pterosaur, *Pterodactylus*, pointed outward and presumably could be used for spearing small fish which swam just beneath the surface. A Cretaceous pterosaur, *Pteranodon*, had a wingspread of 27 ft (9 m) (Fig. 12.53).

Birds

The earliest known birds date from the Jurassic. *Archaeopteryx*, whose remains have been found in the lithographic limestones near Solenhofen, Germany, had a skeleton more similar to that of a reptile than to that of a modern bird (Fig. 12.54). However, the presence of impressions of feathers associated with the skeleton shows that *Archaeopteryx* was a bird.

By Cretaceous time, the breastbones of birds had been considerably enlarged. This adaptation

permitted the attachment of powerful muscles for working the wings. In what appears to be a reversal of an evolutionary trend, *Hesperornis* lost the power of flight and became adapted to an aquatic habitat. Its long legs and sharp teeth indicate that this bird dove to appreciable depths for fish.

Mammals

The record of the early evolution of the mammals is fragmentary due to the scarcity of fossil remains, which are mostly comprised of teeth and fragments of jaws. The remains that have been found suggest that the Mesozoic mammals were relatively small and few in number. However, the brain cavities of the mammals were large when compared with those of the reptiles, and presumably they were more intelligent. It was prob-

Figure 12.50   Reconstruction of the skeleton of the horned dinosaur *Triceratops prorsus* from the Upper Cretaceous Lance Formation in Niobrara County, Wyoming. (Courtesy of the Smithsonian Institution.)

ably this factor which enabled them to survive under the domination of the dinosaurs.

The earliest fossil mammals have been found in deposits of latest Triassic age, and by the Jurassic five orders had evolved (Fig. 12.55). Each order is characterized by the shape of its teeth. Of these, the pantotheres were the basic stock from which the important Cenozoic mammals evolved. The teeth of the pantotheres were sharp and included well-developed canines. Their diet probably consisted of insects, worms, grubs, and very small reptiles.

Plants

The record of Mesozoic flora has been reconstructed from macrofossils and assemblages of

Figure 12.51   Skeleton of an ichthyosaur. (Courtesy of the Field Museum of Natural History, Chicago.)

Figure 12.52   Restoration of a Jurassic seascape with plesiosaurs on the left and fishlike ichthyosaurs on the right. (From a painting by Charles R. Knight, courtesy of the Field Museum of Natural History, Chicago.)

spores and pollen. However, since it is not always possible to relate the spores and pollen to the plants that produced them, such reconstructions are subject to a degree of uncertainty. This is a less serious problem in Upper Cretaceous deposits because many Late Cretaceous plants closely resembled modern forms.

Mesozoic floras differ markedly from those of the Paleozoic. Cycadeoides and cycads achieved prominence during the Jurassic and Early Cretaceous. These plants looked like palms and grew more than 40 ft (13 m) high (Fig. 12.54). In the southern continents, the flora included ginkgoes and seed ferns in addition to cycadeoides. Seed ferns such as *Sphenopteris* and *Glossopteris* declined in importance through the Mesozoic.

Among the gymnosperms, conifers became increasingly important during the Mesozoic. This is

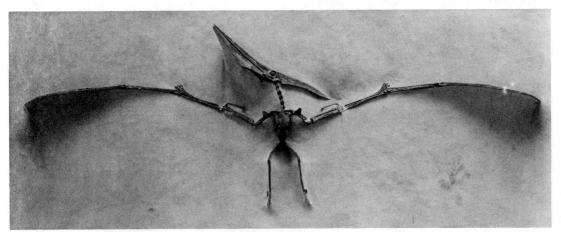

Figure 12.53   Restored skeleton of the flying reptile *Pteranodon ingens* from the Upper Cretaceous chalk deposits of western Kansas. (Courtesy of the Smithsonian Institution.)

Figure 12.54 Restoration of a Jurassic landscape showing the small dinosaur *Ornitholestes* in the foreground at the left, several birds of the genus *Archaeopteryx*, and a number of pterosaurs in the air and on the trees. (From a painting by Charles R. Knight, courtesy of the Field Museum of Natural History, Chicago.)

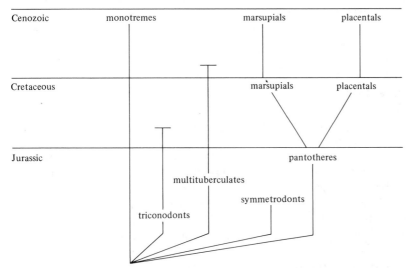

Figure 12.55   Evolution of the mammals. (Courtesy of E. H. Colbert, *in* Ref. 47.)

especially true of the cedar-, pine-, and firlike forms. Ginkgoes, which were also widespread, occur in the Arctic and central regions of North America, in Europe, and in Asia.

The most significant development in plant life in the Mesozoic was the appearance of flowering plants, the angiosperms. These are divided into two subclasses: the monocotyledons, which are characterized by an uncovered seed or single seed-leaf, a lack of bark and pith, and simple leaf veinations, and the more advanced dicotyledons, which have a covered or double seed-leaf, bark, pith, and complex leaf veinations. Pollen from the monocotyledons appears in rocks as old as Jurassic and definite palm stems have been found in Middle Jurassic deposits (48). The explosive development of the angiosperms occurred during the Late Cretaceous. Only about 5% of the Early Cretaceous vascular flora was comprised of angiosperms, but by the end of the Cretaceous approximately 95% of the vascular flora consisted of angiosperms. The dicotyledons, which first appeared near the boundary between the Early and Late Cretaceous, had become the dominant angiosperms by the end of the period. Cretaceous monocotyledons include lilies and palms; among the Cretaceous dicotyledons are oak, birch, fig, willow, and a variety of shrubs.

Extinctions

A large number of important groups became extinct near the end of the Mesozoic. These include the ammonoids, dinosaurs, pterosaurs, plesiosaurs, ichthyosaurs, and mosasaurs. Of the reptiles, only the lizards, snakes, turtles, and crocodiles survived. Among the plants, the cycadeoides became extinct at the end of the Mesozoic.

The theories proposed for the mass extinctions are similar to those which have been suggested to explain earlier mass extinctions. These include worldwide catastrophes, climatic changes, changes in sea level, and diseases. As for catastrophic explanations, there is no indication from the rock record that any catastrophe occurred near the end of the Mesozoic. In the sequences that span the Cretaceous-Tertiary boundary, there is only a steady decline in certain groups. On the other hand, there is a well-documented record of a cooling of the climate during the Cretaceous (Fig. 14.26). Reconstructions of paleogeography show that this cooling was accompanied by a gradual withdrawal of the seas from the continents. The cooling of the climate and the withdrawal of the seas may have had a profound effect on the dinosaurs and related groups. Since withdrawal was probably associated with regional uplift, swamps, which were the principal habitat of the herbivorous dinosaurs, would have been drained, thus eliminating these forms. Consequently, the carnivorous dinosaurs which preyed on them may have become extinct for lack of an adequate food supply. Other dinosaurs, which had become very specialized during the Jurassic and Cretaceous, may not have been able to adapt to rapid changes in climate. Since dinosaurs were cold blooded, the decrease in the mean temperature of their habitats may have been a significant factor in their extinction.

Another possible cause for the extinction of some groups of reptiles is the rise of the mammals. Mesozoic mammals were not large enough to compete directly with reptiles for food and living space, but they may have preyed on reptile eggs and young. Modern reptiles leave their eggs virtually unprotected, and it is not illogical to suppose that Mesozoic reptiles did the same. However, it is unlikely that the mammals were the major cause for the extinctions, since at that time there were still relatively few mammals.

Disease among the dinosaurs and related groups has been suggested as a cause of mass extinctions, but no evidence of this exists in the fossil record. Furthermore, diseases generally affect one or at most only a few species, not a group as large and diverse as the dinosaurs.

In the case of the marine reptiles and the ammonoids, the restriction of the epicontinental seas during the Cretaceous may have been an important factor in their extinction. A narrowing of these seas would have greatly increased the competition for food among the ichthyosaurs, plesiosaurs, mosasaurs, and ammonoids. The fact remains that no single cause has yet been suggested which can account for the variety of extinctions and for the environments in which they occurred.

MESOZOIC CLIMATE

Triassic

It may be inferred from the pattern of paleoclimatic indicators that the climate during the Triassic was relatively warm and dry (Fig. 12.1). No glacial deposits of this age have been confirmed, and dune sandstones and red beds are common at low as well as high paleolatitudes. Coals are more restricted and are less important than coals of other periods.

Jurassic

The distribution of Jurassic paleoclimatic indicators suggests that the Jurassic climate was warm, but not as dry as that of the Triassic (Fig. 12.2). Reefs, evaporites, and red beds are generally confined to a zone within 30° of the paleoequator. Temperate or subtropical coals are widespread from 30 to 70° paleolatitude, and no Jurassic glacial deposits have been reported.

Cretaceous

The Cretaceous climate was initially warm and rather moist (Fig. 12.3). The flora was remarkably uniform throughout the world during the Early Cretaceous, with subtropical forms in latitudes up to 70° from the paleoequator. Ferns and cycads from high latitudes are similar to those growing today in the subtropical South American rain forests. Cretaceous coals are found on almost every continent at paleolatitudes ranging from 0 to almost 90°. The climatic deterioration in the Late Cretaceous continued into the Tertiary.

SUMMARY

At the beginning of the Mesozoic, all the continents were part of one landmass, Pangaea. The breakup of the continents probably began during the Late Triassic with the separation of North America from Gondwanaland. The separations of North America from Greenland, South America from Africa, and Antarctica and Australia from Africa and India probably began during the Late Jurassic. Finally, the separation of Greenland from Europe and the separation of Antarctica from Australia were probably initiated during the Late Cretaceous or early Tertiary.

The separation of North America and Gondwanaland resulted in the formation of a series of fault basins approximately parallel to the line of separation. As these continents continued to separate, geosynclines developed along the margins of the newly created Atlantic Ocean and Gulf of Mexico. The separation also resulted in the folding, faulting, uplift, and intrusion of the deposits of the Cordilleran Geosyncline during the Late Triassic and again during the Late Jurassic. The mountain range that formed during the latter deformation furnished sediments for the Coast Range Geosyncline, which replaced the Cordilleran Geosyncline.

Geosynclinal deposition occurred on the Pacific and Tethyan margins of the continents as oceanic plates moved under these margins. As the Eurasian and African plates moved together, a series of geanticlinal ridges was produced within the northern Tethyan Geosyncline. Thick sequences of clastics resulted from the erosion of these uplifts.

The Mesozoic saw the rise of the reptiles, which dominated almost all the available terrestrial environments and the oceans and air as well. Some of the terrestrial dinosaurs reached gigantic proportions, and the plesiosaurs and mosasaurs in the marine realm were almost as impressive. The mammals originated in the Late Triassic, but evolved very slowly. The extinctions at the end of the Mesozoic include the dinosaurs, pterosaurs, plesiosaurs, ichthyosaurs, mosasaurs, and ammonoids. These extinctions may have been related to an uplift of the land and cooling of the climate, which destroyed many habitats and resulted in increased competition for food and living space.

Horsetails, seed ferns, and conifers were the important floras of the Early and Middle Triassic, but from Late Triassic through Early Cretaceous, floras were dominated by ferns, cycadeoides, cycads, ginkgoes, and conifers. Angiosperms, which first appeared during the Jurassic, expanded rapidly into most terrestrial environments during the Late Cretaceous.

Mesozoic climates were generally warmer than those of either the Paleozoic or the Cenozoic. The Triassic was unusually arid, but the climate more moist during the Jurassic and Cretaceous.

REFERENCES CITED

1. J. D. Phillips and D. Forsyth, 1972 Plate tectonics, paleomagnetism and the opening of the Atlantic: *Geol. Soc. Amer. Bull.*, v. 82, p. 1579.

2. G. P. Erickson and J. L. Kulp, 1961, Potassium-argon measurements on the Palisades Sill, New Jersey: *Geol. Soc. Amer. Bull.*, v. 72, p. 649.

   R. L. Armstrong and J. Bescanon, 1969, Upper Triassic time scale (abstr.): *Trans. Amer. Geophy. Union*, v. 50, p. 329.

3. A. A. Abdel-Monem and J. L. Kulp, 1968, Paleogeography and the source of sediments of the Triassic basin, New Jersey, by K-Ar dating: *Geol. Soc. Amer. Bull.*, v. 79, p. 1231.

   G. de V. Klein, 1969, Deposition of Triassic sedimentary rocks in separate basins, eastern North America: *Geol. Soc. Amer. Bull.*, v. 80, p. 1825.

4. W. H. Kanes and J. P. Conolly, 1971, Rifting of Europe from Africa: Triassic-Jurassic history of the Moroccan Atlas Mountains: *Abstracts with Programs*, v. 3, no. 7, Geological Society of America, p. 617.

5. U. Jux, 1961, *The palynologic age of diapiric and bedded salt in the Gulf Coastal Province:* Louisiana Geological Survey Bull. 38.

6. A. J. Eardley, 1962, *Structural Geology of North America:* Harper and Row, New York.

7. C. L. Drake, M. Ewing, and G. H. Sutton, 1959, Continental margins and geosynclines: The east coast of North America North of Cape Hatteras, *in Physics and Chemistry of the Earth*, v. 3: Pergamon Press, New York, p. 110.

8. N. D. Newell, 1955, Bahamian Platforms, *in* A. Poldervaart, ed., *Crust of the Earth:* Geological Society of America Special Paper 62, p. 303.

9. H. Gill, L. A. Sirkin, and J. A. Doyle, 1969, Cretaceous deltas in the New Jersey Coastal Plain, *in Abstracts with Programs for 1969, Part 7:* Geological Society of America, p. 79.

   J. P. Owens, J. P. Minard, and N. F. Sohl, 1968, Cretaceous deltas in the northern New Jersey coastal plain, in R. Finks, ed., *Guidebook to Field Excursions*, NYS Geol. Assoc. 10th Ann. Mtg., Queens, N.Y.

10. C. Schuchert, 1968, *Historical geology of the Antillean-Caribbean Region:* Hafner, New York.

11. P. H. Mattson, 1960, Geology of the Mayaguez area, Puerto Rico: *Geol. Soc. Amer. Bull.,* v. 71, p. 319.

12. A. A. Meyerhoff, K. M. Khudoley, and C. W. Hatten, 1969, Geologic significance of radiometric dates from Cuba: *Amer. Assoc. Petrol. Geol. Bull.,* v. 53, p. 2494.

13. C. K. Seyfert, 1968, Continental accretion during the Paleozoic and Mesozoic in the Klamath Mountains of northern California (abstr.): *Trans. Amer. Geophys. Union,* v. 49. p. 326.

    W. Hamilton, 1969, Mesozoic California and the underflow of Pacific mantle: *Geol. Soc. Amer. Bull.,* v. 80, p. 2409.

14. C. H. Stephens, 1969, Middle to Late Triassic deformation in Inyo, White and northern Argus Mountains, California, *in Abstracts with Programs for 1969, Part 5:* Geological Society of America, p. 78.

15. M. D. Crittenden, Jr., 1969, Interaction between Sevier Orogenic Belt and Uinta structures near Salt Lake City, Utah, *in Abstracts with Programs for 1969, Part 5:* Geological Society of America, p. 18.

16. E. H. McKee and D. B. Nash, 1967, Potassium-argon ages of granitic rocks in the Inyo Batholith, east-central California: *Geol. Soc. Amer. Bull.,* v. 78, p. 669.

    F. W. McDowell and J. L. Kulp, 1969, Potassium-argon dating of the Idaho Batholith: *Geol. Soc. Amer. Bull.,* v. 80, p. 2379.

    G. H. Curtis, J. F. Everden, and J. I. Lipson, 1958, *Age determinations of some granitic rocks in California by the potassium-argon method:* California Division Mines Special Rep. 54.

17. E. H. Bailey, M. C. Blake, Jr., and D. L. Jones, 1970, *On-land Mesozoic oceanic crust in California Coast Ranges:* U.S. Geological Survey Prof. Paper 700-C, p. C-70.

18. E. H. Bailey, W. P. Irwin, and D. L. Jones, 1964, *Franciscan and related rocks and their significance in the geology of western California:* California Division of Mines and Geology Bull. 183.

    E. H. Bailey and M. C. Blake, 1970, Late Mesozoic tectonic development of western California: *Geotectonics,* no. 3, p. 148; no. 4, p. 225.

19. J. Suppe, 1969, Times of metamorphism in the Franciscan terrain of the Northern Coast Ranges, California: *Geol. Soc. Amer. Bull.,* v. 80, p. 135.

20. S. H. Knight, 1953, Summary of Cenozoic history of the Medicine Bow Mountains, Wyoming: *Wyo. Geol. Assoc. Guidebook, Eighth Ann. Field Conf.,* p. 65.

21. R. Trumpy, 1960, Paleotectonic evolution of the central and western Alps: *Geol. Soc. Amer. Bull.,* v. 71, p. 843.

22. P. G. Temple and L. J. Perry, 1962, Geology and oil occurrence, southeast Turkey: *Bull. Amer. Assoc. Petrol. Geol.,* v. 46, p. 1596.

23. R. Brinkmann, 1969, *Geologic Evolution of Europe,* 2nd. ed., transl., J. E. Sanders: Hafner, New York.

24. R. G. Coleman, 1971, Plate tectonic emplacement of upper mantle peridotites along continental edges: *Jr. Geophys. Res.,* v. 76, p. 1212.

25. J. Aubouin, 1965, *Geosynclines:* Elsevier, New York.

26. R. W. Van Bemmelen, 1954, *Mountain Building:* Martinus Nijhoff, The Hague, Holland.

27. Y. Kawano and Y. Ueda, 1967, Periods of the igneous activities of the granitic rocks in Japan by K-Ar dating method, *in* T. Matsumoto, ed., *Age and Nature of the Circum-Pacific Orogenesis: Tectonophysics,* v. 4, p. 523.

28. C. A. Fleming, 1962, New Zealand biogeography; A paleontologist's approach: *Tuatara,* v. 10, p. 63.

29. C. A. Landis and D. S. Coombs, 1967, Metamorphic belts and orgenesis in southern New Zealand, *in* T. Matsumoto, ed., *Age and Nature of the Circum-Pacific Orogenesis: Tectonophysics,* v. 4, p. 501.

30. H. Burgl, 1967, The orogenesis in the Andean system of Colombia, *in* T. Matsumoto, ed., *Age and Nature of the Circum-Pacific Orogenesis: Tectonophysics,* v. 4, p. 429.

31. R. E. Clemons and L. E. Long, 1971, Petrologic and Rb-Sr Isotopic Study of the Chiquimula Pluton, Southeastern Guatemala: *Geol. Soc. Amer. Bull.,* v. 82, p. 2729.

32. M. Halpern, 1967, Rubidium-strontium isotopic age measurements of plutonic igneous rocks in eastern Ellsworthland and northern

Antarctica Peninsula, Antarctica: *Jr. Geophys. Res.,* v. 72, p. 5133.

33. I. McDougall and N. R. Ruegg, 1966, Potassium-argon dates on the Serra Geral Formation: *Geochim. Cosmochim. Acta,* v. 30, p. 191.

    G. Amarl, V. G. Cordani, K. Kawashita, and J. H. Reynolds, 1966, Potassium-argon dates on basaltic rocks from Southern Brazil: *Geochim. Cosmochim. Acta,* v. 30, p. 159.

    K. M. Creer, J. A. Miller, and A. G. Smith, 1965, Radiometric age of the Serra Geral Formation: *Nature,* v. 207, p. 282.

34. I. McDougall, 1963, Potassium-argon age measurements on dolerites from Antarctica and South Africa: *Jr. Geophys. Res.,* v. 68, p.1535.

35. I. McDougall, 1961, Determination of the age of a basic igneous intrusion by the potassium-argon method: *Nature,* v. 190, p. 1184.

36. D. A. Valencio and J. F. Vilas, 1969, Age of the separation of South America and Africa: *Nature,* v. 223, p. 1353.

37. M. Ewing et al., 1969, Site 4, *in Initial reports of the deep sea drilling project:* v. 1, National Science Foundation, p. 179.

    A. S. Laughton et al., 1970, Deep Sea Drilling Project: Leg 12: *Geotimes,* v. 15, no. 9, p. 10.

38. The Scientific staff of Leg 15, 1971, Deep sea drilling project: Leg 15, *Geotimes,* v. 16. no. 4, p. 12.

39. G. L. Johnson and J. A. Pew, 1968, Extension of the Mid-Labrador Sea Ridge: *Nature,* v. 217, p. 1033.

40. J. Hospers and S. I. Van Andel, 1969, Paleomagnetism and tectonics: *Earth Sci. Reviews,* v. 5, p. 5.

41. S. W. Carey, 1958, The tectonic approach to continental drift, *in* S. W. Carey, ed., *Continental Drift: A symposium:* Univ. of Tasmania, p. 177.

42. E. J. W. Jones and J. I. Ewing, 1969, Age of the Bay of Biscay: Evidence from seismic profiles and bottom samples: *Science,* v. 166, p. 102.

43. C. A. Burk et al., 1969, Deep-sea drilling into the Challenger Knoll, central Gulf of Mexico: *Amer. Assoc. Petrol. Geol. Bull.,* v. 53, p. 1338.

44. A. G. Fischer et al., 1970, Geologic history of the western north Pacific: *Science,* v. 168, p. 1211, May 29.

    J. B. Davis and E. E. Bray, 1969, Analyses of oil and cap rock from Challenger (Sigsbee) Knoll, in *Initial reports of the deep sea drilling project:* v. 1, National Science Foundation, p. 415.

45. C. Teichert, 1967, Major features of cephalopod evolution, in C. Teichert and E. L. Yochelson, eds., *Essays in Paleontology and Stratigraphy:* Special Paper 2, Department of Geology, University of Kansas, p. 162.

    B. Kurten, 1969, Continental Drift and evolution: *Sci. Amer.,* v. 220, no. 3, p. 54.

46. A. S. Romer, 1968, *The Procession of Life:* World, New York.

47. E. H. Colbert, 1961, *Evolution of the Vertebrates:* Science Editions, New York.

48. W. D. Tidwell, S. R. Rushforth, and J. L. Reveal, 1970, *Paloxylon simperi* and *Palmoxylon pristima:* Two pre-Cretaceous angiosperms from Utah: *Science,* v. 168, p. 835.

13
The Tertiary Period

By the early Tertiary, the continents had achieved their present outlines, but the Atlantic and Indian oceans had not yet reached their present widths (Fig. 13.1). As the continents moved away from the mid-oceanic ridges, the Atlantic and Indian oceans increased in size, with a consequent reduction in the size of the Pacific Ocean basin. The Tethyan Geosyncline was transformed into a mountain range as Africa and India collided with Europe and Asia. Erosion of these mountains resulted in a great flood of clastic sediment, which was mostly deposited in adjacent lowlands. Geosynclinal deposition continued on the Pacific margins of the continents as oceanic plates moved under continental plates.

The Tertiary saw the rise of the mammals, which occupied the various ecological niches once dominated by the dinosaurs. The anthropoids, a group of primates that includes monkeys, apes, and man, appear in rocks of Oligocene age, and the remains of what are probably man's direct ancestors occur in Miocene and Pliocene strata.

Gradual cooling of the climate during the Tertiary may have been related to such factors as mountain building, epirogenic uplifts of the continents, and migration of the continents into the polar regions. Glaciation in Antarctica has been traced to the early Tertiary, when the Antarctic continent gained its polar position.

NORTH AMERICA

The coastal plains and continental shelves of North America received thick sequences of sediments throughout the Tertiary, particularly in the Coast

Range, Gulf Coast, and East Coast geosynclines. The Arctic Slope Basin may also prove to be of geosynclinal proportions when the results of recent oil exploration in this area have been compiled. Epicontinental seas were restricted during the Tertiary, and consequently marine Tertiary deposits are rather limited in the continental interior.

Coast Range Geosyncline

Marine and nonmarine sediments were deposited during the Tertiary in a series of basins along the Pacific margin of North America. The extent of these basins was limited by highlands which had been uplifted in pre-Tertiary orogenies.

Central and Southern California. The seas began a gradual transgression in the Paleocene following the regression at the end of the Cretaceous. They reached their maximum extent during the Eocene and again in the Miocene. Marine deposits are largely clastics derived from the Mesocordilleran Geanticline and from local uplifts. The thickness of these deposits, which were laid down in several basins, often exceeds 15,000 ft (5000 m). Siliceous deposits, such as cherts and diatomites, are common within the basins and may have been derived in part from weathering of volcanic ash. Late Miocene folding caused a partial withdrawal of the seas, and by the end of the Pliocene they were restricted to a narrow strip along the margin of the continent.

Washington, Oregon, and Northern California. Eugeosynclinal volcanic and sedimentary rocks of

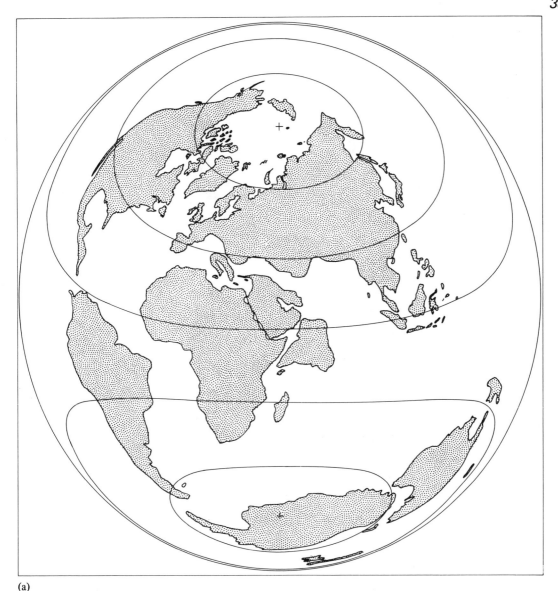

(a)

Figure 13.1 Reconstructions of the continents for the: (a) earliest Tertiary, (b) middle Tertiary, (c) present. (Modified after Phillips and Forsyth, Ref. 1.)

Tertiary age occur in western Washington, western Oregon, and northwestern California. Tuffs and rhyolitic basaltic flows are commonly interbedded with marine clastics and nonmarine coal-bearing strata. Some of the basalts have pillow structures typical of subaqueous extrusions. However, other volcanic rocks were extruded subaerially and,

according to paleogeographic reconstructions, they originated in a chain of volcanic islands (2).

A narrow band of late Tertiary andesites extends from northern California through central Oregon to central Washington. These rocks may have been formed as a result of the underflow of the Pacific Plate along a zone of subduction. This

Fig. 13.1

(b)

mechanism may also account for folding and faulting of many Tertiary sequences in this region.

Alaska. Tertiary deposits of eugeosynclinal character are widely exposed in southern Alaska. These deposits include volcanic flows, agglomerates, breccias, and tuffs interbedded with clastic sedimentary rocks. The eugeosyncline extended from the mainland of Alaska westward through the Aleutian Islands. The western Aleutians are bordered on both the north and south by deep ocean and are underlain by oceanic crust. This relationship provides a unique opportunity for tracing a eugeosyncline from a continent into an ocean basin. Early Tertiary plant fossils have been found in sedimentary rocks of Amchitka Island, near the

382

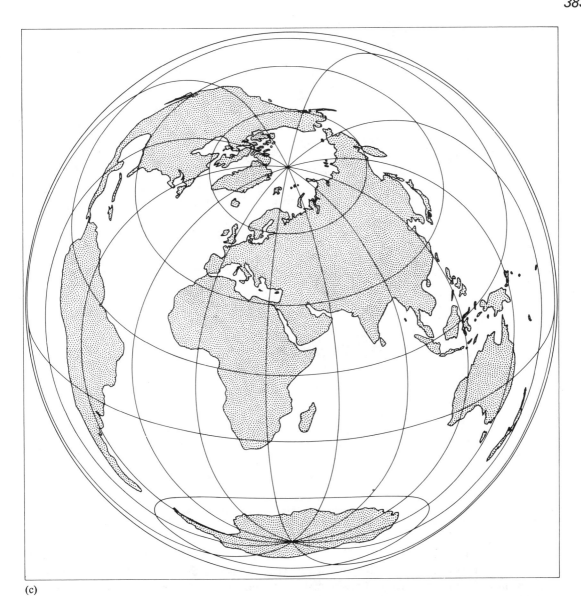

(c)

center of the Aleutian chain. Therefore the Aleutian Arc must have originated during or before the early Tertiary.

Columbia Plateau

A very large part of the Pacific Northwest is covered by basalt flows of late Tertiary and Quaternary age (Fig. 13.2). The numerous flows of the Columbia Plateau and the Snake River Plain cover an area of approximately 200,000 square miles (340,000 km^2) and are as much as 10,000 ft (3000 m) thick. The thickness of individual flows averages a few tens of feet (Fig. 13.3). The lavas were erupted in a very fluid state and in some cases traveled distances of more than 100 miles (170 km).

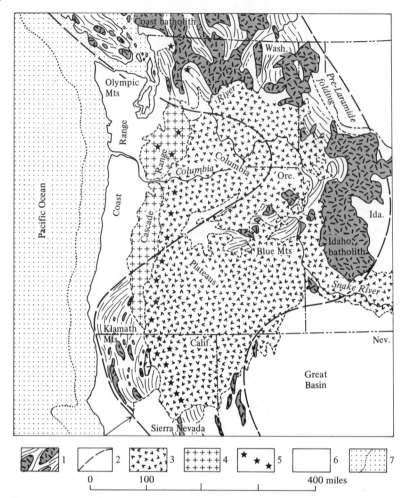

Figure 13.2   Map of the Pacific Northwest showing the distribution of plateau basalts and andesitic volcanoes of Tertiary and Quaternary age. (After P. B. King, Ref. 3.)

Basin and Range Province

Extensive volcanic and plutonic activity occurred in the Basin and Range province of Nevada, eastern California, and western Utah during the middle and late Tertiary. In Miocene and later time, this entire region was uplifted and faulted. The faults, which are dominantly normal faults, are thought to have been produced when the earth's crust became extended as it was arched upward. It has been suggested that the uplift resulted from the overriding of a portion of the East Pacific Rise by North America. The pattern of magnetic anomalies in the floor of the northeastern Pacific Ocean indicates that a section of the East Pacific Rise now lies under eastern Nevada and western Utah. Heating and expansion of the mantle under the Basin and Range Province and would have resulted in a large-scale arching of the overlying crust. Faulting produced a series of elongated ridges in this province (Fig. 13.4). The vertical displacements on these faults often exceed 10,000 ft (3000 m). The valleys between the ridges were filled with thousands of feet of sediments as the ridges were uplifted.

Figure 13.3 Tertiary volcanics of the Snake River Plain in Idaho. Several different lava flows can be seen in the walls of the Snake River Canyon near Idaho Falls.

Central and Southern Rockies

During the early Tertiary, a series of uplifts developed in Montana, Wyoming, South Dakota, Colorado, Utah, and New Mexico. These uplifts are generally anticlinal and are bordered by steeply dipping reverse faults on one or both sides (Figs. 13.5 and 13.6). Erosion of these uplifts has supplied up to 10,000 ft (3000 m) of shale, sandstone, and conglomerate to adjacent basins (Fig. 13.7). In certain basins, freshwater lakes were formed. Shales deposited in these lakes contain a rich fauna of fossil fish and insects. In the Green River Basin of southwestern Wyoming, northwestern Colorado, and eastern Utah, oil shales constitute a very large potential reserve of gas and oil.

In the Bighorn Basin of northwestern Wyoming, Paleozoic sedimentary rocks overlie Tertiary deposits (Fig. 13.8). These Paleozoic rocks apparently slid downhill under the influence of gravity. The source of the slide block was probably an uplifted area in the Bighorn Mountains to the west (4). Apparently the zone of detachment was a shale at the base of a thick limestone sequence.

Figure 13.4 Fault scarp bordering elongate ridge in Basin and Range Province. (Photo by Dennis Burke, courtesy of the U.S. Geological Survey.)

Figure 13.5 Large anticline in southern Utah produced during the early Tertiary deformation of the region. (U.S. Bureau of Reclamation.)

Figure 13.6 Folded Paleozoic sedimentary rocks on the east flank of the Bighorn Mountains, Wyoming. The folding and faulting occurred during the early Tertiary.

Figure 13.7 Eocene lake deposits of Bryce Canyon National Park have been eroded into pillars and spires by the action of wind and water. (Union Pacific Railroad Photo.)

Figure 13.8 Heart Mountain north of Cody, Wyoming, is capped by Paleozoic limestones. These are separated from the underlying Tertiary sediments by a gravity thrust fault.

Figure 13.9   Flat-lying late Tertiary volcanic rocks exposed on Carter Mountain along the South Fork of the Shoshone River west of Cody, Wyoming.

Figure 13.10   Folded Paleozoic limestones in Banff National Park, Canada. The folding occurred during the early Tertiary.

In the vicinity of Yellowstone National Park, Wyoming, volcanic conglomerates, tuffs, and breccias of middle and late Tertiary age overlie early Tertiary deposits (Fig. 13.9). In one area, volcanic deposits repeatedly covered forests, and many trees were fossilized in upright positions.

Northern Rocky Mountains
Deposits in the northern Cordilleran Geosyncline were intensely folded and faulted during the early Tertiary (Fig. 13.10). Seismic surveys and deep drilling indicate that the dip of thrust faults decreases with depth and that the basement is not involved in the deformation (Fig. 13.11). This type of deformation contrasts with that of the central and southern Rockies, where the basement has been displaced along steeply dipping reverse faults.

The Great Plains
Continental clastics were deposited over a large area east of the Rocky Mountains during the early and middle Tertiary. In the Badlands of South Dakota, numerous fossilized remains of early mammals have been recovered from one such deposit, the Brule Clay of Oligocene age (Fig. 13.12). Tuffs interbedded with these sediments were derived from volcanoes lying to the west. A sea covered much of the southeastern part of the United States in early Tertiary time, but receded southward when the Great Plains were uplifted during the later part of the middle Tertiary. Much of the early Tertiary section was eroded from the Great Plains at this time.

East Coast and Gulf Coast Geosynclines
Thick sequences of Tertiary sediments are exposed on the Atlantic and Gulf Coastal Plains. Seismic surveys and drilling indicate that deposits of this age also underlie the continental shelf of the Atlantic Ocean and the Gulf of Mexico. The sediments are mostly sands, clays, and muds that reach a maximum thickness of 20,000 ft (6000 m) in the Mississippi River delta. Most of these deposits are marine in origin, but continental sediments some-times occur on the landward side of the geosyncline. Limestones containing reef-building corals occur in Florida and the Bahamas. More than 8000 ft (2500 m) of Tertiary carbonates have been penetrated in drilling operations on the Bahama Banks.

In Paleocene time, the seas transgressed from the Gulf Coastal region as far north as southern Illinois. The position of the shoreline fluctuated considerably throughout the remainder of the Tertiary as the seas slowly withdrew from the continent. During periods of regression, deltaic sands overlapped marine shales. The resulting interfingering of sands and shales created stratigraphic traps in which important reserves of oil and gas are found. By the end of the Tertiary, the seas were limited to the outer margin of the continent.

The Greater Antilles
Volcanism and plutonism were widespread during the early Tertiary in Cuba, Hispaniola, and Puerto Rico. In Puerto Rico, volcanic and sedimentary rocks of Paleocene through Middle Eocene age conformably overlie rocks of Cretaceous (Maestrichtian age (6). The Tertiary deposits are intruded by granite batholiths that have been radiometrically dated at 41 million years (Late Eocene) (7). Volcanic activity ceased in the Greater Antilles in the Middle Eocene. It is interesting to note, however, that volcanism in the Lesser Antilles began about the same time. Possibly the direction of sea-floor spreading changed during the Middle Eocene. Oligocene and younger deposits are relatively undeformed in the Greater Antilles.

The Lesser Antilles
The Lesser Antilles are a chain of islands extending southward from the eastern end of the Greater Antilles. Volcanic activity has been continuous here from the Middle Eocene to the present time. Coral reefs that once fringed subaerial volcanoes have contributed to the formation of limestones associated with Tertiary volcanics. Clastic sediments were derived largely from volcanics, but

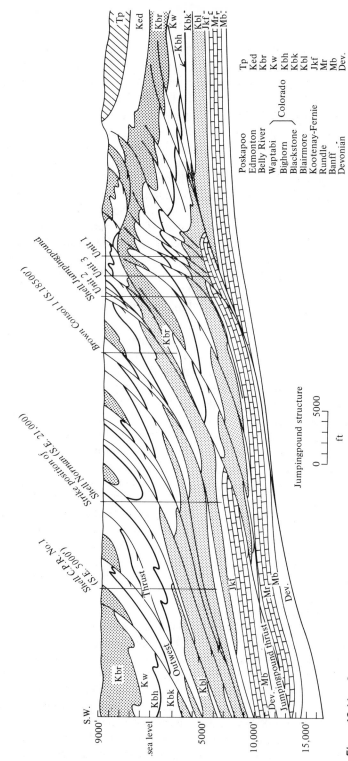

Figure 13.11 Structural cross section across the Jumpingpond gas field in western Alberta, Canada. Note that the thrust faults flatten with depth. (After Fox, Ref. 5. Reproduced with the permission of The American Association of Petroleum Geologists.)

390

Figure 13.12   Oligocene sediments of continental origin in the Badlands National Monument of South Dakota.

some of the clastics in Barbados (for example, those of the New Scotland Formation) may have been derived from deep ocean deposits in the vicinity of northern South America. These deposits are quite similar to sediments cored from the Atlantic Ocean basin east of the island.

EURASIA

Tertiary geosynclinal deposits are widely exposed in the northern Tethyan, Indonesian, and Pacific Ocean geosynclines. Marine sediments are also present near the outer borders of the continental interior, and continental clastics occur in regions adjacent to mountain belts.

Northern Tethyan Geosyncline

During the early Tertiary, a series of geanticlinal ridges in the Northern Tethyan Geosyncline contributed a thick deposit of marine clastics (flysch) to the geosyncline. Late in the Oligocene, the deposits of the Alpine region were intensely folded and uplifted. The folds, which are termed *nappes*,

are typically recumbent and are cut by thrust faults (Fig. 13.13). The mountains that resulted from this deformation were rapidly eroded, and a thick layer of continental clastics (*molasse*) was deposited in the Peri-Alpine Depression to the north. The deposits range in age from Late Oligocene to Late Miocene. By Late Miocene time, the zone of deformation had spread northward and the deposits of the Peri-Alpine Depression were folded. The Jura Mountains and the Carpathians were formed at this time. Vertical uplift without major faulting or folding continued throughout the remainder of the Tertiary.

The cause of the Tertiary deformation within the northern Tethyan Geosyncline may have been compression between the Eurasia Plate and the Africa and India plates.

North Atlantic Margin

Lavas were erupted during the early Tertiary in Ireland, Scotland, the Inner Hebrides, the Faeroe Islands, Spitsbergen, Greenland, and Baffin Island (Fig. 13.14). Columnar jointed lavas of this age

(a)

(b)

Figure 13.13 Folded and faulted sedimentary rocks in the Alps: (a) Near Brienz, Switzerland; (b) the Matterhorn, Switzerland, where sequences have been repeated several times by folding. (Courtesy of CIBA-Geigy.)

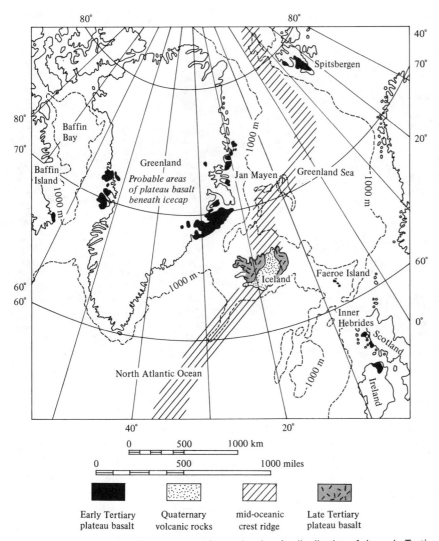

Figure 13.14 Map of the North Atlantic Ocean showing the distribution of the early Tertiary plateau basalts. (After King, Ref. 8.)

form the Giant's Causeway in Ireland (Fig. 13.15). In Scotland, the ages of lavas have been paleontologically determined as Paleocene and Eocene. Radiometric dating indicates that most of the lavas from the North Atlantic borders are approximately the same age (50-60 million years) (9).

The Tertiary volcanism in the North Atlantic region may have accompanied the initial separation of Europe and Greenland. The width of the magnetic anomalies south of Ireland indicates that the half-rate of sea-floor spreading in this area has been approximately 1 cm/year for the past 4 million years. Since the width of the Atlantic Ocean basin at that latitude is approximately 685 miles, (1100 km), the separation of Greenland and Europe began 50 to 60 million years ago, if the rate of sea-floor spreading has been constant.

Continental Interior of Europe

At the beginning of the Tertiary, central Europe

Figure 13.15   Columnar basalts of the Giant's Causeway, County Antrim, North Ireland. (Crown Copyright Geological Survey photo. Reproduced by permission of the Controller, Her Brittanic Majesty's Stationery Office.)

was entirely above sea level. The seas transgressed onto the continental interior from the west during the Early Paleocene, and by the Late Oligocene they extended as far east as southern Russia. A gradual regression began during the Miocene, and by the end of the Pliocene the seas had withdrawn completely from the continental interior.

Tertiary deformation in the continental interior includes faulting in central Europe and folding in the Pyrenees. Fault basins (grabens) developed during the Oligocene and received thick sequences of marine and continental sediments. In southern France, volcanoes erupted along faults bordering the basins.

The Pyrenees Mountains separating France and Spain were folded at the close of Eocene time. This deformation was probably caused by a movement of the Iberian peninsula toward Europe by about 30 miles (50 km) (10).

Indonesia Geosyncline

Thick deltaic, estuarine, and shallow marine sequences were deposited in the Indonesia Geosyncline in Paleocene and Eocene time. In the Late Miocene, intense folding affected the Lesser Sunda Islands, southern Java, and southern Sumatra. This deformation may have been caused by compression between the India Plate and the Eurasia Plate. These plates had been moving toward each other since the beginning of separation of Australia and Antarctica. Because marine Pliocene sediments are restricted to the outer rims of the present islands, the seas probably had regressed from the islands by Pliocene time.

Pacific Ocean Geosyncline

The Tertiary history of the Pacific Ocean Geosyncline includes several deformational episodes, repeated transgression and regression of the sea, and deposition of thick marine and continental strata. Tertiary volcanic rocks are abundant in the southern part of the geosyncline and coals are common in nonmarine sequences. Deformation during the Tertiary may have been more or less continuous, but episodes of especially intense folding occurred at the end of the Eocene, in the Late Oligocene, between the Middle and Late Miocene, and between the Late Pliocene and Early Pleistocene. The continuing volcanism and tectonism suggest that throughout the Tertiary the western Pacific was bordered by active island arcs adjacent to zones of subduction.

Continental Interior of Asia

The northern and central regions of the continental interior of Asia were generally above sea level during the Tertiary, but southern Asia was covered by a shallow sea during the Paleocene and Eocene. Coal-bearing molasse of Miocene and Pliocene age forms a clastic wedge that was derived from highlands to the south.

AUSTRALIA

Marine Tertiary deposits are restricted to basins near the outer margins of Australia. These deposits are dominantly limestone, but minor clastics are also present. The sea transgressed into the marginal basins during the Paleocene and reached its maximum extent during the Late Eocene and Early Miocene. The seas regressed slowly from the continent throughout the remainder of the Tertiary.

Coal-bearing terrestrial clastics of Tertiary age occur in basins near the center of the continent.

Tertiary volcanic rocks crop out in a broad, discontinuous belt along the eastern and southern border of Australia. The volcanics in the south range in age from Paleocene to Eocene and may have been erupted when Australia and Antarctica

Figure 13.16 Reverse fault offsetting Carboniferous sedimentary rocks, Lebung Pass, central Himalayas. (From Geological Survey of India, Memoir 23.)

began to drift apart. Based on the age of magnetic anomalies, we infer that the separation of these two continents began approximately 50 to 60 million years ago (Paleocene to Eocene).

NEW ZEALAND

New Zealand was subjected to several strong orogenic episodes during the Tertiary. These deformations culminated during the Late Pliocene and Early Pleistocene. At the beginning of the Tertiary, New Zealand was emergent, but its surface had been reduced to low relief. Transgression of the seas began during the Eocene and reached a maximum in Middle or Late Oligocene time. The seas began to recede before the end of the Oligocene, and no subsequent incursions occurred for the duration of the Tertiary.

Tertiary sedimentary rocks are dominantly marine clastics that are interbedded with limestone and coal. The clastic deposits were eroded from geanticlinal ridges that bordered New Zealand. Felsic to mafic volcanics are abundant in Tertiary sequence, and therefore it is likely that New Zealand was an active island arc at that time.

INDIA

Southern Tethyan Geosyncline

Early Tertiary deposits of the Southern Tethyan Geosyncline are dominantly limestones and marine clastics. The first major phase of deformation in the geosyncline occurred during the Oligocene and included intense folding and intrusion of granitic rocks. The Murree Series of Miocene age contains more than 8000 ft (2500 m) of nonmarine clastics eroded from highlands to the north.

A second major deformational episode occurred during the Middle Miocene, when the deposits of the Southern Tethyan Geosyncline were again folded and faulted and great thrust sheets or nappes moved southward from the core of the Himalayan Orogenic Belt (Fig. 13.16). The Siwalik "System," a sequence of molasse-type sediments,

was deposited during this orogenic episode. These sediments form a clastic wedge 15,000 ft (5000 m) thick, ranging from Middle Miocene to Early Pleistocene in age. As in the Northern Tethyan Geosyncline, the Tertiary deformations in the Southern Tethyan Geosyncline were probably the result of a collision between the India and Eurasia plates.

Continental Interior

Volcanism was widespread in the western continental interior of India during the Early Tertiary. Basalts of the Deccan Traps overlie Paleocene beds and are themselves overlain by Eocene strata (11). Potassium-argon dates range from 42 to 65 million years (Paleocene to Late Eocene) (12). Numerous large dikes and massive intrusions occur along the western border of India, particularly in the area north of Bombay. Since the traps here reach a thickness of about 10,000 ft (3000 m), this region was probably the center from which the basalts erupted. The Deccan Traps may have been associated with the initial separation of India and Africa. The oldest deposits recovered by the *Glomar Challenger* in the Arabian Sea west of India are of Paleocene age (13), which is in agreement with an early Tertiary age of separation.

Marine Tertiary sediments are restricted to the outer margins of the continental interior of India. Continental deposits eroded from the Himalayas occur in a trough just south of the mountains.

MIDDLE EAST

During the Tertiary, erosion of geanticlines on the northern border of the Southern Tethyan Geosyncline supplied thick sequences of clastic sediments to the geosyncline. These clastics grade southward into carbonates. The continental clastics of Miocene age are very thick and closely resemble the alpine molasse (Fig. 13.17). The principal folding in the central regions of the geosyncline also probably occurred in the Miocene (Fig. 13.18).

Figure 13.17   Miocene clastics of the Hofuf Formation in northwestern Saudi Arabia. (Courtesy of Aramco.)

AFRICA

Southern Tethyan Geosyncline

Africa and Europe were connected at Gibraltar during the early Tertiary, but the Tethys Sea separated the central and eastern portion of northern Africa from Europe. At this time, Italy may have been part of an island arc bordering Africa. If so, extensive deformation in Italy during the Tertiary would probably have been related to a collision of Italy with southern Europe.

In western Italy, Eocene and older rocks, which are recumbently folded, are unconformably overlain by Oligocene beds; therefore, folding probably occurred near the end of the Eocene. The folding resulted in the formation of a series of geanticlinal ridges, which furnished thick sequences of clastics during the Oligocene in western Italy. In Miocene time, new ridges were uplifted to the west of the older ridges, and large blocks slid from these ridges into soft muds to the east. This resulted in a chaotic mass of exotic blocks in a shaly matrix, known as the *argille scagliose*. This unit, some of which is similar in many respects to the wildflysch of the Alps, contains blocks measuring thousands of feet across (Fig. 13.19).

Limestone and minor clastics were deposited during the Paleocene and Early Eocene in northwestern Africa. However, the uplift of a geanticlinal ridge in the geosyncline resulted in the deposition of a thick sequence of clastic sediments during the Late Eocene. Deformation also occurred at the end of the Oligocene and at the end of the Miocene, a time when volcanism was widespread and granitic batholiths were intruded. The extensive deformation, volcanism, and plutonism indicates that this area was an active island arc during the Tertiary.

Figure 13.18   The sedimentary rocks of the Zagros Mountains in southwestern Iran were folded during the Tertiary. (Aerofilms Limited, courtesy of the British Petroleum Co.)

Figure 13.19   Exotic limestone mass in *argille scagliose* near Passo Della Cisa, Italy. Notice the recumbent folds, which may have been produced by drag during sliding. (Photo courtesy of B. M. Page.)

Figure 13.20　The Sphinx and the Great Pyramid, near Cairo, Egypt. These structures are built largely of nummulite-bearing Eocene limestone. (Photo by Eleanor Catena.)

Continental Interior

Marine Tertiary deposits are widespread in northern and central Africa. In Egypt and Libya, the Tertiary sequence is more than 10,000 ft (3000 m) thick. Marine deposits are dominantly limestone with minor quartz sandstone and shale (Fig. 13.20). These deposits are essentially undeformed, in contrast to the intensely folded and faulted Tertiary strata to the north, to the east, and to the west (Fig. 13.21). Apparently this area was protected from compressional forces that developed during the joining of the Africa and Eurasia plates.

A series of rift zones cutting through the east-

Figure 13.21　Flat-lying limestones, shales, and sandstones of Paleocene to Eocene age behind the Temple of Hatsheput, Luxor, Egypt. (Photograph By Eleanor Catena.)

ern half of the continent are among most conspicuous geologic features in Africa. The rift valleys often contain elongated lakes, such as Lake Nyasa and Lake Tanganyika, and are bordered by volcanoes, most notably Mount Kilimanjaro and Mount Kenya. The rift valleys are the result of normal faulting that occurred in late Tertiary and Quaternary time. They intersect the mid-oceanic ridge system at the southern end of the Red Sea, but the blocks on either side of the rift valleys do not appear to be moving apart nearly as fast as the sea-floor moves away from mid-oceanic ridges.

The Red Sea and the Gulf of Aden are believed to have formed during the separation of Saudi Arabia from Africa (Fig. 13.22). Normal faults, which commonly border these bodies of water, were probably developed during the initial phases of separation. The width of the magnetic anomalies in the Red Sea indicates that the separation of Saudi Arabia and Africa began between 5 and 15 million years ago. The oldest sediments recovered by the *Glomar Challenger* from the Red Sea are of Late Miocene age, 7 to 12 million years old (14).

SOUTH AMERICA

Andean Region

Continued uplift in the Andean Geosyncline during the Tertiary caused the seas to become even

Figure 13.22   Satellite photograph of the Gulf of Aden and the southern end of the Red Sea. (Courtesy of Monem Abdel-Gawad, NASA.)

(a)

(b)

Figure 13.23 Sedimentary rocks in the Andes
Mountains, folded during the early Tertiary: (a)
Chevron fold in beds of the Nantoco Formation
of Cretaceous age south of Copiapo, Chile. (Photo
by Robert Dingman, U.S. Geological Survey.) (b)
Syncline in Middle Cretaceous limestone near Lima
in western Peru. (Photo by C. R. Peterson,
La Oroya, Peru.)

more restricted than they had been at the end of
the Cretaceous. Marine deposition in Paleocene
time was limited to a rather narrow coastal strip
along the westernmost margin of South America.
In Eocene time, however, the seas transgressed
through a gap in the Andean highlands and formed
a shallow epicontinental sea east of the highlands.

Folding and volcanism were widespread at the
end of the Middle Eocene and at the end of the
Early Oligocene. Radiometric dating indicates that

granitic rocks were emplaced between about 30
and 55 million years ago (15).

Deformation was almost continuous during
the Miocene, and an especially intense deformation occurred at the beginning of the Late Miocene, when the geosynclinal deposits were folded
and faulted (Fig. 13.23). Many geologists believe
that this activity represents the principal phase of
the Andean Orogeny. During the Pliocene, the
highlands of the Andean region were eroded to
low relief. Renewed folding, faulting, and uplift
during the Late Pliocene were responsible for the
present elevation of the Andes. The Tertiary
deformations in the Andean region were probably
caused by the continued movement of an oceanic
plate under South America.

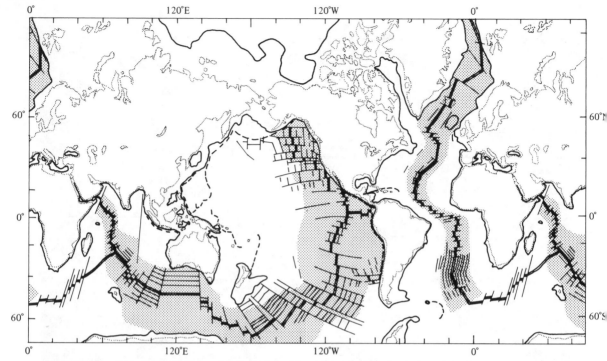

Figure 13.24   Map of the world showing the new sea floor added since the beginning of the Tertiary.
(Reproduced from F. J. Vine, Ref. 16.)

Continental Interior

Marine transgressions in South America during the Tertiary were confined to the margins of the continent. Clastics in this region were derived both from the craton and from the eastern slope of the Andes.

ANTARCTICA

The Tertiary history of Antarctica is very similar to that of South America. Sedimentary and volcanic rocks of geosynclinal character were deposited in a relatively narrow belt extending along the Pacific border from South Victoria Land to the Palmer Peninsula. As in South America, orogenic and volcanic activities during the Tertiary were probably associated with the eastward movement of an oceanic plate under the continent.

The Scotia Island Arc, located between Antarctica and South America, is similar in configuration to the Lesser Antilles Island Arc in the Caribbean region. Both are convex toward the east and both are underlain by a westward-dipping Benioff Zone. It is likely that the Tertiary volcanic rocks of the Scotia Arc resulted from a westward movement of an oceanic plate beneath the arc.

OCEAN BASINS

The Atlantic and Indian oceans increased in width as the continents bordering them continued to separate. A very large area of new sea-floor was created at the crests of the Mid-Atlantic and Indian Ocean ridges during the Tertiary (Fig. 13.24). New sea-floor was also created in the Pacific Ocean basin, but this basin decreased in area as the continents separated.

The oldest known sediments in the North Atlantic between Greenland and Norway are of Paleocene age (17). These deposits were taken about two-thirds of the distance from the crest of the Mid-Atlantic Ridge and the continental margin and, therefore, approximately two-thirds of the opening of this part of the North Atlantic occurred during Tertiary time.

Deep drilling and piston coring have recovered Tertiary sediments from all of the oceans of the world. The sediments include siliceous and calcareous oozes, "red" clays, and turbidites. The siliceous oozes consist of the remains of radiolarians, diatoms, silicoflagellates, and sponge spicules (Fig. 13.25). Calcareous oozes contain skeletons of foraminiferans, coccoliths, and discoasters. The so-called red clays, found in the middle latitudes in areas of low organic productivity, are generally chocolate brown and only rarely red. They consist of wind-blown dust, volcanic ash, organic debris, meteoric dust, and the products of submarine weathering. Turbidites of

Tertiary age are common in the deep ocean adjacent to the continents, but some are found thousands of miles from land (18).

The age of the oldest sediments in any region generally increases with increasing distance from mid-oceanic ridges (Fig. 13.26). In the equatorial Pacific, no Paleocene sediments have been found within 4000 miles (6500 km) of the crest of the East Pacific Rise, and no Eocene sediments have been found within 1800 miles (3000 km) of the crest of the rise. No Miocene or Pliocene sediments have been found within 200 miles (350 km) and 60 miles (100 km), respectively, of the crest of the rise. A similar pattern has been observed in the Atlantic and Indian oceans.

Tertiary volcanics, which are widespread in ocean basins, are mainly basalts. The age of the volcanics in the ocean basins generally increases with distance from mid-oceanic ridges. However,

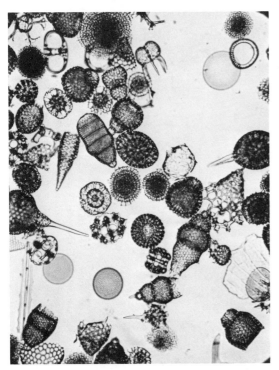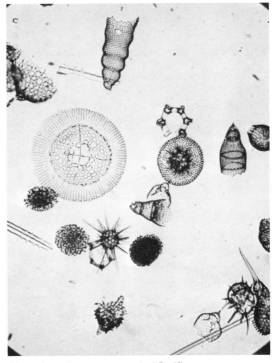

Figure 13.25   Lower and Middle Miocene radiolarian assemblages from the western tropical Pacific recovered by the drilling ship *Glomar Challenger.* (Photos courtesy of Annika Sanfilippo.)

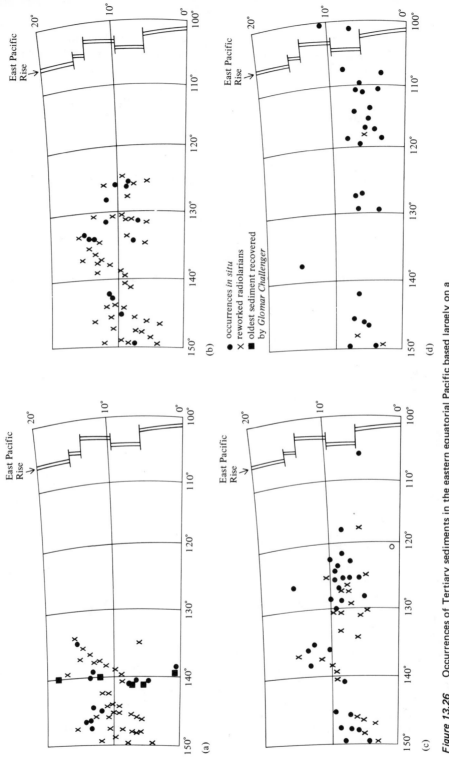

Figure 13.26 Occurrences of Tertiary sediments in the eastern equatorial Pacific based largely on a study of their radiolarians: (a) Eocene; (b) Oligocene; (c) Miocene; (d) Pliocene. (Modified after Riedel, Ref. 19. Copyright © 1967 by the American Association for the Advancement of Science, with additional data from Burckle et al., Ref. 20. Copyright © 1967 by the American Association for the Advancement of Science.)

● occurrences *in situ*
× reworked radiolarians
■ oldest sediment recovered by *Glomar Challenger*

in some areas, young volcanics occur far from these ridges. The Hawaiian Islands, for example, are composed of Late Tertiary and Quaternary volcanics, but they are thousands of miles from the East Pacific Rise. Why they are located so far from a ridge crest is something of a mystery.

Magnetic Anomalies and Sea-Floor Spreading

The trends of magnetic anomalies and transform faults may be used to determine the direction of sea-floor spreading. Magnetic anomalies are generally oriented perpendicular to the direction of sea-floor spreading at the time of their formation.

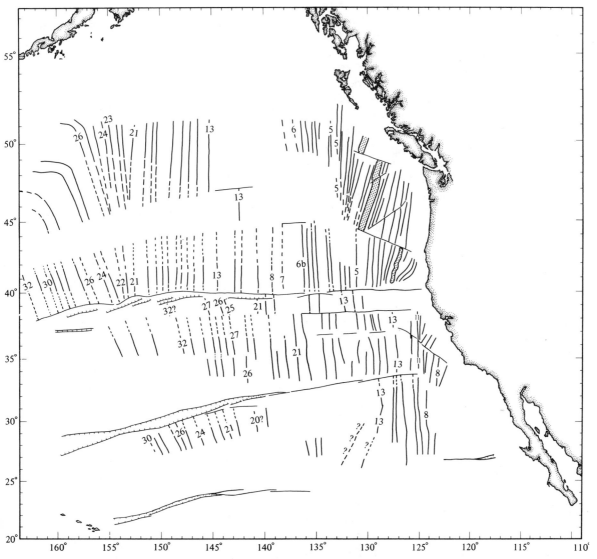

Figure 13.27   Fracture zones and magnetic anomalies in the northeastern Pacific Ocean Basin. The numbers of the anomalies are from Heirtzler et al. (22). Anomalies 22 to 25 are thought to have formed 10 to 15 million years ago. Changes in the trend of magnetic anomalies indicate changes in the direction of sea-floor spreading. (From Menard and Atwater, Ref. 21.)

Transform faults are almost always parallel to the direction of sea-floor spreading, especially when two transform faults are parallel to each other.

Pacific Ocean. Magnetic anomalies near the crest of the Juan de Fuca and Gorda ridges (west of Washington and Oregon) trend approximately N30°E and transform faults offsetting the ridges trend about N60°W (Fig. 13.27). These trends indicate that during the late Tertiary and Quaternary, the Pacific Plate was moving approximately N60°W relative to the small oceanic plate to the east of the Juan de Fuca and Gorda ridges. The trends of the magnetic anomalies and transform faults near the Gulf of California are similar to those of the Juan de Fuca and Gorda ridges. Thus the Pacific Plate is moving approximately N60°W relative to the Cocos Plate east of the East Pacific Rise. In the South Pacific, transform faults near the crest of the East Pacific Rise also trend about N60°W. Evidently the Pacific Plate has been moving N60°W relative to the Southeast Pacific Plate to the east.

In the northern Pacific, anomalies inferred to be between 10 and 50 million years old trend approximately north-south and are offset by inactive transform faults trending approximately east-west (Fig. 13.27). Therefore, it is likely that the direction of the sea-floor spreading during the Pliocene differed from that during the Eocene, Oligocene, and Miocene. Since anomalies inferred to have formed between 50 and 75 million years ago trend in a northwesterly direction, the direction of the sea-floor spreading may have changed again during the early Tertiary (Fig. 13.27).

Atlantic Ocean. In the northern North Atlantic, magnetic anomalies formed within the last 10 million years trend northeasterly, and transform faults offsetting them trend in a northwesterly direction. North America appears to have been rotating relative to Europe about a pole of spreading in northeastern Siberia. Transform faults in the equatorial Atlantic indicate a westerly movement of North America relative to Africa during the late Tertiary and Quaternary. However,

a change in the trend of transform faults in the vicinity of Anomaly 5 suggests a change in the direction of sea-floor spreading approximately 10 million years ago (23).

Indian Ocean. In the North Indian Ocean and the Red Sea, transform faults trend northeasterly and magnetic anomalies trent northwesterly. The India Plate appears to have been rotating relative to the Africa Plate about a pole of spreading located in Libya (24). In the South Indian Ocean, transform faults trend north-south and magnetic anomalies trend east-west. At the present time, Australia is moving northward relative to Antarctica.

TERTIARY LIFE

The extinction of the dinosaurs at the end of the Cretaceous left the emerging Tertiary plains relatively unoccupied. Paleocene mammals were small and not very abundant, but mammals increased greatly in size and number during the Eocene and Oligocene. With the disappearance of most of the marine reptiles, the bony fish became quite numerous. In the absence of flying reptiles, the bird population grew tremendously. The changes in the marine invertebrate communities were equally important. Lyell's subdivision of the Tertiary based on the percentages of modern species provides an indication of the gradual modernization of these faunas (Table 13.1).

Protozoans

A major expansion of the Foraminifera, particularly the larger varieties, occurred during the Tertiary. The large, disk-shaped nummulites and orbitoids are extremely abundant in many Eocene and Oligocene limestones of the Tethyan, Caribbean, and Indo-Pacific regions. The small planktonic Foraminifera are important in correlating Tertiary sequences because of their widespread distribution and rapid evolution. Their ecological zonation has been useful in determining changes in

Table 13.1 Subdivision of Cenozoic based on percentage of modern species

Epoch		Modern species (%)
Pleistocene	} Neogene	90-100
Pliocene		50-90
Miocene		20-40
Oligocene	} Paleogene	10-15
Eocene		1-5
Paleocene		0

water depth within depositional basins. For example, Foraminifera have indicated that the San Joaquin Valley of California went from a marine basin more than 6000 ft (2000 m) deep during the Oligocene to a shallow neritic basin in Pliocene time.

Invertebrates

Invertebrate faunas of the Tertiary were dominated by the pelecypods and the gastropods. The pelecypods *Venericardia* and *Ostrea* are important worldwide index fossils in Eocene and younger sequences; *Pecten* and *Venus* are important in Miocene and younger deposits. Gastropods, such as *Turritella*, were abundant in Tertiary shelf faunas. Other important invertebrates include corals, echinoids, crinoids, starfish, and bryozoans.

Fish, Amphibians, and Reptiles

During the Tertiary, the teleosts (bony fish) achieved a position of dominance among the marine and freshwater vertebrates. Numerous well-preserved fossil fish have been recovered from shales deposited in intermontane lakes in the Rocky Mountains (Fig. 13.28). Sharks were prominent in Tertiary oceans, and shark teeth are common fossils, especially in deposits of Miocene age. Some of these teeth are more than 6 in. (15 cm) long (Fig. 13.29). The shark that possessed these teeth may have been longer than 70 ft (23 m).

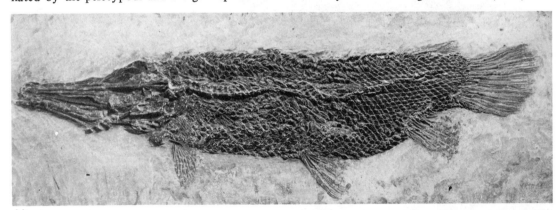

(a)

(b)

Figure 13.28 Fossil fish from the Eocene Green River Shale of southwestern Wyoming. [(a) Courtesy of the Buffalo Museum of Science; (b) Courtesy of the Field Museum of Natural History, Chicago.]

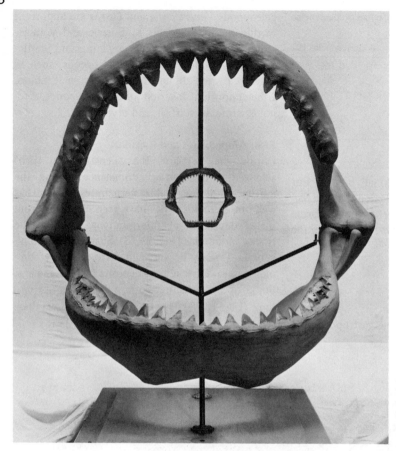

Figure 13.29 Reconstruction of the jaws of a very large Miocene shark. The jaws of a large modern shark appear small in comparison. (Courtesy of the Field Museum of Natural History, Chicago.)

The scarcity of fossil amphibians and reptiles suggests that these creatures were greatly reduced in numbers during the Tertiary. Tertiary amphibians include frogs, salamanders, and toads. Turtles, crocodiles, lizards, and snakes were the only reptiles to survive the extinctions at the end of the Mesozoic.

Birds

One of the interesting evolutionary trends among the birds during the Tertiary was the development of giant flightless varieties. In the absence of competition from large carnivores, many birds lost the power of flight. Their great size enabled them to prey on numerous Tertiary mammals. *Diatryma*, which lived in Wyoming during the Eocene, occupied an ecological niche left by the carnivorous dinosaurs. It was a stocky bird, nearly 7 feet tall (2.1 m) and it had a massive head (Fig. 13.30). These giant birds became extinct when large carnivorous placental mammals appeared. Other flightless giants evolved in South America, Australia, New Zealand, and Madagascar before the arrival of large carnivores. However, only the rheas of South America, the ostriches of Africa, and the emus and cassawaries on the islands near Australia are alive today. Perhaps these were able to survive because they are fast runners.

The majority of fossil birds of the Tertiary are quite modern in aspect, and most orders of modern birds had appeared by the Eocene. However, at least one variety of toothed birds lived during the Cenozoic. The partial remains of a giant toothed bird were recently found in beds of Pliocene age in California. The wing span of this bird has been estimated at 16 ft (4.9 m).

Mammals

Tertiary mammals are dominantly marsupials and placentals; both these orders stem from the pantotheres. Born alive, marsupials are tiny and extremely immature at birth, and they must be transferred immediately to the mother's pouch for continued nourishment and development. The young of placental mammals are born in a more advanced state. Their delayed birth is made possible by the development of a highly efficient nutrient connection, the placenta, between the mother and the fetus. As a result of this delayed birth, the young have a much better chance of survival. Many placental mammals are relatively independent only a few hours after birth.

In Eurasia, Africa, and North America, placental mammals replaced the marsupials during the early Tertiary. However, South America and Australia became isolated from the other continents before the evolution of the placental mammals. Consequently, marsupials thrived on these continents during the Tertiary.

Paleocene mammals were of small to medium size and had five toes. The increase in the size of many mammals in Eocene time is reminiscent of the trend toward giantism in the dinosaurs. This is a common trend, since metabolic efficiency generally increases with the size of an animal. Another important evolutionary development among the mammals was the increase in brain size. The brain of the lower mammals is rather small and smooth, whereas the more advanced mammals have a larger brain with a highly involuted surface.

Archaic Hoofed Mammals. Primitive mammals ranged from very small carnivores to large browsing forms. The condylarths and amblypods

Figure 13.30   The large flightless bird *Diatryma* lived during the Eocene in Wyoming. It grew up to 7 ft tall and had a skull as large as the skull of a small horse. (Courtesy of the Field Museum of Natural History, Chicago.)

Figure 13.31   Reconstruction of an Early Tertiary landscape, Wyoming, showing two condylarths (*Phenacodus*). (Courtesy of the American Museum of Natural History.)

Figure 13.32   Reconstruction of an Eocene landscape in Wyoming. Two uintatheres watch a group of tiny, primitive horses (*Hyracotherium*). (Courtesy of the Field Museum of Natural History, Chicago.)

were the most important of the Paleocene herbivores, but both became extinct during the Eocene. The small and relatively agile condylarths were characterized by five toes and a long tail (Fig. 13.31). The amblypods were more stocky and had sharp canine teeth. Uintatheres were strange-looking amblypods with three pairs of horns (Fig. 13.32).

Odd-toed Hoofed Mammals. The odd-toed hoofed mammals have a large central toe which bears most or all of the weight. This group includes the horse, the rhinoceros, and the extinct titanotheres and chalicotheres. The evolutionary history of the horse is documented by an extensive fossil record (Fig. 13.33). The remains of the oldest known horse *Hyracotherium* (formerly *Eohippus*) have

been found in late Paleocene and Eocene strata (Fig. 13.33). These horses were the size of a small dog and had four toes on each forefoot and three toes on each hind foot. Their teeth exhibit the characteristics of browsing animals. During the middle Tertiary, horses increased in size, and their toes decreased in number. In the Miocene, most horses changed their feeding habits from browsing to grazing as meadows and plains became more extensive. One-toed horses similar to the modern horse appeared in the Pliocene.

Titanotheres were large mammals similar to the rhinoceros (Fig. 13.34). They were typified by the brontotheres, which were as much as 8 ft (2.4 m) high at the shoulder. Titanotheres had four forefoot digits and three hind-foot digits, stout limbs, and bony-knob horns. They were prevalent

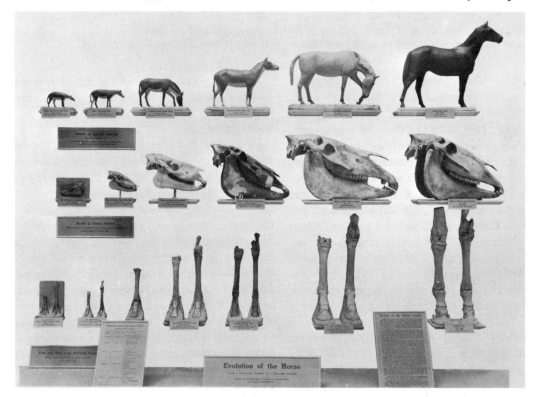

Figure 13.33 Stages in the evolution of the horse from the Eocene to the present: Top row, changes in the relative size of the body of the horse; middle row, changes in the size of the skull; bottom row, changes in the size and configuration of the fore and hind legs of the horse. (Courtesy of the Field Museum of Natural History, Chicago.)

Figure 13.34   Reconstruction of a group of titanotheres (***Brontops robustus***) from Early Oligocene deposits of North America. (Courtesy of the Field Museum of Natural History, Chicago.)

in western North America during the early Tertiary but became extinct in the Oligocene.

Chalicotheres had clawlike feet and a horselike body (Fig. 13.35). These strange-looking mammals probably used their claws for digging up roots and tubers. Chalicotheres appeared during the Eocene and became extinct in the Pleistocene.

Several groups of rhinoceroses appeared during the Eocene. The early forms were small and hornless. Some were adapted for running; others were semiaquatic. Very large rhinoceroses developed in Late Oligocene time. One of these giants, *Baluchitherium,* was 18 ft (3.5 m) high at the shoulder and was probably the largest land mammal that ever lived.

Even-toed Hoofed Mammals. The oreodonts thrived between Late Eocene and the Early Pleistocene and were common during the Oligocene in western North America (Fig. 13.35). They are related to the modern camel. The first true camels appeared in the Eocene in North America as small four-toes forms. As in the evolution of

the horse, the camels increased in size as their toes decreased in number. During the Pliocene, camels migrated from North America to Asia, Africa, and South America.

Pigs and peccaries appeared in the Eocene, and giant pigs evolved during the Oligocene and Miocene (Fig. 13.35). The skull of one specimen is nearly 3 ft (1 m) long.

Elephants, Deer, and Cattle. The proboscideans first appeared during the Eocene. Typical of the early elephants was *Paleomastodon,* a small, short-tusked form which lived in northern Africa and southern Asia. In Miocene time, elephants migrated to Europe and North America. Miocene elephants included semiaquatic shovel-tusked forms (Fig. 13.36). Elephants with large upper tusks appeared during the Pliocene. Deer and cattle first appeared in Eurasia in Oligocene time. These forms did not migrate to North America until the Pleistocene.

Primates. With the exception of man, apes, and

Figure 13.35 Restoration of a Miocene landscape in western Nebraska with wild hogs (in the foreground) and a chalicothere browsing on the lower branches of a tree. At the far left is a herd of oreodonts. (From a painting by Charles R. Knight, courtesy of the Field Museum of Natural History, Chicago.)

Figure 13.36 Restoration of the shovel-tusked elephant *Platybelodon grangeri,* which lived during the Late Tertiary in North America and Asia. (From a painting by J. C. Hansen, courtesy of the Field Museum of Natural History, Chicago.)

baboons, all primates are tree dwellers, as most probably were their ancestors. The anatomy of the primates represents an adaptation to this environment. Tree dwellers must be agile and able to grasp branches as they move from tree to tree. Stereoscopic vision, which is a benefit of the forward eye orientation of the primates, provides depth perception.

The main primate groups of the early Tertiary were the lemurs, lorises, and tarsioids (Fig. 13.37). Primates have been found in strata as old as late Paleozoic (26), but the anthropoids, which include monkeys, apes, and man, did not appear until the Oligocene. The remains of *Parapithecus* (a monkeylike form) and *Aegyptopithecus* (an apelike form) have been found in the Fayum Depression of Egypt (Fig. 13.38). *Propliopithecus,* whose remains have been retrieved from Oligocene deposits of Africa, may be an ancestor of man.

The similarity of its dentition to that of modern man provides the basis for this suggestion.

Fossil remains of *Dryopithecus* have been found in Early Miocene deposits in Europe, Asia, and Africa (Fig. 13.39). The canine teeth of this primate are larger than those of the hominids. It has been suggested that the modern chimpanzees and gorillas evolved from *Dryopithecus* or a closely related form.

The oldest fossil hominids have been found in Early Miocene deposits in Kenya, Africa (26). These have been assigned to the genus *Kenyapithecus.* Remains of *Ramapithecus,* a similar hominid, have been found in India, Africa, and Asia in strata of Late Miocene and Early Pliocene age. From fragments of teeth and jaws, it was determined that *Ramapithecus* was probably an erect biped about 3 ft (1 m) tall. The small size of the incisors and canines implies a dependence on

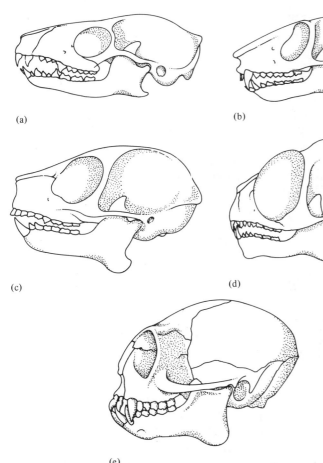

(a)

(b)

(c)

(d)

(e)

Figure 13.37 Skulls of some of the lower primates, somewhat enlarged from the natural size: (a) Tree-shrew, (b) mouse lemur, (c) an Eocene tarsioid, (d) the modern tarsier, (e) marmoset. The arrangement of the skulls illustrates some of the evolutionary changes that occurred in the development of the higher primates from the primitive primates. Two of the important developments were the shortening of the skull and the movement of the eyes to the front of the skull, which permitted stereoscopic vision. (From Clark, Ref. 25.)

the use of the hands for grasping and tearing vegetation. *Ramapithecus* may have used simple tools such as sticks and rocks, but there is no evidence of this in the fossil record. The similarity of the teeth of *Ramapithecus* to those of modern man indicates that this hominid could be a direct ancestor of man. The tool-making australopithecines of the Late Pliocene are discussed in Chapter 14.

Other Mammals. The creodonts were the dominant carnivores of the early Tertiary. These include catlike forms, such as *Oxyaena,* and doglike forms. Creodonts expanded greatly in size and numbers in Miocene time, but all had become ex-

tinct by the end of the Pliocene. The aquatic cetaceans, which appeared in the Middle Eocene, include carnivores such as porpoises and herbivores such as whales. Seals, sea lions, and walruses did not appear until Miocene time.

The rodents, including squirrels, rats, mice, and porcupines, emerged first in the Paleocene, and the lagomorphs (rabbits) appeared during the Eocene. The edentates, which include anteaters, sloths, glyptodonts, and armadillos, arose in Paleocene time and were an important part of the Late Pliocene fauna of South America.

Mammalian Faunas and Continental Drift
According to Björn Kürten, the evolution of the

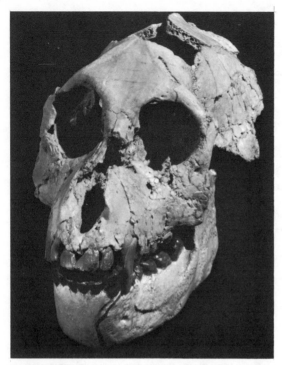

Figure 13.38 Skull of *Aegyptopithecus*, an Oligocene primate, probably a close relative to the ancestors of man. (Photo courtesy of E. L. Simons.)

mammals was significantly influenced by the displacements between continents (27). During the early stages of mammalian evolution, the continents were close enough to permit migration between the continents, and primitive mammals had gained worldwide distribution by the Early Cretaceous. In the Late Cretaceous, however, the continents had separated sufficiently to prevent many faunal migrations. Each continent produced several different orders of mammals, and during the early Tertiary there were seven isolated or semi-isolated faunal provinces.

After the separation of Australia and Antarctica in the early Tertiary, Australia was completely isolated from the other continents. This isolation resulted in the adaptive radiation of such primitive mammals as the monotremes (spiny anteater and platypus) and the marsupials (kangaroos, wombats, and bandicoots).

The presence of certain fossil edentates and notoungulates in early Tertiary deposits of both North and South America indicates that these continents were probably connected by a land bridge at that time. However, during most of the Tertiary, South America was isolated from the other continents. About six orders of mammals evolved in South America during the Tertiary. These include the archaic placental mammals and several orders of marsupials. Adaptive radiation resulted in the development of a wide variety of mammals which occupied a number of different environments. Many of the marsupials of South America and Australia resemble the more advanced placental mammals. Carnivorous marsupials of South America included foxes, wolves, and cats. *Thylacosmilus* had very large canine teeth which were remarkably similar to those of the sabertooth cat, *Smilodon*, a placental mammal which lived during the Pleistocene. The rodents migrated to South America during the middle Tertiary. Since monkeys originated in Africa and are not found in North America, it is presumed that they somehow were able to cross the Atlantic Ocean to South America. Rodents probably migrated from Eurasia by way of North and Central America across a narrow body of water to South America.

Africa was isolated from the other continents during the early Tertiary. The Tethys Sea was still an open ocean and a shallow sea separated Spain and Africa. Between four and six orders of mammals evolved in Africa during this episode of isolation. These mammals include the elephants, conies, and aardvarks. Approximately 16 orders originated in Eurasia and North America. Among them are the insectivores, bats, primates, cats, dogs, bears, hoofed mammals, rodents, and rabbits.

Several continents that were isolated during the early Tertiary were rejoined in the middle and late Tertiary. Africa was connected to Europe in the Oligocene as a result of its northward drift. Faunal migrations between Africa and Europe resulted in the dispersal of many forms, including elephants and mastodons. South America re-

Figure 13.39 Restoration of the Miocene primate *Dryopithecus* (*Proconsul*). (Courtesy of the British Museum of Natural History.)

mained isolated until the Late Pliocene, when a land bridge was established between North and South America. Ground sloths, porcupines, and armadillos migrated from South America northward, and horses, camels, deer, dogs, cats, and bears migrated southward. The archaic placental mammals and marsupials of South America, which could not compete with the more advanced placental mammals, became extinct. Of the mammals native to South America, only a few edentates survived. In the late Tertiary, the westward movement of North America resulted in the for-

mation of a land bridge between Alaska and Siberia. This route permitted an intermixing of North American and Eurasian faunas.

By the end of the Tertiary, the number of faunal provinces had been reduced from seven to four and the number of mammalian orders declined from 30 to 18. Only Australia retained its primitive fauna until the introduction of advanced mammals by European settlers.

Plants

Many families of modern plants have existed since

the early Tertiary. Paleocene and Eocene temperate forests included poplar, birch, cedar, and alder. By the middle Tertiary, angiosperms had become adapted to a wide variety of different environments ranging from the hot, humid tropics to the colder polar regions. A number of modern genera may be traced to the middle Tertiary. Temperate forests of the Oligocene included oak, beech, chestnut, and conifers in the lowland regions. In the Miocene, forests diminished somewhat in importance, and grasslands reached the height of their development. As the climate became cooler in the middle and late Tertiary, a variety of different plant assemblages developed. Temperate and cool-temperate forests expanded, and the ranges of subtropical plants became more restricted. Boreal forests were dominated by conifers such as spruce, pine, and fir, and shrub-herb plants.

CLIMATE

The worldwide cooling that began during the Late Cretaceous was interrupted by a warming trend between the Paleocene and the Eocene. During this warm interval, tropical and subtropical climates were widespread. Palm trees grew as far north as Germany and the states of Washington and Alaska, and corals lived 10 to 20° north of their present ranges (28). A climatic cooling which began during the early Oligocene resulted in a northward shifting of climatic belts (Fig. 13.40). The fossil plant assemblage from a lignite in Vermont indicates that the climate in that area during the late Oligocene was still significantly warmer than that of today. The depositional environment was probably similar to that of the swamps in the southeastern United States (29).

Plant fossils from the Vienna Basin indicate that the climate of central Europe changed from subtropical-humid during the Early Miocene, to subtropical-dry in the Middle Miocene, to warm-dry in the Late Miocene. Studies of mollusks in the circum-Pacific region indicate that there was a general warming trend in this area from the Early to the Middle Miocene (30). However, the climate of the Pliocene was in general cooler and drier than the Miocene.

Glaciation

A study of cores taken in the deep ocean near

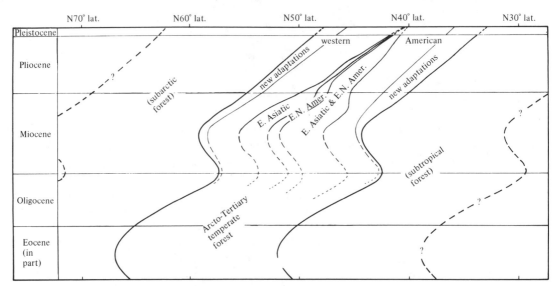

Figure 13.40   Generalized indication of the migration of forests during the Tertiary in western North America. (After Dorf, Ref. 28.)

Figure 13.41 Taylor Valley, Antarctica, looking eastward toward McMurdo Sound. Circles mark basalts overlying Late Pliocene glacial deposits. (Courtesy of the U.S. Department of the Interior.)

Antarctica indicates that an ice cap existed there as early as Eocene time (31). Rafted debris has been found in cores from the north Pacific and Bering Sea in deposits as old as Late Miocene (32). Glaciation of Late Miocene and Pliocene age has been reported from the St. Elias Mountains in southern Alaska, but this may have been an alpine glaciation (33). Glacial deposits in the Jones Mountains of Antarctica are overlain by lava flows which are probably between 7 and 10 million years old (Late Miocene) (34). The extent of these

deposits suggests that they were the result of continental glaciation.

Glacial deposits of Late Pliocene age are widespread. Basaltic scoria overlying glacial sediments in the Taylor Valley in Antarctica has been dated at approximately 2.6 million years. Since these deposits are above the level of the present valley glacier, continental glaciation must have begun more than 2.6 million years ago (35) (Fig. 13.41).

In the Sierra Nevada Mountains of California, till is underlain and overlain by volcanic rocks that have been dated at 3.0 and 2.7 million years, respectively (36). In Iceland, glaciation began at about 3.0 million years ago, and glacial deposits more than 2 million years old have been reported from the Soviet Union and South America (37).

SUMMARY

With the continued separation of the continents during the Tertiary, the Atlantic and Indian oceans widened while the Pacific Ocean decreased in size. Geosynclines bordered all three oceans and the Tethys Sea as well.

Extensive tectonic activity occurred in the geosynclines bordering the Pacific Ocean and the Tethys Sea as oceanic plates moved under the continental margins. In the Alpine and Himalayan regions, there was intense deformation during the Oligocene and Miocene as a result of compression between major continental plates.

The seas transgressed into the interiors of the continents during the Paleocene and Eocene but were less extensive than they had been in the Mesozoic. An irregular withdrawal of the seas began in the Miocene, and by the end of the Tertiary, the seas were confined to the outer margins of the continents.

Mammals expanded rapidly in size, number, and variety during the Tertiary, coming to inhabit most of the available environments by the end of Eocene time. The Miocene primates *Kenyapithecus* and *Ramapithecus* are probably ancestors of man, and as the Tertiary ended, *Australopithecus* was making primitive tools.

Following a warming trend during the Paleocene and Eocene, the climate began a general cooling trend that ended with glaciation in many mountainous and polar areas in the late Tertiary. The Tertiary glaciations were a prelude to the widespread glaciation that followed in the Pleistocene.

REFERENCES CITED

1. J. D. Phillips and D. Forsyth, 1972, Plate tectonics, paleomagnetism, and the opening of the Atlantic: *Geol. Soc. Amer. Bull.*, v. 83, p. 1579.

2. P. D. Snaveley, 1968, Tholeiitic and alkalic basalts of the Eocene Siletz River volcanics, Oregon Coast Range: *Amer. Jr. Sci.*, v. 266, p. 454.

3. P. B. King, 1959, *The Evolution of North America:* Princeton University Princeton, N.J.

4. W. G. Pierce, 1957, Hart Mountain and South Fork Detachment Thrusts of Wyoming: *Amer. Assoc. Petrol. Geol. Bull.*, v. 41, p. 591.

5. F. G. Fox, 1959, Structure and accumulation of hydrocarbons in Southern Foothills, Alberta, Canada: *Amer. Assoc. Petrol. Geol. Bull.*, v. 43, p. 992.

6. P. H. Mattson, 1966, Unconformity between Cretaceous and Eocene rocks in central Puerto Rico: *Trans. Third Caribbean Conference, Kingston, Jamaica, 1962,* p. 49.

7. A. H. Barabas, 1971, K-Ar dating of igneous events and porphyry copper mineralization in west central Puerto Rico: *Abstracts with Programs,* v. 3, no. 7, Geological Society of America, p. 498.

8. P. B. King, 1969, *The Tectonics of North America — A discussion to Accompany the Tectonic Map of North America, Scale 1 : 5,000,000:* U.S. Geological Survey Prof. Paper 628.

9. S. Moorbath and H. Welke, 1969, Lead isotope studies of igneous rocks from the Isle of Skye, Northwest Scotland: *Earth Planet. Sci. Lett.*, v. 5, p. 217.

10. X. Le Pichon, 1972, Discussion of paper by P. R. Voght, R. H. Higgs, and G. L. Johnson, 'Hypotheses on the Origin of the Mediterranean Basin: Magnetic data': *Jr. Geophys. Res.*, v. 77, p. 391.

11. D. N. Wadia, 1966, *Geology of India:* St. Martin's Press, New York.

12. P. Wellman and M. W. McElhinny, 1970, K-Ar Age of the Deccan Traps, India: *Nature,* v. 227, p. 595.

13. Scientific Staff for Leg 23, 1972, Deep Sea Drilling Project in the Arabian Sea: *Geotimes,* v. 17, no. 7, p. 22.

Scientific Staff for Leg 24, 1972, Deep Sea Drilling Project: Leg 24, *Geotimes,* v. 17, no. 9, p. 17.

14. Scientific Staff for Leg 23, 1972, Deep Sea Drilling Project in the Red Sea: *Geotimes,* v. 17, no. 7, p. 24.

15. S. Quirt et al., 1971, Potassium-argon ages of porphyry copper deposits in northern and central Chile: *Abstracts with Programs,* v. 3, no. 7, Geological Society of America, p. 676.

16. F. J. Vine, 1970, The geophysical year: *Nature,* v. 227, p. 1013.

17. T. Saito, L. H. Burkle, and D. R. Horn, 1967, Paleocene core from the Norwegian Basin: *Nature,* v. 216, p. 357.

18. M. Ewing, R. Houta, and J. Ewing, 1969, South Pacific sediment distribution: *Jr. Geophys. Res.,* v. 74, p. 2477.

19. W. R. Riedel, 1967, Radiolarian evidence consistent with spreading of the Pacific floor: *Science,* v. 157, p. 540, Aug. 4.

20. L. H. Burckle et al., 1967, Tertiary sediment from the East Pacific Rise: *Science,* v. 157, p. 537, Aug. 4.

21. H. W. Menard and T. Atwater, 1968, Changes in direction of sea-floor spreading: *Nature,* v. 219, p. 463.

22. J. R. Heirtzler et al., 1968, Marine magnetic anomalies, geomagnetic field reversals, and motions of the ocean floor and continents: *Jr. Geophys. Res.,* v. 73, p. 2119.

23. P. J. Fox, W. C. Pitman, III, and F. Shepard, 1969, Crustal plates in the central Atlantic: Evidence for at least two poles of rotation: *Science,* v. 165, p. 487.

24. W. J. Morgan, 1968, Rises, trenches, great faults, and crustal blocks: *Jr. Geophys. Res.,* v. 73, p. 1959.

25. W. E. Le Gros Clark, 1965, *History of the Primates:* University of Chicago Press, Chicago.

26. L. S. B. Leakey, 1967, An Early Miocene member of Hominidae: *Nature,* v. 213, p. 155.

27. B. Kürten, 1969, Continental drift and evolution: *Sci. Amer.,* v. 220, no. 14, p. 54.

28. E. Dorf, 1964, The use of fossil plants in palaeoclimatic interpretations, *in* A. E. M. Nairn, ed., *Problems in Palaeoclimatology:* Wiley-Interscience, New York, p. 13.

29. J. A. Wolfe and E. S. Barghoorn, 1960, Genetic change in Tertiary floras in relation to age: *Amer. Jr. Sci.,* v. 258-A, p. 388.

30. W. O. Addicott, 1966, Tertiary climatic change in the marginal northeastern Pacific Ocean: *Science,* v. 165, p. 583.

31. S. V. Margolis and J. P. Kennett, 1970, Antarctic glaciation during the Tertiary recorded in sub-Antarctic deep-sea cores: *Science,* v. 170, p. 1085.

32. Scientific staff for Leg 19, 1971, Deep Sea Drilling Project: Leg 19: *Geotimes,* v. 16, no. 11, p. 12.

33. G. H. Denton and R. L. Armstrong, 1969, Miocene-Pliocene glaciations in southern Alaska: *Amer. Jr. Sci.,* v. 267, p. 1121.

34. R. H. Rutherford et al., in press, Tertiary glaciation in the Jones Mountains: *SCAR Symposium on Antarctic Geology, 1970.*

35. R. L. Armstrong, W. Hamilton, and G. H. Denton, 1968, Glaciation in Taylor Valley, Antarctica, older than 2.7 million years: *Science,* v. 159, p. 187.

36. R. R. Curry, 1966, Glaciation about 3,000,000 years ago in the Sierra Nevada: *Science,* v. 154, p. 770.

37. J. H. Mercer, 1969, Glaciation in southern Argentina more than two million years ago: *Science,* v. 164, p. 823.

The Quaternary Period

*I*n the late 1800s geologists were puzzled by the "drift" that covers much of northern and central Europe. They observed that the random mounds and ridges were commonly comprised of unsorted sediments quite unlike ordinary waterlain deposits. Some cobbles and boulders in the drift differed considerably from the underlying bedrock but could be matched with distant outcrops. It was suggested that these deposits had been carried by icebergs or by the waters of the biblical flood, but these explanations failed to account for many of the surface features.

Those who believed that the flood or icebergs were not adequate depositional mechanisms sought an alternative explanation. James Hutton and Ventez-Sitten were among the first scientists to propose that glaciers were the source of the drift in the vicinity of the Alps. Ventez-Sitten, a Swiss civil engineer, demonstrated that the drift that was being deposited by modern glaciers was identical to drift found well beyond the present extent of those glaciers. In 1832 A. Bernhardi proposed that continental glaciers from the north had at one time extended into central Europe and were responsible for the lowland drift (1). The great naturalist Louis Agassiz later demonstrated that continental glaciers had at one time also covered much of North America. He further suggested that glaciation in both Europe and North America occurred during a "great ice period." By the middle of the nineteenth century, Agassiz had become a leading proponent of the glacial theory.

QUATERNARY CHRONOLOGY

The French geologist J. Desnoyers applied the term Quaternary to the soils and gravels overlying Tertiary deposits. Subsequently, the Quaternary was divided into two epochs, the Pleistocene, which includes the "ice ages," and the Holocene or Recent, which comprises much of postglacial time. An early attempt to date the beginning of the Pleistocene was based on the assumption that approximately 250,000 years were required for continental ice sheets to form, advance, and recede. The existence of four distinct glacial episodes provided an age of one million years for the beginning of the Pleistocene and the onset of the ice ages. Although the currently accepted figure for the beginning of the epoch is 1.8 million years, the original figure was a surprisingly accurate estimate.

RECONSTRUCTION OF QUATERNARY CLIMATES

The shell chemistry of fossil Foraminifera (forams) depends on the temperature of the ocean in which they lived and constructed their calcium carbonate tests. Harold Urey found that when seawater evaporates, the water vapor is slightly enriched in the lighter oxygen isotope ^{16}O and a slight concentration of the heavier isotope ^{18}O remains in the seawater (2). Since the rate of evaporation generally increases with temperature, the amount of ^{18}O enrichment depends on the temperature of the sea-

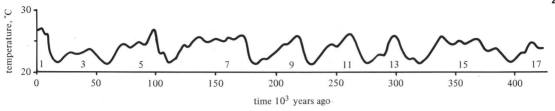

Figure 14.1 Paleotemperatures of the Caribbean as determined by oxygen isotope studies. (After Emiliani, Ref. 3. Copyright © 1966 by the University of Chicago.)

water. When compared with modern shell chemistry and seawater temperatures, the ratio of ^{18}O to ^{16}O of the tests of fossil planktonic forams provides the temperature of the surface water of the ocean in which they lived (Fig. 14.1).

The relative abundance of certain forams is also a function of the temperature of the seawater in which they lived. *Globigerina pachyderma*, for example, is restricted to cold climates, whereas *Globorotalia menardii* is found principally in warm climates (Fig. 14.2a). Another species, *Globorotalia truncatulinoides* (Fig. 14.2b), has a unique characteristic of coiling to the left in cold water and to the right in warm water. An abundance of left over right coiled tests in mid-latitude ocean sediments might indicate that glacial conditions had once occurred there (Fig. 14.3).

Terrestrial climates may also be inferred from the study of the remains of fossil plants, such as leaves, pollen, and spores, and fossil insects. Pollen and spores in ancient lake sediments provide an

(a)

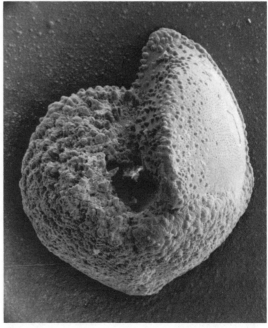

(b)

Figure 14.2 Pleistocene foraminiferans: (a) *Globorotalia menardii*; (b) *Globorotalia truncatulinoides*, right coiling. (Photo courtesy of J. P. Kennett.)

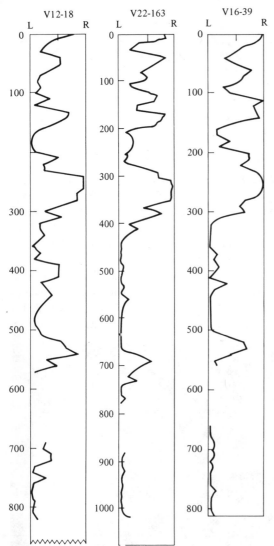

Figure 14.3 Correlation of three cores from the South Atlantic as defined by changes in the direction of coiling of *Globorotalia truncatulinoides.* The scale runs from 100% left coiling (left) to 100% right coiling (right). Numbers to the left of the columns are depths in the cores, in centimeters. The percentage of right-coiling forms increases with increasing temperature of the surface waters. Warmer surface waters occurred during interglacial and interstadial periods. (From Ericson and Wollin, Ref. 4. Copyright © 1968 by the American Association for the Advancement of Science.)

indication of the sequences of vegetation that once grew near the lake; insect remains give an approxi-

mation of temperature at the time of deposition. Both varieties of fossils are now considered to be indispensable in climate reconstructions (Fig. 14.4). A composite of the percentages of spores and pollen in a core may reveal the vegetational history of a region (Fig. 14.5).

QUATERNARY STRATIGRAPHY

Pliocene-Pleistocene Boundary

The boundary between the Pliocene and the Pleistocene is generally placed at the beginning of widespread glaciation. However, since continental glaciation may have begun as early as 60 million years ago (see Chapter 13), it is not possible to define the Pleistocene as the time of the ice ages.

The type section for the basal Pleistocene is located in southern Italy. Here the boundary between the Pliocene and Pleistocene has been placed at the first appearance of coldwater invertebrates, such as the pelecypod *Arctica islandica* and the foram *Hyalinea baltica*. This boundary also marks the extinction of most of the discoasters (microalgae) and the first appearance of large numbers of *Globorotalia truncatulinoides* (6). From studies of cores taken in the South Atlantic, David Ericson and Goesta Wollin found that these floral and faunal changes occurred at approximately the beginning of the Olduvai (Gilsá) Event, approximately 1.8 million years ago (7) (Fig. 7.27). The basal Pleistocene marine beds in southern Italy have been assigned to the Calabrian Stage (Fig. 14.6). Sedimentary units that have been classically assigned to the basal or Villafranchian Stage of the terrestrial Pleistocene are found in northern Italy along the Po River. The fauna of the Upper Villafranchian includes one-toed horses, the first true elephants, and distinctive species of dogs, carnivores, rodents, deer, and other vertebrates. Marine fossils found interbedded with terrestrial deposits of the Upper Villafranchian Stage are quite similar to those found in deposits of the Calabrian Stage. In the Olduvai Gorge in Tanzania, volcanic rocks interbedded with sediments con-

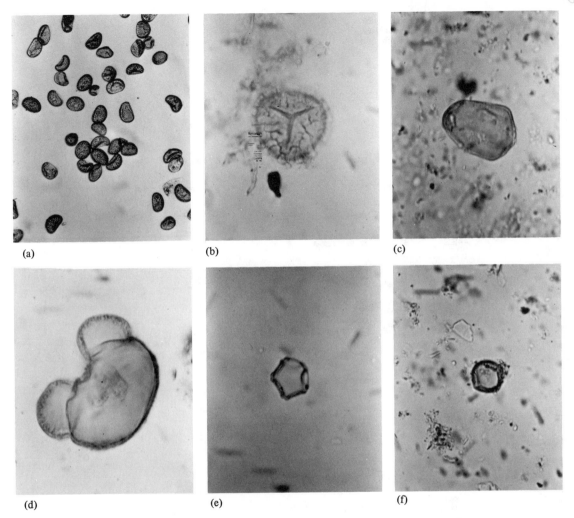

Figure 14.4   Photomicrographs of Quaternary spores and pollens: (a) fern spores; (b) *Lycopodium*; (c) Cyperaceae (sedge); (d) *Picea*; (e) *Alunus*; (f) Rosaceae.

taining an Upper Villafranchian fauna have been dated at approximately 1.75 million years old (8). This age indicates that the Pleistocene began more than 1.75 million years ago.

Pleistocene-Holocene Boundary

Although no universal criteria for the boundary between the Pleistocene and the Holocene have been established, the dissipation of the continental ice sheets provides acceptable time planes. The glaciers receded from North America about 5000 years ago and from Europe 8000 years ago (9). Stratigraphers variously place the Pleistocene-Holocene boundary at either the midpoint of worldwide warming of the oceans or at the midpoint in the rise of sea level. The midpoint in the warming of the oceans occurred between 11,000 and 12,000 years ago. At this time there was an increase in the abundance of *Globorotalia menardii*, several species of planktonic radiolarians, and warm-water coccoliths in the mid-

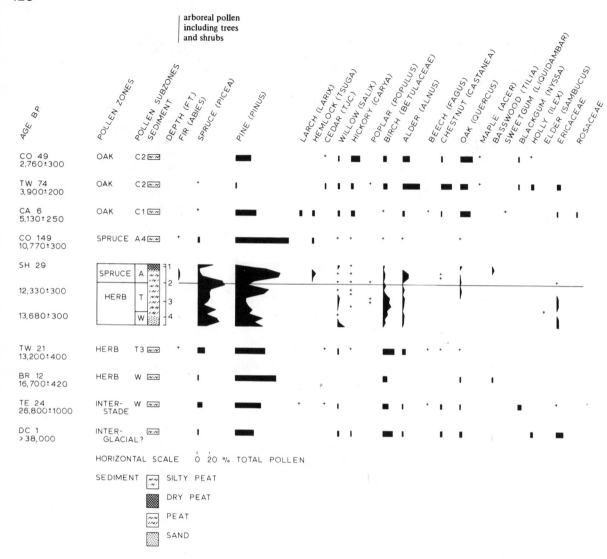

Figure 14.5 Radiocarbon-dated pollen diagram of Late Pleistocene deposits from New York and New Jersey. The decrease in the percentage of spruce and pine pollen from 16,700 years B.P. to the present is an indication of a gradual warming of the climate during that interval. (From Sirkin et al., Ref. 5.)

latitudes. The midpoint in the rise of sea level occurred about 13,500 years ago, when ice still covered much of North America (Fig. 14.7).

DISTRIBUTION OF PLEISTOCENE GLACIERS

Modern glaciers cover only about 10% of the earth's land surface, and most of these are located in Antarctica, Greenland, Iceland, and mountainous terrains (Fig. 14.8). Pleistocene glaciers covered more than three times that area or about 17 million square miles (44 million km^2) based on the outer limits of glacial deposits (Fig. 14.9, Table 14.1). Seismic studies and test drilling in Green-

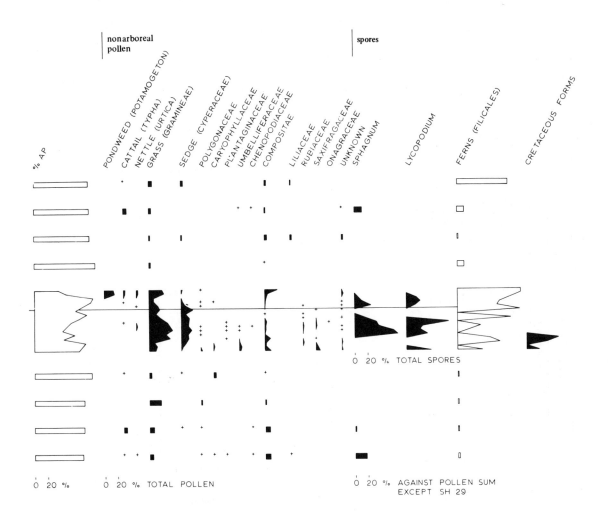

nonarboreal pollen   spores

0 20 % TOTAL SPORES

0 20 %   0 20 % TOTAL POLLEN   0 20 % AGAINST POLLEN SUM EXCEPT SH 29

land and Antarctica indicate that the ice reaches thicknesses of 10,000 and 13,000 ft (3000 and 4000 m), respectively (11). The North American and Scandinavian ice sheets were probably of comparable thickness. It has been estimated that the Pleistocene ice sheets had a volume of 18.4 million cubic miles (76.7 million km³). A corresponding volume of seawater converted to glacial ice would require a drop in sea level of about 644 ft (197 m) (1). However, loading of the crust would result in a flow of mantle from when the ice that would tend to cause a rise in sea level. The actual lowering of sea level would probably be only about 431 ft (133 m). At the present time, about 6.3 million

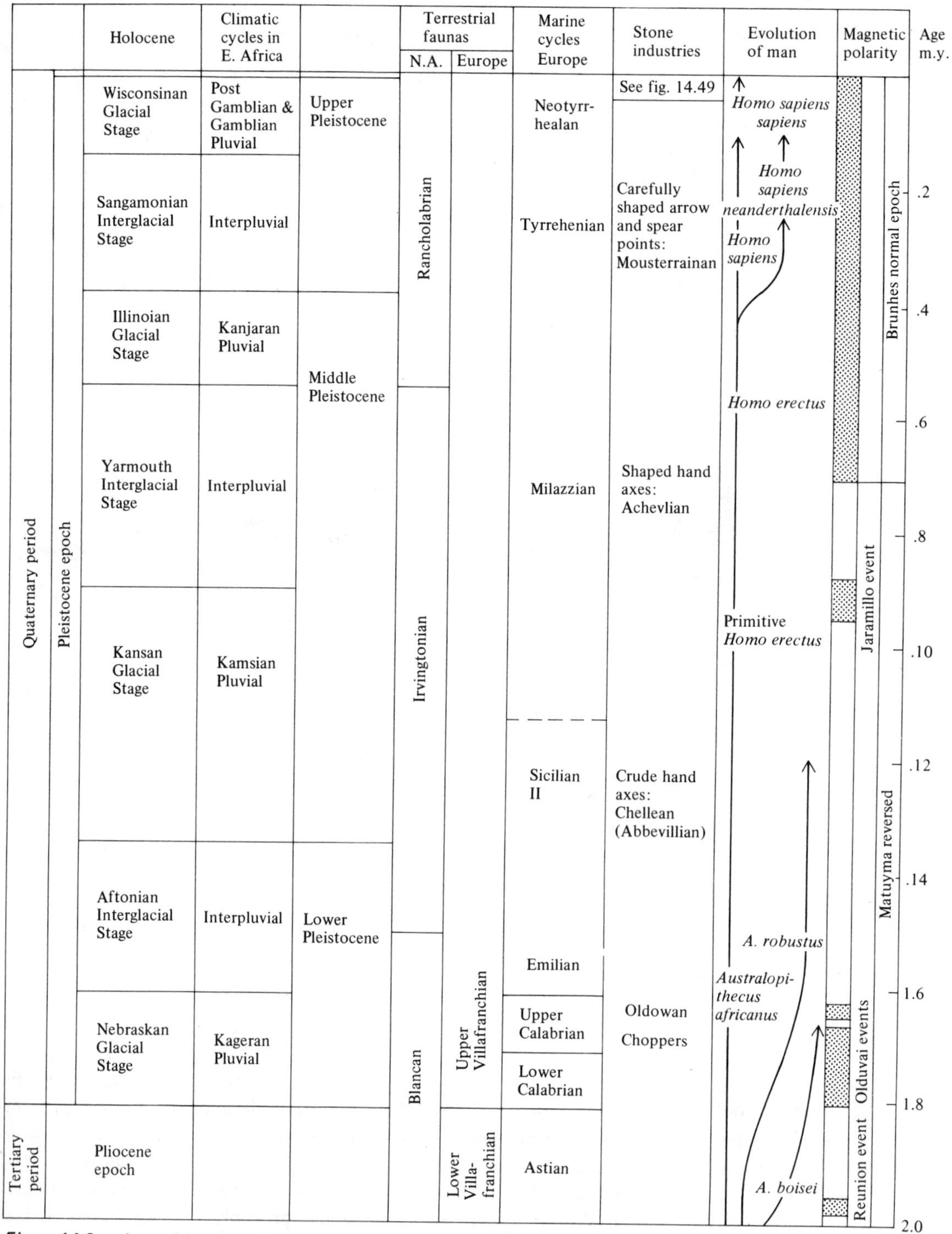

		Holocene	Climatic cycles in E. Africa		Terrestrial faunas		Marine cycles Europe	Stone industries	Evolution of man	Magnetic polarity	Age m.y.
					N.A.	Europe					
Quaternary period	Pleistocene epoch	Wisconsinan Glacial Stage	Post Gamblian & Gamblian Pluvial	Upper Pleistocene	Rancholabrian		Neotyrr-healan	See fig. 14.49	Homo sapiens sapiens ↑	Brunhes normal epoch	
		Sangamonian Interglacial Stage	Interpluvial				Tyrrehenian	Carefully shaped arrow and spear points: Mousterrainan	Homo sapiens neanderthalensis ↑ Homo sapiens		.2
		Illinoian Glacial Stage	Kanjaran Pluvial	Middle Pleistocene			Milazzian	Shaped hand axes: Achevlian	Homo erectus		.4
		Yarmouth Interglacial Stage	Interpluvial		Irvingtonian						.6
		Kansan Glacial Stage	Kamsian Pluvial						Primitive Homo erectus	Jaramillo event	.8
							Sicilian II	Crude hand axes: Chellean (Abbevillian)	↑	Matuyma reversed	.10
		Aftonian Interglacial Stage	Interpluvial	Lower Pleistocene	Blancan	Upper Villafranchian	Emilian		A. robustus Australopithecus africanus		.12
							Upper Calabrian	Oldowan Choppers		Reunion event Olduvai events	.14
		Nebraskan Glacial Stage	Kageran Pluvial				Lower Calabrian				1.6
Tertiary period		Pliocene epoch				Lower Villa-franchian	Astian		A. boisei		1.8
											2.0

Figure 14.6   A provisional stratigraphic chart comparing climatic cycles, marine cycles, stone industries, and the ranges of the ancestors of man.

Table 14.1   Approximate land areas covered by Pleistocene glaciers[a]

Continent	Square miles	Square kilometers
North America, Greenland, Hawaii	7,128,227	18,512,391
Europe and Iceland	2,779,404	7,208,145
Asia	1,523,505	3,951,000
South America	335,472	870,000
Africa	732	1,900
Australia, New Zealand, and New Guinea	11,580	30,000
Antarctica	5,325,136	13,810,000
Total	17,105,056	44,383,436

[a]From Flint (1).

cubic miles (26.25 million km^3) of ice occupies the continents. If it all melted, this volume of ice would raise the sea level about 210 ft (65 m) (1).

MULTIPLE GLACIATIONS

Evidence of multiple glaciations has been found in North America, Europe, and Asia. The most convincing evidence is the presence of till sheets or outwash gravels separated by paleosols, gumbotils (weathered tills), or marine sediments. In the Alps, five distinct outwash deposits, each separated from its neighbors by paleosols, indicate that five separate episodes of glaciation occurred in that area (Table 14.2). Four and possibly five glacial stages, each represented by till, outwash, and moraines, have been identified in northwestern Europe and have been correlated with the Alpine sequences. Interglacials there are represented by paleosols, unconformities, or fossiliferous marine and terrestrial deposits.

T. C. Chamberlain and F. Leverett were among the first to record evidence of multiple

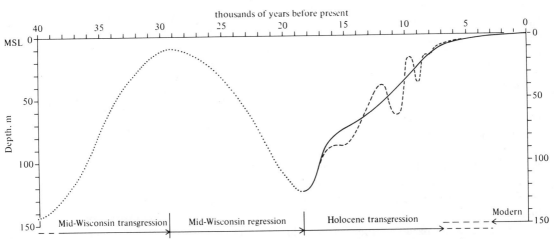

Figure 14.7   Variation in sea level throughout the past 40,000 years based on radiocarbon dates of fresh and salt marsh peats and shallow marine fossils plotted against depth below sea level where they were found. (From Curray, Ref. 10. Copyright © 1965 by Princeton University Press, reprinted with permission.)

Table 14.2   Suggested correlations of North American and European glacial and interglacial stages[a]

Midcontinent of North America	Rocky Mountains of North America	European Alps	Northern Europe
Wisconsinan — Late Interstadial	Pinedale	Würm — Late	Weichselian
			Eemian
Wisconsinan — Middle Interstadial	Bull Lake — Late Stage	Würm — Early	Warthian
		Riss-Würm	*Treene*
Wisconsinan — Early	Bull Lake — Early Stage	Riss	Saalian
Sangamonian		*Mindel-Riss*	*Holstein*
Illinoian	Sacajawea Ridge	Mindel	Elsterian
Yarmouthian		*Günz-Mindel*	*Cromerian*
			Beestonian
Kansan	Cedar Point	Günz	Menapian
Aftonian		*Donau-Günz*	*Waalian*
Nebraskan	Washakie Point	Donau	Eburdnian

[a] "Interglacials" are italicized.

glaciations in North America. In the Great Plains region, they found lithologically distinct tills separated by gumbotils. Four major glacial advances have since been documented in central North America (Fig. 14.10).

Pluvial and Interpluvial Features

Striking fluctuations in climate occurred during the Pleistocene in areas distant from the glaciers. Many regions which today have warm-dry and cool-dry climates experienced cooler pluvial (rainy) climates during the galcial stages. Large, deep lakes formed in the western United States, in parts of the Sahara and Gobi deserts, and in central Australia (Fig. 14.11). These lakes left wave-cut cliffs and beach deposits far up on the modern hillsides. Wave-cut terraces occur as much as 300 ft above Utah's Great Salt Lake, a remnant of Lake Bonneville (Fig. 14.12). Cores taken from the Great Salt Lake reveal several cycles of lacustrine silts, shoreline sands, and salt (12). The presence of shoreline sands below the present lake level indicates that the climates during interpluvial intervals corresponding to interglacial intervals may have been even more arid than the present climate.

Pluvial climates resulted from changes in atmospheric wind and pressure patterns because continental glaciers were present and ocean temperatures were cooler. Today most precipitation in

the mid-latitudes is associated with the activity of frontal cyclonic storms occurring mainly between 35 and 50° latitude. During glaciations, however, the frontal zone moved southward and brought an increase in precipitation to the arid regions. Arid belts may have shifted toward the equator at such times (13).

Evidence of Multiple Glaciations from the Ocean Basins

Considerable information about multiple glaciations has been derived from analysis of sediments from the deep ocean basins. Although the glacial record on land generally features numerous unconformities, ocean cores commonly contain a complete Pleistocene record. Ericson and Wollin have identified at least four major cold episodes, separated by three warmer episodes, in sediments dating back approximately 2.0 million years (Fig. 14.13). They assume that these four cold episodes correlate with the four major glacial advances in Europe and North America.

PRE-WISCONSINAN STRATIGRAPHY

Interpretation of the Pre-Wisconsinan segment of Pleistocene history has been complicated by the erosion and deposition of the Wisconsinan Ice Sheets and by the fact that deposits of this age are out of the range of radiocarbon dating.

(a)

Figure 14.8 Active glaciers in:
(a) Southern Victoria Land,
Antarctica (U.S. Navy Photo-
graph, courtesy of the U.S. Geo-
logical Survey.); (b) the Alps
(Courtesy of Francois Arbey.);
(c) Bylot Island, just north of
Baffin Island (Courtesy of the
National Air Photo Library,
Department of Energy, Mines
and Resources, Canada); (d) the
Bagley Ice Field, South Alaska.
(Courtesy of British Petroleum
Co.)

(b)

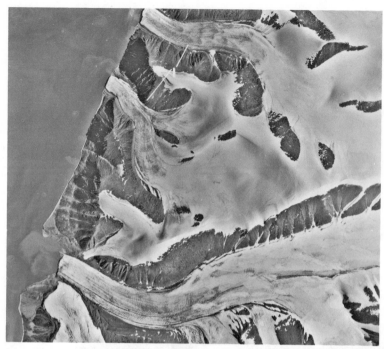

(c)

(d)

Figure 14.9   Extent of Pleistocene glaciation (shaded area) in the northern hemisphere. (After Flint, Ref. 1. Copyright © 1970 by John Wiley, New York.)

Nebraskan Glacial Stage

In Nebraska, two tills have been designated as Neraskan in age, and therefore it is likely that there was more than one separate glacial advance during that interval (1). The age of the Nebraskan has been the subject of much recent debate. Oxygen-isotope studies by Emiliani reveal eight cooler than average intervals in the last 425,000 years (Fig. 14.1). Emiliani has suggested that the Nebraskan occurred within this time interval. However, Ericson and Wollin have suggested that the cold interval which began in the Early Pleistocene (about 1.8 million years ago) is the Nebraskan Glacial Stage.

Nebraskan glacial deposits contain a late Blancan mammalian fauna, and the boundary between the Blancan and Irvingtonian has been radiometrically dated at about 1.5 million years old (1). It therefore appears that the Nebraskan is more than 1.5 million years old.

In Africa, there were four major pluvial stages during the Pleistocene. These are thought to correlate with the four major glacial advances in North America (Fig. 14.6). The oldest of the pluvial stages, the Kageran Pluvial, has been radiometrically dated at about 1.75 million years old (8). If the Nebraskan Glacial Stage correlates with the Kageran Pluvial, it would be of about the same age.

Aftonian Interglacial Stage

A deep soil developed in Nebraskan glacial deposits during the Aftonian Interglacial Stage. Rainfall during this interval was probably comparable to that of today, but warmer temperatures resulted in a dryer climate (1).

Volcanic ash underlying deposits of the Aftonian Interglacial Stage and overlying deposits of Lower Pliocene age has been dated at 1.9 million years (18). The Aftonian deposits are separated from the ash by a long weathering interval, so that the Aftonian began sometime after 1.9 million years ago. Ericson and Wollin propose that the Aftonian Interglacial Stage lasted from about 1.6 to 1.35 million years ago. Aftonian deposits contain an Irvingtonian mammalian fauna, and since the Irvingtonian began about 1.5 million years ago, the radiometric data are consistent with the date proposed by Ericson and Wollin for the beginning of the Aftonian Stage. However, deposits in Nebraska which are thought to be Aftonian in age contain a reversely magnetized volcanic ash dated at 1.2 million years (19). Thus, Ericson and Wollin's date for the end of the Aftonian may be too great.

Kansan Glacial Stage

In Nebraska, there are three tills of Kansan age (1).

Evidently there were at least three separate glacial advances during the Kansan Glacial Stage. Till of probable Kansan age is overlain by normally magnetized volcanic ash dated at 600,000 years (19). This date indicates that the Kansan till is older than 600,000 years.

Yarmouthian Interglacial Stage

A deep soil developed in Kansan drift before the Illinoian Glacial Stage. The climate during the Yarmouthian Stage was probably somewhat warmer and dryer than it is today (1).

Illinoian Glacial Stage

There are three distinct Illinoian tills in Illinois and Indiana. Since they differ little in their degree of weathering, it is likely that they represent three closely spaced glacial advances. River terrace deposits in Europe correlated with the Illinoian

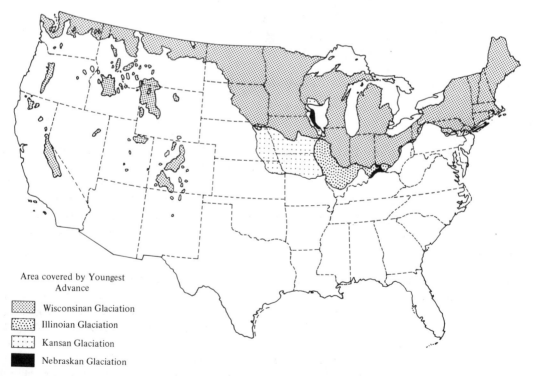

Area covered by Youngest
Advance

Wisconsinan Glaciation
Illinoian Glaciation
Kansan Glaciation
Nebraskan Glaciation

Figure 14.10   Borders of the Nebraskan, Kansan, Illinoian, and Wisconsinan ice sheets in the central United States based on the extent of the glacial drift of different ages. Where older drift is buried beneath younger drift, the borders are not shown. (From Denny, Ref. 11.)

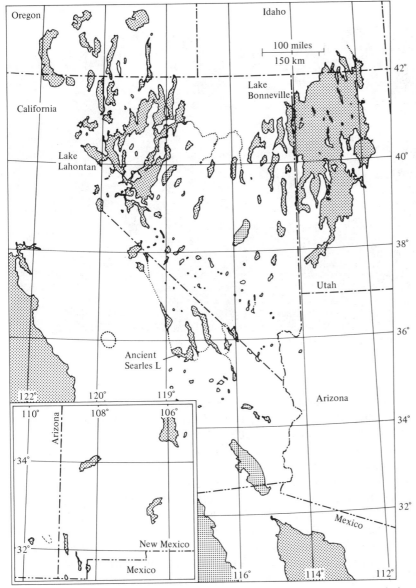

Figure 14.11 Lakes of the Great Basin in the western part of the United States during the Pleisto-cene at the time of maximum glaciation. (From Flint, Ref. 1. Copyright © 1971 by John Wiley, New York.)

Glacial Stage have been dated at approximately 400,000 years (1).

Sangamonian Interglacial Stage

The Sangamonian Interglacial Stage is widely rep-resented by both sediments and a deep soil devel-oped in Illinoian glacial deposits. This soil is much thicker and more mature than the soil that has developed since the retreat of the last Wisconsinan Ice Sheet (1). Determination of the age of the

Figure 14.12   Wave-cut terraces of the ancient Lake Bonneville north of Salt Lake City, Utah. (Courtesy of the Utah State Historical Society.)

Sangamonian is a matter of considerable controversy among Pleistocene stratigraphers. Some have suggested that the warm interval which occurred between about 60,000 and 100,000 years ago correlates with the Sangamonian Stage (Fig. 14.14). However, it is not likely that a soil as deep and mature as that which formed during the Sangamonian could have formed in only 40,000 years. It is more likely that the warm interval 60,000 to 100,000 years ago is an interstadial of the Wisconsinan Glacial Stage, and that the Sangamonian lasted from about 375,000 to 120,000 years B.P. (Fig. 14.13).

WISCONSINAN GLACIAL STAGE

Deposits of the Wisconsinan Stage have been extensively studied throughout the world. There were apparently three major glacial advances during the Wisconsinan. This has allowed the Wisconsinan to be subdivided into three stadials and two interstadials (Fig. 14.14).

Early Wisconsinan Stadial

The Early Wisconsinan Stadial lasted between 120,000 and 100,000 years ago. Glacial climates during this interval are indicated by studies of foraminiferans and oxygen isotopes (Fig. 14.14). Elevated marine terraces and coral reefs between 100,000 and 75,000 years old have been found in the Bahamas, Barbados, and the Mediterranean region. These terraces and coral reefs may have formed during the interstadial separating the Early and Middle Wisconsinan Stadials. Their elevation indicates that sea level was probably as high or higher than it is today, and therefore the continental ice sheets had presumably disappeared completely from North America (except for Greenland) and Eurasia during this interstadial.

Middle Wisconsinan Stadial

The Middle Wisconsinan Stadial lasted from about 60,000 to 44,000 years B.P. In central North America, there were at least three pulsations

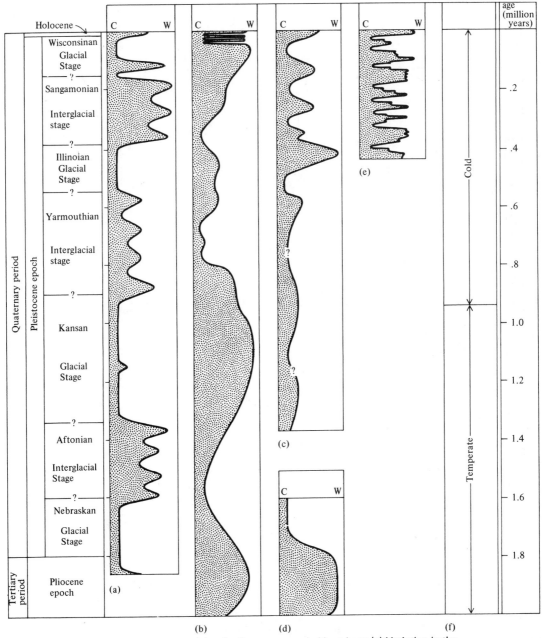

Figure 14.13 Generalized climate curves for the Quaternary period based on: (a) Variation in the frequency of *Globorotalia menardii* (After Ericson and Wollin, Ref. 4. Copyright © 1968 for the American Association for the Advancement of Science.); (b) planktonic foraminiferans from the northern Gulf of Mexico (After J. H. Beard, Esso Production Research Co., Ref. 14.); (c) planktonic foraminiferans from the southern oceans (After Kennett, Ref. 15.); (d) planktonic foraminiferans from the Mediterannean, the Caribbean, and the Gulf of Mexico (After Lamb, Ref. 16.); (e) paleo-temperatures of the Caribbean as determined by oxygen isotope studies (After Emiliani, Ref. 3. Copyright © 1966 by the University of Chicago.); (f) Antarctica radiolarians (After Bandy and Casey, Ref. 17.). The time scale at the left has been inferred from the climatic curves: C, cold; W, warm.

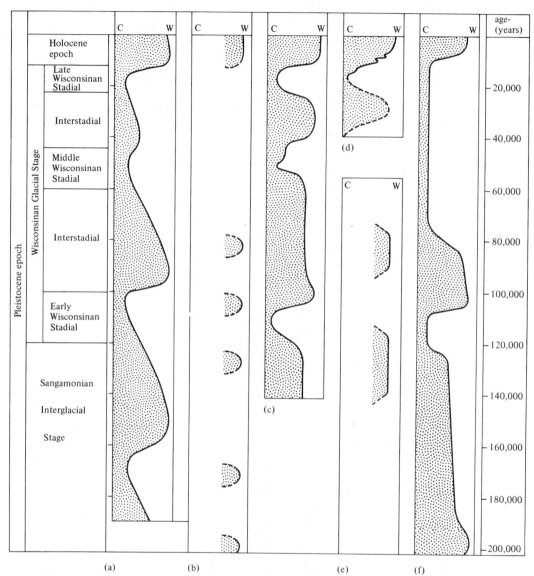

Figure 14.14   Generalized climatic curves for the Late Pleistocene and Holocene epochs based on:
(a) Paleotemperatures of the Caribbean as determined by oxygen isotope studies (After Emiliani,
Ref. 3. Copyright © 1966 by the University of Chicago.); (b) the study of raised coral reefs from the
Barbados Islands (After Mesolella, Ref. 20. Copyright © 1969 by the University of Chicago.); (c) the
relative abundance of *Globigerina rubescens* from the northern Indian Ocean (After Frerichs, Ref. 21.
Copyright © 1968 by the American Association for the Advancement of Science.); (d) sea-level curves
from the Atlantic shelf of the United States (After Curray, Ref. 10. Copyright © 1965 by Princeton
University Press.); (e) marine deposits in North Africa (Data from Stearns and Thurber, Ref. 22.);
(f) variations in the frequency of *Globorotalia menardii* (After Ericson and Wollin, Ref. 4. Copyright ©
1968 by the American Association for the Advancement of Science.)

during the advance of the ice sheet (Fig. 14.15). It is not known how far north the ice retreated between pulsations, but the lack of marine terraces 60,000 to 44,000 years old above the present sea level indicates that the continental ice sheets of the Northern Hemisphere probably did not melt completely during the retreats. After advancing, the ice remained at its maximum extent for only a short period of time (Fig. 14.15). In northern Europe, drift assigned to the early Würm rests on Wärthe drift. Since the Wärthe drift is not significantly weathered, these two glacials are probably relatively close in age and are separated by an interstadial rather than an interglacial. The Wärthe drift was probably deposited during the Middle Wisconsinan Stadial.

Warming and partial deglaciation occurred between about 44,000 and 28,000 years B.P. during a second interstadial. This interval includes the Farmdalian Substage in central North America and the Paudorf Interstadial in Europe. Deposits of this age in southern England include the "Arctic Plant Bed" which contains fossil mosses and dwarf birches along with the bones of rhinoceroses and mammoths. These and similar deposits have been dated at between 29,000 and 32,000 years B.P. by the radiocarbon method (1).

Late Wisconsinan Stadial
The glacial advance which began about 28,000 years ago reached its maximum at about 19,000 years B.P. The rate of advance in the mid-continent has been calculated to be between 80 and 125 ft (35 and 105 m) per year (23). The southernmost margin of the ice is marked by a complex of end moraines and outwash deposits. The ice began to retreat shortly after it reached its maximum extent. Minor fluctuations in climate caused readvances of the ice front over recently exposed terrain (Fig. 14.15). The melting of the ice left exposed such glacial features as *roches moutonnées*, U-shaped valleys, end moraines, and streamline topography (Fig. 14.16).

The Great Lakes Region. As the Late Wisconsinan ice sheet receded, the lobate margin of the ice

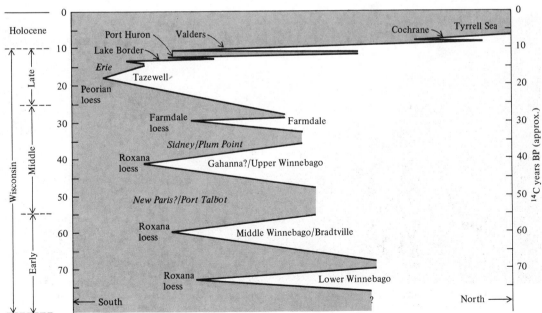

Figure 14.15 Diagrammatic sketch showing the advance and retreat of the ice sheet during Middle and Late Wisconsinan time in central North America. Nonglacial intervals are in italics. (From Flint, Ref. 1. Copyright © 1971 by John Wiley, New York.)

(a)

(b)   *Figure 14.16*   Erosional and depositional features produced by Pleistocene glaciers: (a) Small tarn in a cirque in the Klamath Mountains of northern California — Notice the striated roche moutonnée in the foreground; (b) U-shaped valley along the "Million Dollar Highway" in western Colorado; (c) recessional moraines east of Hudson Bay, Canada; (d) deeply scoured topography in the Northwest Territories, Canada. [(c) and (d) courtesy of the National Air Photo Library, Department of Energy, Mines and Resources, Canada.]

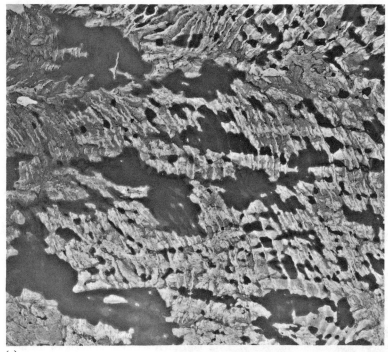

(c)

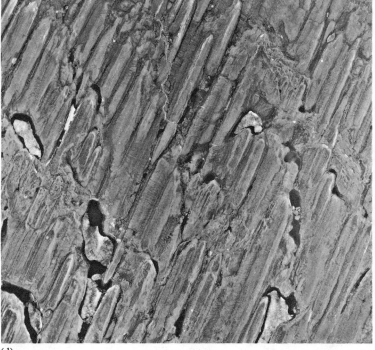

(d)

created three major lakes – Lake Chicago in the Lake Michigan basin, Lake Saginaw in the Lake Huron basin, and Lake Maumee in the Lake Erie basin; these were followed by a succession of intermediate lake stages (Fig. 14.17). Drainage from these lakes was at first southward into the Gulf of Mexico and the Atlantic Ocean. A readvance of the ice about 13,000 years B.P. resulted in a change in lake levels and lake outlets (Fig. 14.17b). About 12,000 years B.P., before extensive crustal rebound, a marine incursion into the St. Lawrence basin created the Champlain Sea. At about the same time, the sea also invaded the lower Hudson Valley. The marine phase ended when rebound elevated the crust north of the glacial terminus. The ice that had retreated rapidly between 12,750 and 12,000 years B.P. readvanced and extended as far south as southern Wisconsin. This advance covered a forest at Two Creeks, Wisconsin, where wood from beneath the till has been dated at about 11,850 years B. P. (1).

Glacial Lake Agassiz formed west of the Great Lakes as the ice receded from that region. This lake covered more than 73,000 square miles (200,000 km²) and lasted from about 12,500 to 7,500 years B.P. (1). Drainage from this lake, which was first southward into the Gulf of Mexico shifted eastward into the Great Lakes and finally northward into Hudson Bay.

Northeastern United States. The retreat of the Late Wisconsinan ice sheet from northeastern United States probably began between 17,000 and 19,000 years ago. The rate of retreat has been determined in the Hudson Valley from radiocarbon dates and pollen stratigraphy in lake and bog sediments (24). The age of the tundra or herb pollen zone in a given bog provides an approximate age for the beginning of deglaciation in a specific region. The retreat of the ice was slow at first, but by 15,000 B.P. the ice front had retreated into southern New York and southern New England (Fig. 14.18). The ice margin reached the northern Hudson region by about 13,000 years B.P., and the St. Lawrence lowlands by about 12,600 years B.P.

Europe. The maximum advance of the Scandinavian ice sheet during the Late Wisconsinan occurred between about 17,000 and 20,000 years B.P. At this time, the southern margin of the ice sheet extended from the continental shelf surrounding the British Isles, across central Europe to northwestern Siberia (Fig. 14.9). Regardless of whether the ice sheet in the British Isles coalesced with the main ice sheet at this time, these two ice sheets had separated soon after the onset of deglaciation. By 13,000 years B.P. the ice still covered most of the Scandinavian Peninsula (Fig. 14.19). As the ice continued to retreat, a proglacial lake, the Baltic Ice Lake developed south of the ice margin (Fig. 14.20). Approximately 9000 years B.P. sea level rose sufficiently to form an inland sea, the Yoldia Sea, which was named after the Arctic mollusk *Yoldia arctica*, a common fossil found in the marine sediments of the Yoldia Sea. Postglacial rebound isolated the Baltic basin from the sea and created the Ancylus Lake, named for *Ancylus fluviatus*, a freshwater mollusk of the lake sediments. Ancylus Lake persisted through the melting of the last remnants of the Scandinavian ice sheet, but about 6000 years B.P. the sea again occupied the basin. The Littorina Sea, named after the marine snail *Littorina littorea*, had decreased in size as the Baltic Shield continued to be uplifted.

THE HOLOCENE

If we define the Holocene as beginning about 11,000 years ago, sea level has risen 150 ft (50 m) during that time. As a result, much of the continental shelves was flooded. Holocene sediments have accumulated in river deltas, flood plains, and beaches. In most areas these deposits are relatively thin owing to the short interval of deposition. However, in many estuaries where Pleistocene erosion has deepened the channels, Holocene deposits may exceed 100 ft (30 m).

Palynological studies of Holocene sediments have indicated that 6000 to 4000 years ago climates were generally warmer than those of today. This interval has been referred to as the

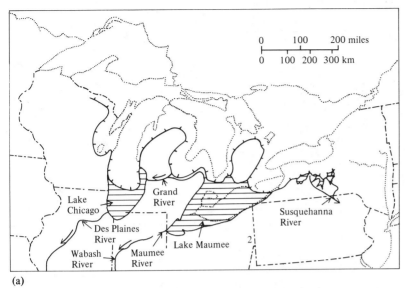

(a)

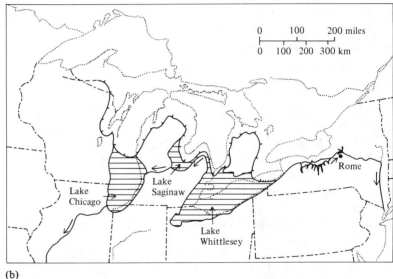

(b)

climatic optimum. Since then climates have gradually cooled and several episodes of glacial expansion occurred. The midpoints of "little ice ages" have been reported at 3575, 3200, 2800, and 250 years B.P.

POSTGLACIAL REBOUND

The weight of glacial ice depressed the earth's crust and caused a slow flow of the mantle from beneath the ice-loaded crust. The amount of crustal depression during Pleistocene glaciations was approximately 1 ft for each 4 ft of ice. When the ice melted, the crust slowly rose as the mantle material flowed back under the deglaciated crust. This uplift is still going on in northern North America, Scandinavia, and Asia, where the rate of uplift may exceed one ft (30 cm) per century (Figs. 14.21 and 14.22).

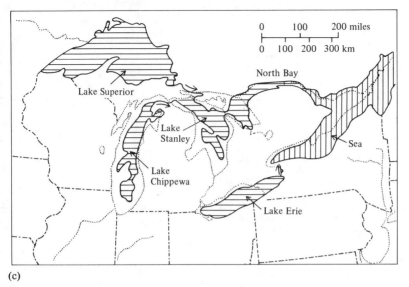

(c)

Figure 14.17 Evolution of the Great Lakes: (a) At the time of Lake Maumee, (ca. 13,750 B.P.), (b) at the time of the Port Huron readvance (ca. 13,000 years B.P.), (c) at the time of Lake Chippewa. (After Flint, Ref. 1. Copyright © 1971 by John Wiley, New York.)

TECTONISM AND VOLCANISM

The tectonic and volcanic history of the Quaternary is a result of movements of crustal and oceanic plates which began in Tertiary time. Many mountains and plateaus were uplifted without appreciable folding. The Colorado Plateau, for example, was uplifted nearly 5000 ft (1500 km) during the late Tertiary and early Quaternary. Uplift of the continental interior was in part responsible for the lowering of sea level during the Quaternary. In the Basin and Range Province, normal faulting and volcanism, which began in the Tertiary, continued into the Quaternary.

Folding occurred along the Pacific margins of the continents as crustal plates continued to move away from the Mid-Atlantic and Mid-Indian Ocean ridges. Strike-slip faulting was particularly prevalent in these regions. The San Andreas Fault System in California is one of the most thoroughly studied strike-slip fault systems in the world (Fig. 14.23). Over its entire length this fault network forms the boundary between sections of the America Plate and the Pacific Plate (Fig. 14.24). In the region of the San Andreas Fault, the America Plate is moving southwestward and the Pacific Plate is moving northwestward relative to Eurasia. The combination of these motions produces a right-lateral displacement along the fault (Fig. 14.24). The Queen Charlotte and Merriweather faults to the north may be of the same origin (Fig. 14.24). Other major strike-slip faults active during the Quaternary include the Philippine Fault, the Alpine Fault in New Zealand, and the Atacama Fault in South America. The Alpine Fault marks the junction between the Pacific Plate and India Plate (Fig. 1.10).

The prevalence of earthquakes and active volcanoes in the "ring of fire" around the Pacific Ocean basin indicates that oceanic plates are moving under the Pacific margins of the continents at the present time (Fig. 14.25). However, volcanism in the western United States appears to have been less intense in the Late Pleistocene than during the late Tertiary and Early Pleistocene. Currently, the only active volcano in the contiguous United States is Mount Lassen, although Mount Rainier and Mount Shasta have hot regions at depth. Volcanism in the African Rift System produced such peaks as Mount Kenya and the extensive ash deposits in the Olduvai Gorge. In the Atlantic

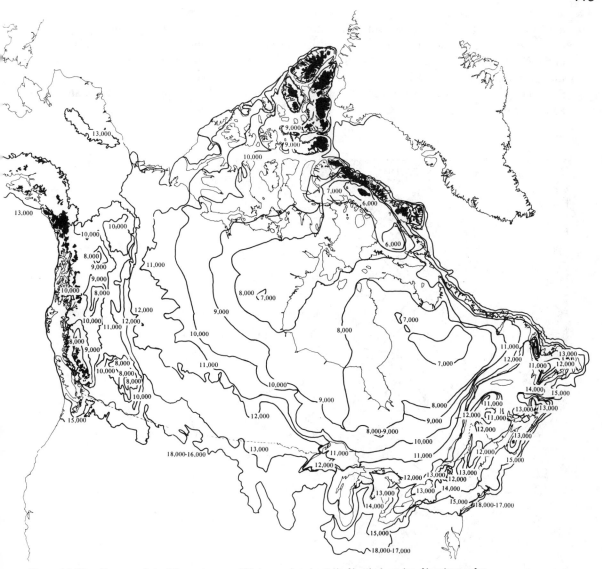

Figure 14.18 Retreat of the Wisconsinan and Holocene ice sheets in North America. Numbers refer to the age of the ice margin along the adjacent line. (After a map published by the Geological Survey of Canada.)

Ocean basin, continued volcanism produced new volcanic islands such as Tristan da Cunha and, most recently, Surtsey on the crest of the Mid-Atlantic Ridge. Widening of the North Atlantic has occurred at a rate of about 1 in. (2.5 cm) per year, with a total widening of about 39 miles (48 km) since the beginning of the Pleistocene. The South Atlantic has widened approximately 48 miles (80 km), and the southern Indian Ocean has widened about 72 miles (120 km) in that time.

CAUSES OF THE ICE AGES

Numerous theories have been proposed to explain

why the earth has experienced repeated glaciations during much of the late Cenozoic. An explanation of Pleistocene ice ages should first account for the initial onset of cold climates and then explain why the glaciations have been cyclical. Many geologists feel, too, that an acceptable theory should be applicable to prior ice ages.

Initiation of the Ice Ages

Quaternary glaciations are generally presumed to have resulted from a gradual cooling of the entire earth which began during the Eocene (Fig. 14.26). This climatic change may have been initiated by the uplift of the continents and associated withdrawal of the seas from the continents. Since air

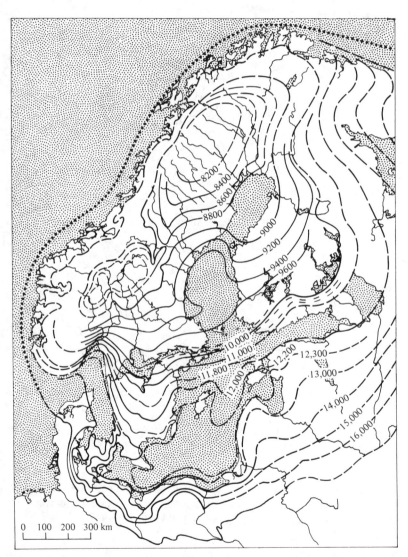

0 100 200 300 km

Figure 14.19   Retreat of the Wisconsinan and Holocene ice sheets in Scandinavia as determined by varve counts and [14] C dating. (After De Geer, Ref. 25.)

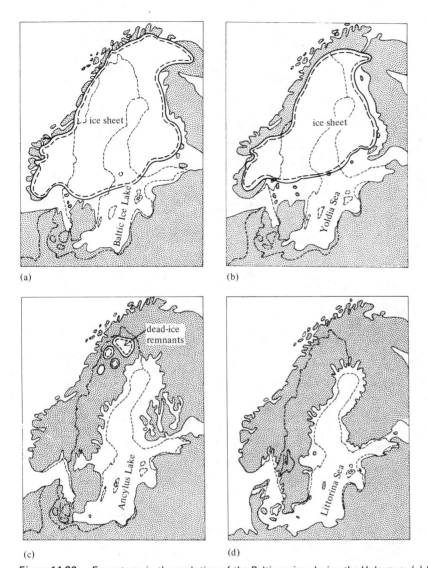

Figure 14.20 Four stages in the evolution of the Baltic region during the Holocene: (a) Baltic Ice Lake (ca. 8500 years B.P.); (b) Yoldia Sea (ca. 7900 years B.P.); (c) Ancylus Lake (ca. 7200 years B.P.); (d) Littorina Sea (ca. 5200 years B.P.). (After Lundqvist, Ref. 26. Copyright © 1965 by John Wiley, New York.)

temperature normally decreases with altitude (at a lapse rate of about 1°F/1000 ft or 1°C/480 m), the continents would have cooled through increased elevation. Furthermore, an increase in land area would also have increased the rate of erosion, the amount of dust in the atmosphere, and consequently the opacity of the atmosphere. Dust would also have provided condensation nuclei for

water vapor, leading to increased cloud cover and precipitation. Similarly, widespread volcanism in the Tertiary would have contributed dust to the atmosphere and may have supplemented this process.

It is likely that polar wandering and continental drift played a significant role in the cooling of the earth during the Tertiary. Throughout the

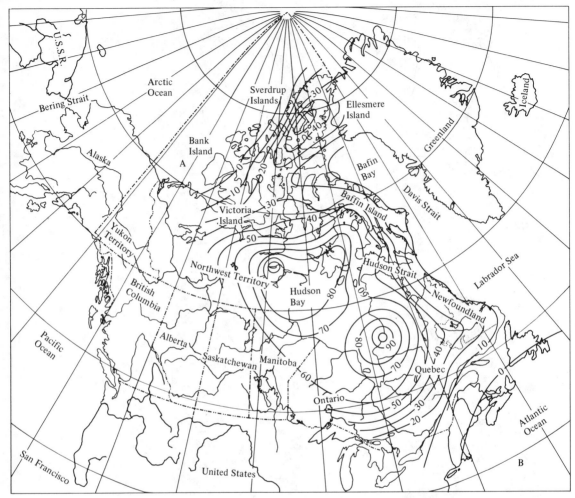

Figure 14.21 Postglacial uplift (in meters) of North America as determined by the elevation of marine deposits 6000 years old. (After Andrews, Ref. 27.)

Paleozoic and during most of the Mesozoic, the North Pole was located in the Pacific Ocean region (Fig. 14.27). In the Tertiary, however, the pole migrated into the Arctic Ocean. If, as proposed in our reconstruction of the continents, the northeastern tip of Siberia was separated from the rest of Siberia prior to the separation of Europe and Greenland, there would have been a free interchange between the waters of the Arctic and Pacific Oceans during the Paleozoic, Mesozoic, and early Tertiary (Fig. 14.27). The gap between the northeastern tip of Siberia and the rest of Siberia narrowed as Greenland drifted away from Europe. By the late Tertiary, the gap had closed; the pole was then in approximately the same position it occupies today. The location of the pole over a body of water with limited circulation would have helped to cool the Arctic region and may have contributed to the initiation of glaciation in that region.

In the southern hemisphere the South Pole was located in the Pacific Ocean throughout most of the Mesozoic. Both the pole and Antarctica migrated toward their present positions and the South Pole reached the continental margin during the Cretaceous (Fig. 14.28). The movement of Antarctica into a polar region could have accelerated climatic cooling.

The closing of the Isthmus of Panama during the Pliocene also may have been a factor influencing the onset of a glacial climate. This closing would have strengthened the Gulf Stream and increased precipitation in the northern hemisphere, thereby contributing to the accumulation of glacial ice.

Episodes of widespread deformation, maximum withdrawal of the seas, and location of a geographic pole on or near a continent are also associated with Precambrian and Paleozoic continental glaciations (Fig. 14.28). Therefore it is likely that these factors are very important in the formation of continental ice sheets.

Causes of Multiple Glaciation

Cyclical glaciation requires controlling factors which are repetitive, such as:

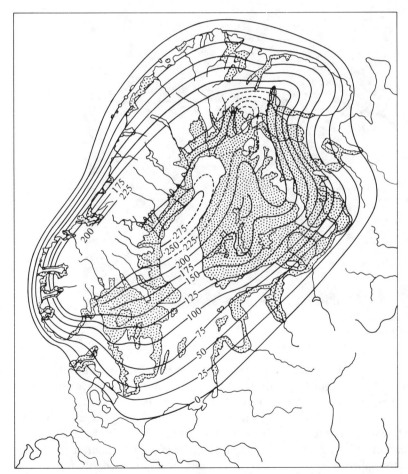

Figure 14.22   Postglacial uplift of Fennoscandia as determined by the highest strand mark of the ocean during the late Quaternary. (After Wright, Ref. 28.)

Figure 14.23   Trace of the San Andreas Fault showing a stream offset in a right-lateral sense. (Photo courtesy of the U. S. Geological Survey.)

1. Astronomical control, which involves variations in the amount of solar radiation which the northern hemisphere receives during a year.
2. Atmospheric control, which depends on variations in the carbon dioxide and water vapor content of the atmosphere.
3. Oceanic control, which involves freezing and thawing of the Arctic Ocean.

Astronomical Control. Glaciations may have been triggered by changes in the amount of solar radiation (insolation) received at various latitudes at different times of the year. Cyclic changes in the shape of the earth's orbit or in the inclination of the earth's axis might bring about such changes. This mechanism was proposed by the Russian meteorologist, M. Milankovitch. Milankovitch postulated that cool summers in the northern hemi-

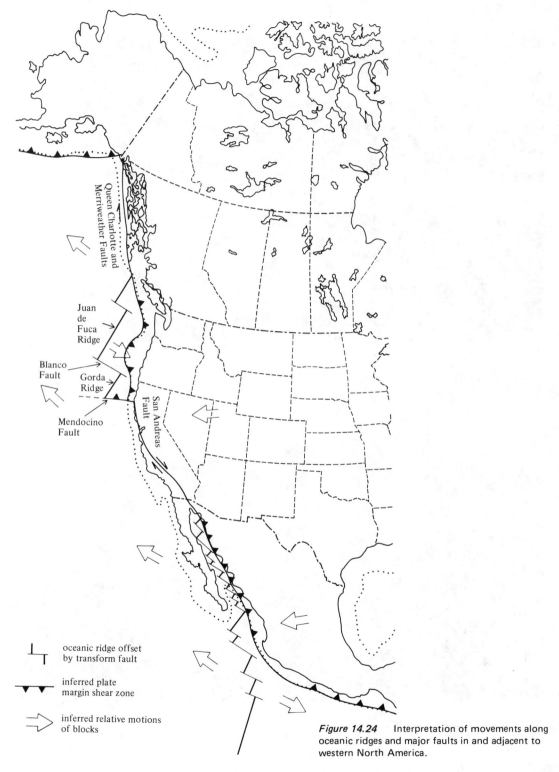

Queen Charlotte and
Merriweather Faults

Juan
de
Fuca
Ridge

Blanco
Fault

Gorda
Ridge

Mendocino
Fault

San Andreas
Fault

⊥⊤ oceanic ridge offset
by transform fault

▼▼ inferred plate
margin shear zone

⇨ inferred relative motions
of blocks

Figure 14.24 Interpretation of movements along
oceanic ridges and major faults in and adjacent to
western North America.

(a)

(b)

Figure 14.25   Quaternary volcanoes: (a) Mount Rainier in the western part of the state of Washington (Union Pacific Railroad Photo.); (b) Mount Fuji, an active volcano (Courtesy of the National Park Association of Japan.)

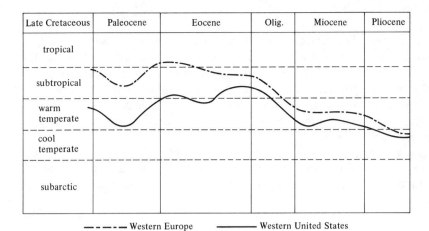

Late Cretaceous	Paleocene	Eocene	Olig.	Miocene	Pliocene
tropical					
subtropical					
warm temperate					
cool temperate					
subarctic					

— · — · — Western Europe   ———— Western United States

Figure 14.26 Curves illustrating climatic changes in western Europe and western United States during the Tertiary. (From Dorf, Ref. 29. Copyright © 1964 John Wiley, London.)

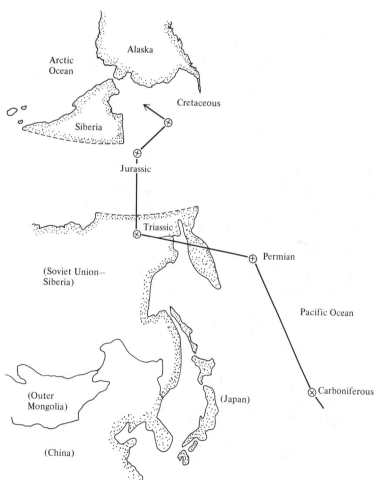

Figure 14.27 (a) Polar wandering curve for the Late Pleistocene and Mesozoic relative to Asia; (b) polar wandering curve for the Late Pleistocene, Mesozoic, and Cenozoic relative to Antarctica.

(a)

453

sphere would favor the accumulation of glacial ice. The earth's axis is now inclined at approximately 23½° to the plane of the earth's orbit around the sun. It is this inclination that produces the seasons. However, the angle of inclination of the axis varies between 21½ and 24½°. This angle fluctuates in a 40,000-year cycle. A greater inclination would produce warmer summers and colder winters; a lesser inclination would produce warmer winters and cooler summers (Fig. 14.29).

The earth's orbit around the sun is an ellipse of slightly varying eccentricity. At minimum eccentricity, the orbit is almost circular; at maximum eccentricity, the ratio of the semiminor axis to the semimajor axis is about 0.998. The eccentricity of the earth's orbit varies in a 92,000-year cycle.

The earth wobbles like a top, and thus the rotational poles "point" in various directions in space (Fig. 14.30). For example, the North Pole

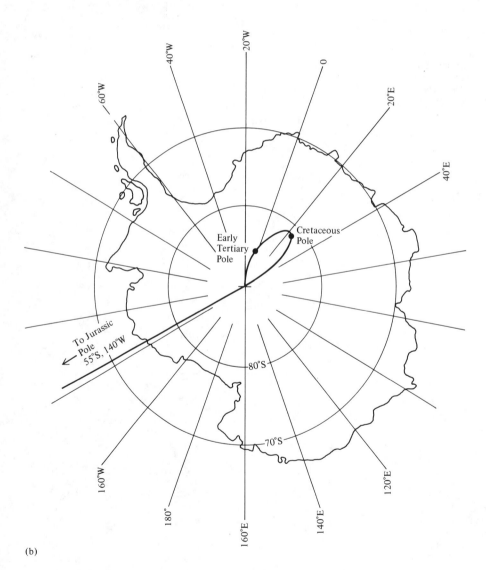

(b)

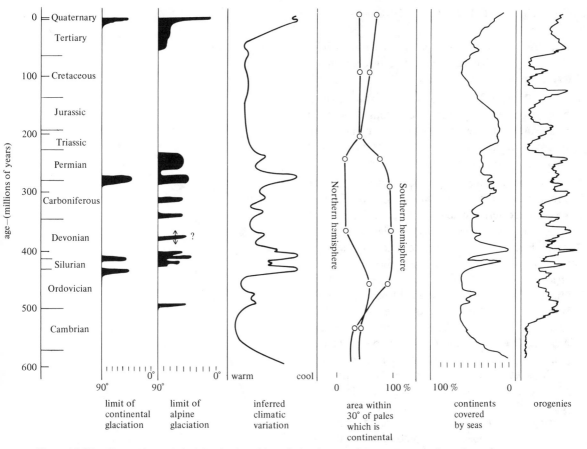

Figure 14.28 Comparison of glacial episodes with periods when continents were near the poles and periods of deformation and uplift.

now points approximately at the North Star, but this has not always been the case. The precession of the equinoxes — the shift in the seasons through time due to this wobble — varies in a 21,000-year cycle. At the present time, the earth's orbit is at about maximum eccentricity, and summer occurs when the earth is farthest from the sun. These conditions are favorable for the accumulation of glacial ice (Fig. 14.30).

Based on these cycles, the variation in summertime insolation for northern hemisphere latitudes may be calculated and plotted against time (Fig. 14.31). In general, there is fairly good correlation between this curve and paleotemperatures based on oxygen isotope studies (Fig. 14.31). The maximum of Late Wisconsinan glaciation occurred about 19,000 years ago, and there is a minimum in the insolation curve at 22,000 years B.P. The mid-

point of the Middle Wisconsinan Stadial occurred at about 50,000 years B.P., but the closest minimum in the insolation curve occurs at about 70,000 years B.P. The maximum advance of the Early Wisconsinan glaciers occurred about 110,000 years B.P., and there was a minimum insolation about 115,000 years B.P. The lag between the low points in the two curves suggests that astronomical variations may trigger long-term temperature declines.

Atmospheric Control. Variations in the carbon dioxide content of the atmosphere have been suggested as a possible cause of multiple glaciations, since carbon dioxide above its present level of 0.03 percent of the atmospheric gases might result in warmer worldwide climates. The increase in carbon dioxide might be caused by forest fires, volcanism, or other imbalance in the carbon cycle. Conversely, a decrease in carbon dioxide might lead to an increase in glacial ice.

Ocean Control. Maurice Ewing and William Donn have suggested that cyclic glaciations depend on temperature changes within the Arctic and North Atlantic oceans (31). With the onset of a glacial climate, this sequence might be established:

1. An ice-free Arctic Ocean would provide abundant moisture for precipitation in the form of snow in northern Canada and northern Eurasia.

2. Snow and ice would accumulate around the Arctic basin until glacial ice became thick enough to flow southward. As the ice sheets grew, an increasing amount of heat would be reflected back into space and the climate would become cooler. When the ice had moved into more southerly latitudes, it would be nourished by moist air from the Atlantic Ocean and the Gulf of Mexico. At this point, the ice sheet would no longer be dependent on the Arctic Ocean for moisture. At a latitude of approximately 40 to 50° north, the rate of advance of the ice would be balanced by the rate of melting, and the ice front would cease to advance.

3. Ocean temperatures would have steadily decreased as the glaciers advanced, because of the influx of meltwater and general cooling of the climate. A decrease in ocean temperatures would lead to a decrease in the rate of evaporation of the ocean water, and a decrease in precipitation in the northerly regions. The Arctic Ocean might freeze over at this time.

4. A decrease in precipitation would cause the ice to recede, and the oceans would remain cold as the glaciers retreated. If the Arctic Ocean had frozen over, the rate of recession would increase as the ice front moved into more northerly latitudes, where moisture from the Arctic, the Atlantic, or the Gulf of Mexico would not be available.

5. The climate would continue to become warmer in the absence of the ice sheets in North America and Eurasia, and the Arctic ice pack would melt. At this point, the cycle would begin again.

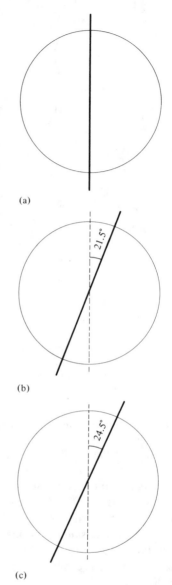

(a)

(b)

(c)

Figure 14.29   Effect of variations in angle of inclination of earth's axis on climate: (a) Hypothetical, inclination of axis vertical, no seasons; (b) cooler-than-average summers, warmer-than-average winters — favorable for glacial ice; (c) warmer-than-average summers, cooler-than-average winters — unfavorable for such accumulation.

PLEISTOCENE LIFE

A striking feature of Pleistocene life is the distinction between cold and warm faunas and floras. Extensive migrations of animal and plant assem-

summer in the
northern hemisphere

winter in the
northern hemisphere

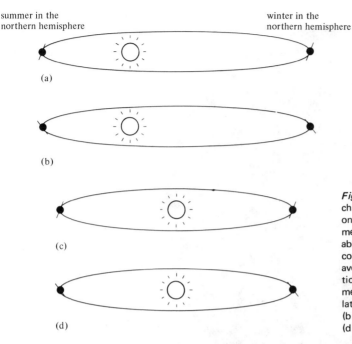

(a)

(b)

(c)

(d)

Figure 14.30 Effect of precession and changes in the ellipticity of the earth's orbit on climate: (a) Warmer-than-average summers, cooler-than-average winters — unfavorable for accumulation of glacial ice; (b) cooler-than-average summers, warmer-than-average winters — favorable for accumulation of glacial ice; (c) and (d) average summers and winters — unfavorable for accumulation of glacial ice. Note that orbits (a) and (b) are more elliptical than average; (c) and (d) are almost circular.

blages occurred in response to alternating glacial and interglacial climates and to the establishment of new connections between continents and islands. Many new forms appeared and many extinctions occurred, especially in formerly isolated groups.

Radiolaria

Radiolaria were abundant during the Pleistocene in both warm and cold waters (Fig. 14.32). Eight species of radiolarians became extinct during the Pleistocene (Fig. 14.33). It has been suggested that the extinctions were in some way caused by reversals in the earth's magnetic field. One species became extinct near the end of the Olduvai Event. Four species became extinct at the end of the Matuyama Reversed Epoch.

The magnetic field of the earth traps most of

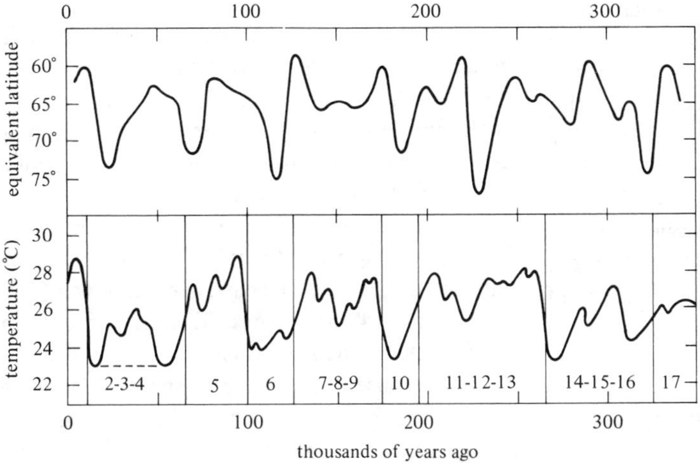

Figure 14.31 Variation of summertime insolation in the northern hemisphere compared with the temperatures of surface water of the Atlantic Ocean (From Broecker, Ref. 30; copyright © 1966 by the American Association for the Advancement of Science.)

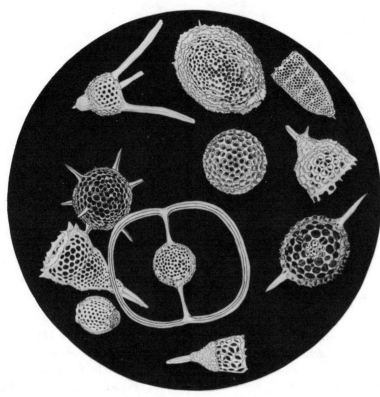

Figure 14.32 Antarctic radiolarians. (From Hays and Opdyke, Ref. 32. Copyright © 1967 by the American Association for the Advancement of Science.)

the radioactive particles and cosmic rays that bombard it, thereby preventing harmful radiation from reaching the earth's surface. During magnetic reversals, which last for about 10,000 years, the earth's magnetic field decreases to about one-quarter of its present strength (33). At such times, there is a significant increase in radiation reaching the earth's surface. The radiation would be especially intense if a large solar flare occurred during intervals of reduced magnetic field (Fig. 14.34). Whether the increases in radiation would produce an appreciable effect on evolution or extinctions has not been established (34). However, experimental results indicate that very low magnetic fields have harmful effects on a large variety of organisms (35).

Terrestrial Faunas

The Lower Pleistocene faunas of southern Europe are distinctly warm-temperature assemblages which included true elephants, rhinoceroses, cattle, and one-toed horses. The contemporaneous cold fauna was comprised mainly of mastodons, woolly mammoths, and woolly rhinoceroses (Figs. 14.35 and 14.36).

The Middle Pleistocene faunas of Europe included warm-climate forms such as straight-tusked elephants, rhinoceroses, hippopotamus, pigs, cattle, horses, bison, elk, deer, bear, a variety of rodents, and man. Cold-climate forms are similar to those of the Early Pleistocene.

The fauna of Europe during the Würm Glacial Stage was comprised of woolly mammoths, woolly rhinoceroses, large cave bears, deer, and man (Figs. 14.37 and 14.38). The cold-climate faunas of North America had many similarities to those of Eurasia because mammals were able to migrate between these continents by way of the Bering Strait during times of lowered sea level. The warm-

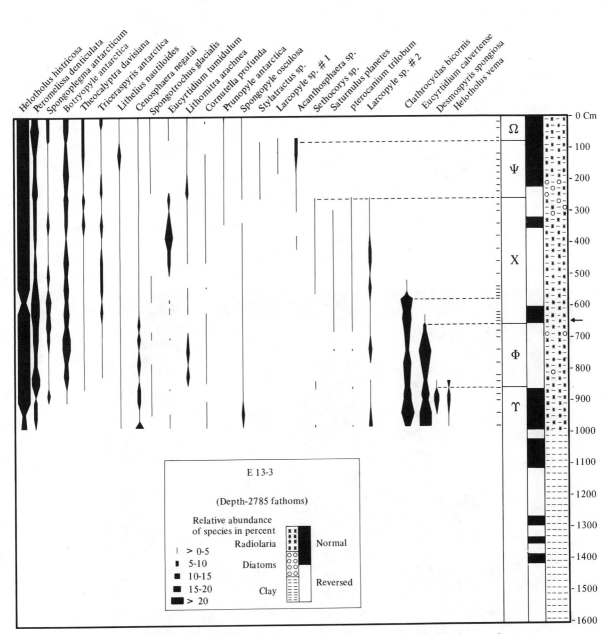

Figure 14.33   Ranges of radiolarian species correlated with magnetic stratigraphy. The black parts of the column at the left are periods of normal polarity and the white parts are periods of reversed polarity. Faunal zones at the left, designated by Greek letters, are based on the extinction or near extinction of certain species of Radiolaria. Notice how these boundaries often occur in the vicinity of magnetic reversals. (From Hays and Opdyke, Ref. 32; copyright © 1967 by the American Association for the Advancement of Science.)

climate faunas of Europe and Africa were also similar and included buffalo, gazelles, and sheep.

A remarkable record of Late Pleistocene warm-climate fauna is preserved in the La Brea tar pits in Los Angeles (Fig. 14.39). The tar pits have yielded very well-preserved bones of thousands of mammals, such as those of the imperial elephant, the saber tooth cat, wolves, lions, horses, bisons, camels, and giant ground sloths (Fig. 14.40). Radiocarbon analysis indicates that the bed (18 ft or 6 m thick) in which these fossils occur ranges in age from 20,000 to 12,000 years B.P. (36).

The connection of South America and North America allowed northward and southward migrations of two distinct faunas which resulted in mass extinctions in both groups. The South American fauna including glyptodonts, camel-like guanacos, the giant ground sloth, armadillos, opossums,

monkeys, giant anteaters, tapirs, rodents, and a marsupial, migrated northward.

Of the large mammals of the Pleistocene, most became extinct between 10,000 and 5,000 years B.P. The list includes the mammoths, mastodons, saber-tooth cats, Irish elk, and giant ground sloths. Horses and camels disappeared from North America at about this time. The bison is one of the few large mammals which persisted in North America. Changes in climate and increased competition due to migrations over land bridges may have caused some extinctions, but these factors were operative throughout the Pleistocene and it is difficult to understand why the extinctions should have been concentrated near the beginning of the Holocene. Man's increased skill as a hunter, as evidenced by the development of more efficient weapons and improved hunting techniques, may

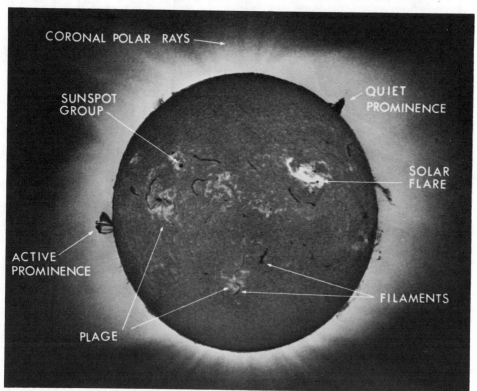

Figure 14.34 Photograph of the sun's surface taken in the light of H-alpha, showing a solar flare. Large quantities of radiation are given off by the sun during flares. (Courtesy of Lockheed Solar Observatory, Lockheed Missiles and Space Co.)

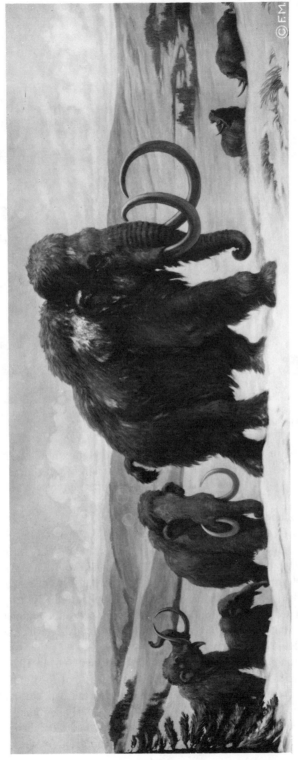

Figure 14.35 Restoration of a Pleistocene landscape in Europe showing a group of woolly mammoths in the foreground and two woolly rhinoceroses behind the mammoths at the right. (From a painting by Charles R. Knight, courtesy of the Field Museum of Natural History, Chicago.)

Figure 14.36   Restoration of a mastodon. The imperial mastodon stood nearly 14 ft (4.5 m) high at the shoulders. (Courtesy of the Field Museum of Natural History, Chicago.)

have been the principal cause of the early Holocene extinctions (37).

Terrestrial Vegetation

Floral assemblages also responded to the alternating glacial and interglacial climates. During glacial advances, the cold, temperate, and warm vegetation migrated toward the equator. When the ice receded, floral assemblages migrated back toward the poles. At the height of the glacial advance during the Late Wisconsinan (about 19,000 years B.P.) in eastern North America, oak forests extended southward to Florida, pine forests grew on the exposed southern continental shelf, spruce forests occupied the eastern and central upland of the United States, and tundra vegetation bordered the ice margin. As the ice receded northward, the climate warmed, and these plant assemblages also migrated northward. During the postglacial warm-

ing trend, vegetation now existing in the southeastern coastal plain of the United States flourished as far north as southern New York.

Man

The most remarkable form of mammalian life to evolve from the Tertiary primates is man. The record of man's evolution is gradually being reconstructed through careful search of likely sites of habitation by interdisciplinary scientific teams. The record is far from complete, but some broad evolutionary trends are evident (Fig. 14.6).

Early Pleistocene Man. In 1924 a miner sent Raymond Dart a box of fossil-bearing breccia from a cave near the railroad station at Taung, South Africa (Fig. 14.41). Among the fossils was an almost complete skull imbedded in the breccia. For many months, Dart worked to free the skull

Figure 14.37   Restoration of a Pleistocene landscape showing two cave bears. (Courtesy of the Field Museum of Natural History, Chicago.)

from the enclosing rock. The skull that emerged was that of a hominid unlike any that had yet been found. It was similar in some respects to the skull of a human, but it also resembled that of an ape (Fig. 14.42). Dart named the group represented by the skull *Australopithecus africanus* (38). From the teeth, he concluded that the individual had died around the age of five or six years. The sharp molars suggested that this hominid was a carnivore. It is interesting to note that among the modern primates, only man eats meat regularly.

Full grown, *Australopithecus africanus* would have weighed 90 lb (41 kg) and would have been about 4 ft (1.3 m) tall (Fig. 14.43). Its brain capacity would have been about 600 cc as an adult, compared with the average capacity of 1350 cc for modern man. However, since *Australopithecus* was relatively small, the ratio of brain weight to body weight was relatively large. This ratio has been estimated at 1 : 42, versus 1 : 47 in modern man. From the position of the opening at the base of the skull and from the shape of the pelvis, it was determined that *Australopithecus* walked upright. The small size of the canines and the shape of the face indicate that tools rather than teeth were used for defense.

Dart believed that *Australopithecus africanus* was a direct ancestor or at least a very close relative of modern man. Many of Dart's contemporaries did not accept this idea because they felt that Asia was the birthplace of man. Until that time the oldest known remains of the ancestors of man had been found in Java and China. But with subsequent discoveries of australopithecines in Africa, the view that man had originated in Asia was finally abandoned.

In 1936 Robert Broom found the remains of a hominid in a cave at Sterkfontein, South Africa. Although he assigned this specimen to a new genus, it is now believed to belong to the same species. In 1948 Broom found the remains of a distinctly different hominid in a cave at Swartkrans, less than a mile from Sterkfontein (39). This hominid was hardier than *Australopithecus africanus*, and Broom named it *Paranthropus robustus*. However, some investigators felt that the specimen was similar enough to *Australopithecus africanus* to be called *Australopithecus robustus*. The robust form averages about 120 lb. (54.5 kg); it has heavy bones, flattened molars, and a thick skull. Like the modern gorilla, *Australopithecus robustus* has a sagittal crest on the top of its skull. The sagittal crest was probably used for the attachment of powerful jaw muscles. The flattened molars and powerful jaws may indicate that the robust form was herbivorous.

The mammalian faunas associated with the australopithecines all lived at approximately the same time (40). The faunas have been correlated with those of the Upper Villafranchian Stage, indicating that the australopithecines of South Africa lived between 1.2 and 1.8 million years ago.

Figure 14.38   The "Irish Elk" pictured here was actually a very large deer with antlers measuring up to 12 ft (4 m) across. (From a painting by Charles R. Knight, courtesy of the Field Museum of Natural History, Chicago.)

Figure 14.39 Restoration of a Pleistocene landscape at Rancho La Brea showing the saber-toothed cat, *Smilodon*, and other animals. (Courtesy of the Field Museum of Natural History, Chicago.)

Figure 14.40   Restoration of a Pleistocene landscape showing large ground sloths and armored armadillos. (From a painting by Charles R. Knight, courtesy of the Field Museum of Natural History, Chicago.)

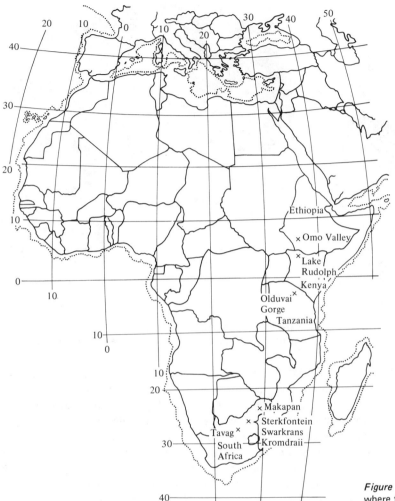

Figure 14.41 Map indicating areas where fossil remains of *Australopithecus* have been found.

Stone tools, mainly choppers and scrapers, have been found at the Swartkrans and Sterkfontein sites (41), but bone tools were also used. Dart noticed that many baboon skulls from the South African caves had a peculiar double depression. He attributed this to a blow from the humerus of an antelope — the distal end of this bone fits perfectly into the depression. The unusual abundance of such "weapons" among the fossils in the caves supports his suggestion. Other bone artifacts in the caves include slender, pointed bones that may have been produced during the extraction of bone marrow.

In another part of Africa, Louis Leakey and his wife Mary have been studying the fossiliferous deposits of Olduvai Gorge in Tanzania. In this 25-mile canyon they found many crude artifacts in the lowest exposed unit, Bed I (42). It was not until 1959 that the remains of a hominid were found. The original discovery of a nearly complete skull (Fig. 14.44) was made by Mary Leakey. Subsequent excavations have resulted in the discovery

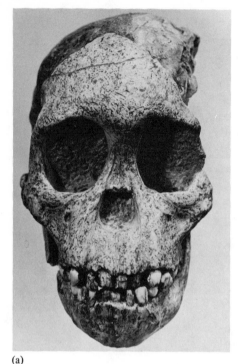

(a)

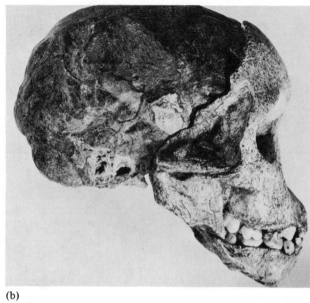

(b)

Figure 14.42   Skull of the "Taung baby" cleaned and described by
Raymond Dart. (Courtesy of P. T. Tobias and A. R. Hughes.)

of numerous bones and artifacts on a "living floor" (Fig. 14.45). The reconstructed skull resembles that of *Australopithecus robustus*, but it is somewhat more massive. Both skulls have a prominent sagittal crest. However, Louis Leakey assigned the skull to a new genus and species, *Zinjanthropus boisei*. Many authorities, however, feel that it is only a new species, *Australopithecus boisei*; others believe the resemblance to *Australopithecus robustus* does not even merit establishment of a new species.

Bed I also yielded the remains of another hominid, which Leakey named *Homo habilis* (43). The cranial volume of this hominid is 673 cc, slightly greater than the average of *Australopithecus africanus*, but well within its range. Whether this hominid is sufficiently different from *Australopithecus africanus* to be assigned to a new genus is debatable. Most workers prefer to restrict the genus *Homo* to those hominids having an average cranial volume of more than 900 cc. Tools found in Bed I at Olduvai Gorge include stone choppers and scrapers as well as polished bone scrapers. It is not clear what species made these very crude tools; they have been assigned to the Oldowan Tool Industry. Bed I, which contains a fauna similar to that of the Upper Villafranchian of southern Europe, an interbedded volcanic tuff, has been dated at approximately 1.75 million years by the potassium-argon method; a basalt flow near the base of Bed I has been dated at 1.8 million years (44).

Australopithecines have also been found near Lake Rudolph in Kenya and in the Omo Valley of Ethiopia (45). The strata in which the fossils occurred have been dated at 2.5 million years (46). Choppers associated with these remains are among the oldest known artifacts (47). In the Omo Basin, teeth and jaw fragments of hominids have been found in deposits as old as 3.1 million years (48). Teeth resembling those of the australopithecines have been found in four different horizons dated at 3.1, 2.4, 2.3, and 2.1 million years. Fragments of the jaw of *Australopithecus (Zinjan-*

Figure 14.43 Restoration of *Australopithecus* as they may have looked 3 or 4 million years ago.
[Courtesy of Zdeněk Burian. From *Prehistoric Man* by Augusta and Burian, Artia, Prague, Czechoslovakia.)

(a) (b)

Figure 14.44 Skulls of *Australopithecus* (*Zinjanthropus*) *boisei from:* (a) Olduvai Gorge, Tanzania;
(b) Lake Rudolph, Kenya. (Courtesy of R. B. Leakey.)

Figure 14.45 Fossiliferous strata near Olduvai Gorge, Tanzania. (Courtesy of R. B. Leakey.)

thropus) boisei have been found in deposits about 2.0 million years old. However, tools have not yet been found associated with these hominid fossils.

Middle Pleistocene Man. A brain case and two fossil teeth were discovered in Bed II at Olduvai Gorge in association with stone tools of the Chellean (Abbevillian) culture (43). The skull, which is relatively large and has heavy brow ridges, may represent a more advanced hominid than *Australopithecus africanus;* indeed, it may belong to a primitive variety of *Homo erectus.* The tools include hand axes, a significant advance over the choppers of the Oldowan culture. Bed II ranges in age from about 1.6 to 1.0 million years.

Eugene Dubois found the remains of a relatively advanced hominid in Middle Pleistocene deposits of Java. The average cranial volume of this hominid, known as the Java man, is between 900 and 1000 cc; the skull is characterized by massive brow ridges (Fig. 14.46). Originally designated *Pithecanthropus erectus*, this post-Villafranchian hominid is now included in the genus *Homo.* Basalt associated with the remains of *Homo erectus* has been dated at about 0.5 million years.

Remains of *Homo erectus* were found by Davidson Black in a cave near Peking, China; nearby was charcoal, which indicates that this early man probably used fire. In addition, many of the skulls have been broken open, suggesting the pos-

Figure 14.46 Restoration of *Homo erectus.* [Courtesy of Zdenêk Burian. From *Prehistoric Man* by Augusta and Burian, Artia, Prague, Czechoslovakia.)

sibility of cannibalism. A jaw fragment unearthed near Heidelberg, Germany, resembles that of *Homo erectus* and may someday prove that this hominid ranged into central Europe.

Late Pleistocene Man. *Homo sapiens* does not appear in the fossil record until about 250,000 years ago. Numerous finds in Europe document the skeletal morphology and habits of these early humans during the Riss-Würm Interglacial Stage and the Würm Glacial Stage. Skull remains of a primitive variety of *Homo sapiens* have been found near Steinheim, Germany, and Swanscombe, England, in deposits more than 200,000 years old.

Remains of a probable subspecies, Neanderthal man (*Homo sapiens neanderthalensis*), were first found in the Neander Valley in Germany. This early human lived from 110,000 to 35,000 years ago in a broad area covering much of Africa, Asia, and Europe. He had a prominent brow ridge, large jaws, and a massive, but not very erect, frame (Fig. 14.47). Neanderthal man had a very large brain

case which averaged about 1500 cc. Tools associated with Neanderthal man include carefully shaped points, hand axes, and needles (Fig. 14.48). Neanderthal man disappeared about 45,000 years ago, perhaps having been assimilated by a more advanced *Homo sapiens.*

Remains of the more modern Cro-Magnon man (*Homo sapiens sapiens*) have been found in Java and Borneo in deposits more than 40,000 years old, but he is best known from numerous sites in Europe and central Asia. He is characterized by reductions in the brow ridge and in the size of the jaw, a higher cranium, and a stature more erect than that of the Neanderthal man (Fig. 14.49). Among his more advanced cultural skills was painting. Beautiful cave drawings in southern Europe were produced between 30,000 and 9000 years ago (49) (Fig. 14.50).

The repeated glaciations resulted in man's migration through Eurasia. Eventually he reached what is now the Bering Strait and crossed into North America. Sparse evidence has come to light for the existence of man in North America prior to 11,000 years B.P. Very crude flints, reported to be artifacts, have been found in alluvial fan deposits in California dated at between 30,000 and 500,000 years B.P. (50). Artifacts have also been reported in deposits which have radiocarbon ages between 24,000 and 13,000 years B.P. (50). In Mexico an obsidian blade found beneath a log was dated at about 23,000 years B.P. In Illinois an artifact has been found in deposits predating the Late Wisconsinan glacial advance and thus perhaps more than 23,000 years old (51). Finally, a skull found near Laguna Beach, California, has been dated by the radiocarbon method at between 18,620 and 15,680 years.

There is abundant radiocarbon data to indicate that large numbers of men migrated to North America about 11,000 years ago. Migration was slowed considerably when a rise in sea level resulted in the flooding of the Bering Strait.

Figure 14.47   Reconstruction of the Neanderthal man. (Courtesy of the Field Museum of Natural History, Chicago.)

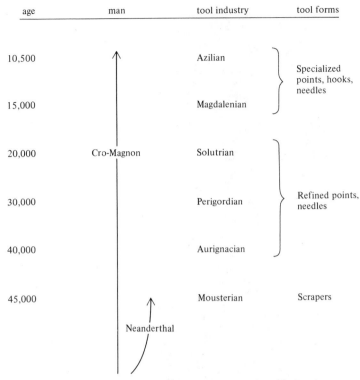

age	man	tool industry	tool forms
10,500		Azilian	Specialized points, hooks, needles
15,000		Magdalenian	
20,000	Cro-Magnon	Solutrian	
30,000		Perigordian	Refined points, needles
40,000		Aurignacian	
45,000		Mousterian	Scrapers
	Neanderthal		

Figure 14.48   Chronology of Late Pleistocene man and tool industries.

SUMMARY

The chronology of the Quaternary has been developed through the use of radiometric dating, magnetic polarity measurements, and paleoclimatic reconstructions. The Pliocene-Pleistocene boundary has been dated at about 1.8 million years at the type locality of the basal Pleistocene. Terrestrial evidence shows that at least four major glacial episodes occurred during the Pleistocene and that each of these had two or more minor advances. A study of oceanic sediments from long cores has also indicated four major cold episodes in the Pleistocene, dated at 1.8 to 1.6, 1.3 to 0.9, 0.55 to 0.38, and 0.12 to 0.01 million years ago. These cold intervals probably correlate with the four major advances of the ice sheets.

During the last cold interval, there were three separate glacial advances, reaching their maximum extent about 110,000, 50,000 and 19,000 years

ago, respectively. The final retreat of ice sheets began shortly after the maximum advance, and by 5000 years B.P. the ice sheets had disappeared from North America and Europe. Tundra vegetation and spruce and pine forests migrated behind the retreating ice.

Postglacial rebound followed the melting of the Late Wisconsinan ice sheets. Tectonism and volcanism continued throughout the Quaternary in the circum-Pacific region and in the boundary regions of the crustal plates. Strike-slip faulting was common in these regions as the Atlantic and Indian oceans widened at rates ranging from 2 to 4 cm per year.

The worldwide cooling that resulted in the Pleistocene glaciers began during the early Tertiary and may have been related to uplift of the continents, polar wandering, and continental drift. Cyclic glaciation has been attributed to changes in

Figure 14.49   Restoration of Cro-Magnon man in central Europe. (From a painting by Maurice Wilson, courtesy of the Trustees of the British Museum of Natural History.)

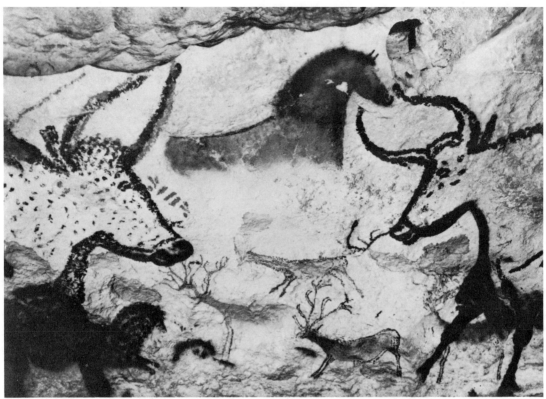

Figure 14.50   Paintings on the wall of a cave near Lascaux, France. (Courtesy of the French Embassy Press and Information Division.)

solar insolation due to changes in the angle and direction of tilt of the earth's axis and changes in the ellipticity of the earth's orbit. Alternatively, cyclic glaciation may have been related to a periodic freezing and thawing of the Arctic Ocean.

Pleistocene marine life was marked by the development of cold and warm water species which migrated in response to changing oceanic temperatures. Terrestrial faunas included mammoths, mastodons, and woolly rhinoceroses in cold climates, and elephants, rhinoceroses, horses, and cattle in warmer climates. The australopithecines which emerged during the latest Tertiary presumably gave rise to the true man, *Homo erectus*, in the Middle Pleistocene. Emerging in the Late Pleistocene, *Homo sapiens* was first represented by Neanderthal man, who appeared about 100,000 years ago. His successor, Cro-Magnon man, appeared about 50,000 years ago.

REFERENCES CITED

1. R. F. Flint, 1971, *Glacial and Quaternary Geology:* Wiley, New York.

2. C. Emiliani, 1958, Ancient temperatures: *Sci. Amer.*, v. 198, no. 2, p. 54.

3. C. Emiliani, 1966, Paleotemperature analysis of Caribbean cores P6304-8 and P6304-9 and a generalized temperature curve for the past 425,000 years: *Jr. Geol.*, v. 74, p. 109.

4. D. B. Ericson and G. Wollin, 1968, Pleistocene climate and chronology in deep-sea sediments: *Science*, v. 162, p. 1227, Dec. 13.

5. L. A. Sirkin, J. P. Owens, J. P. Minard, and M. Rubin, 1970, *Palynology of some upper*

Quaternary peat samples from the New Jersey Coastal Plain: U.S. Geological Survey Prof. Paper 700-D, p. D77.

6. F. T. Banner and W. H. Blow, 1967, The origin, evolution and taxonomy of the foraminiferal genus *Pulleniatina cushman, 1927*: *Micropaleontology,* v. 13, p. 133.

7. D. A. Emilia and D. F. Heinrichs, 1969, Ocean floor spreading: Olduvai and Gilsa Events in the Matuyama Epoch: *Science,* v. 154, p. 349.

8. L. S. B. Leakey, R. Protsch, and R. Berger, 1968, Age of Bed V, Olduvai Gorge, Tanzania: *Science,* v. 162, p. 559.

9. J. C. Frye, H. B. Willman, M. Rubin, and R. F. Black, 1968, *Definition of Wisconsinan Stage:* U.S. Geological Survey Bull. 1274-E, Washington: Government Printing Office.

10. J. R. Curray, 1965, Late Quaternary history, continental shelves of the United States, *in* H. E. Wright and D. G. Frey, eds., *The Quaternary of the United States:* Princeton University Press, Princeton, N.J., p. 723.

11. C. S. Denny, compiler, 1970, Glacial Geology of Coterminous United States, in *The National Atlas of the United States of America:* U.S. Dept. of the Interior, Geological Survey, p. 76.

12. R. B. Morrison and J. C. Frye, 1965, *Correlation of the Middle and Late Quaternary successions of the Lake Lahontan, Lake Bonneville, Rocky Mountains (Wasatch Range), southern Great Plains, and eastern midwest areas — Nevada:* Mackay School of Mines, University of Nevada.

13. R. W. Fairbridge, 1964, African Ice Age aridity, *in* A. E. M. Nairn, ed., *Problems in Paleoclimatology:* Wiley-Interscience, New York, p. 356.

14. J. H. Beard, 1969, Pleistocene paleotemperature record based on planktonic foraminifers, Gulf of Mexico: *Abstracts with Programs for 1969, Part 7,* Geological Society of America, p. 256.

15. J. P. Kennett, 1969, Foraminiferal studies of southern ocean deep-sea cores: *Antarct. Jr.,* v. 4, p. 178.

16. J. L. Lamb, 1969, Planktonic foraminiferal datums and Late Neogene Epoch boundaries in the Mediterranean, Caribbean, and Gulf of Mexico: *Abstracts with Programs for 1969,*

Part 7, Geological Society of America, p. 256.

17. O. L. Bandy and R. E. Casey, 1969, Major late Cenozoic planktonic datum planes, Antarctica to the Tropics: *Antarct. Jr.,* v. 4, p. 170.

18. C. W. Naeser, A. A. Izett, and R. E. Wilcox, 1971, Zircon fission track ages of pearlette-like volcanic ash beds in the Great Plains: *Abstracts with Programs,* v. 3, no. 7, Geological Society of America, p. 657.

19. G. A. Izett, R. E. Wilcox, J. D. Obradovich, and R. L. Reynolds, 1971, Evidence for two pearlette-like ash beds in Nebraska and adjoining areas: *Abstracts with Programs,* v. 3, no. 7, Geological Society of America, p: 265.

20. K. J. Mesolella, R. K. Matthews, W. S. Broecker, and D. L. Thurber, 1969, The astronomical theory of climatic change: Barbados Data: *Jr. Geol.,* v. 77, p. 250.

21. W. E. Frerich, 1968, Pleistocene-Recent boundary and Wisconsin glacial biostratigraphy in the northern Indian Ocean: *Science,* v. 159, p. 1456, March 29.

22. C. E. Stearns and D. L. Thurber, 1965, Th 230-U 234 dates of Late Pleistocene marine fossils from the Mediterranean and Moroccan littorals: *Quaternaria,* v. 7, p. 23.

23. J. P. Kempton and D. L. Gross, 1971, Rate of advance of the Woodfordian (Late Wisconsinan) glacial margin in Illinois: Stratigraphic and radiocarbon evidence: *Geol. Soc. Amer. Bull.,* v. 82, p. 3245.

24. G. G. Connally and L. A. Sirkin, 1972, *The Wisconsinan history of the Hudson-Champlain lobe*: Geol. Soc. Amer. Mem. 136.

25. E. H. De Geer, 1954, Skandinaviens geokronologi: *Geol. Foren. Stockholm Forh.,* v. 76, p. 303.

26. J. Lundqvist, 1965, The Quaternary of Sweden, *in* K. Rankama, ed., *The Quaternary:* Wiley-Interscience, New York, p. 139.

27. J. T. Andrews, 1969, The pattern and interpretation of restrained, post-glacial and residual rebound in the area of the Hudson Bay, *in Earth Science Symposium on Hudson Bay*, Ottawa, 1968: Canada Geological Survey Paper 65-53, p. 49.

28. W. B. Wright, 1937, *The Quaternary Ice Age:* Macmillan, London.

29. E. Dorf, 1964, The use of fossil plants as paleoclimatic indicators, *in* A. E. M. Nairn, ed., *Problems in Paleoclimatology:* Wiley–Interscience, London, p. 13.

30. W. S. Broecker, 1966, Absolute dating and the astronomical theory of glaciation: *Science,* v. 151, p. 299, Jan. 21.

31. W. L. Donn and M. Ewing, 1968, The theory of an ice-free Arctic Ocean: *Meteorol. Monogr.,* v. 8, p. 100.

   W. L. Donn and M. Ewing, 1966, A theory of ice ages III: *Science,* v. 152, p. 1706.

32. J. D. Hays and N. D. Opdyke, 1967, Antarctic radiolaria, magnetic reversals, and climatic change: *Science,* v. 158, p. 1001, Nov. 24.

33. P. J. Smith, 1967, The intensity of the ancient geomagnetic field: A review and analysis: *Geophys. Jr., Roy. Astron. Soc.,* v. 12, p. 321.

34. C. J. Waddington, 1967, Paleomagnetic field reversals and cosmic radiation: *Science,* v. 158, p. 913.

35. I. K. Crain, 1971, Possible direct causal relation between geomagnetic reversals and biological extinctions: *Geol. Soc. Amer. Bull.,* v. 82, p. 2603.

36. T. Y. Ho, L. F. Marcus, and R. Berger, 1969, Radiocarbon dating of petroleum-impregnated bone from tar pits at Rancho La Brea, California: *Science,* v. 164, p. 1051.

37. P. S. Martin and H. E. Wright, Jr., eds., 1967, *Pleistocene Extinctions:* Yale University Press.

38. R. A. Dart, 1925, *Australopithecus africanus:* Man-ape of S. Africa: *Nature,* v. 115, p. 195.

39. R. Broom and J. T. Robinson, 1952, *The Swartkrans ape-men:* Memoir, Transvaal Museum.

40. M. P. Oakley, 1954, Dating of Australopithecinae: *Amer. Jr. Phys. Anthro.,* v. 12, p. 9.

   R. F. Ewer, 1956, The dating of the Australopithecinae: Faunal evidence: *So. Afr. Archaeol. Soc. Bull.,* v. 5, p. 41.

   H. B. S. Cooke, 1963, Pleistocene mammal faunas of Africa, with particular reference to Southern Africa, *in* F. C. Howell and F. Bouliere, eds., *African Ecology and Human Evolution:* Aldine Publishing Co., Chicago.

41. J. T. Robinson, 1970, New finds at the Swartkrans australopithecine site: *Nature,* v. 225, p. 1217.

   R. J. Clarke, F. Clark Howell, and C. K. Brain, 1970, More evidence of an advanced hominid at Swartkrans: *Nature,* v. 225, p. 1219.

   M. D. Leakey, 1970, Stone artifacts from Swartkrans: *Nature,* v. 225, p. 1222.

42. L. S. B. Leakey, 1965, *A Preliminary Report on the Geology and Fauna: Olduvai Gorge 1951-1961:* v. 1, Cambridge, University Press.

   L. S. B. Leakey, 1951, *A Report on the Evolution of the Hand-Axe Culture in Beds I-IV: Olduvai Gorge:* Cambridge, University Press.

43. L. S. B. Leakey, 1961, New finds at Olduvai Gorge: *Nature,* v. 189, p. 649.

44. L. S. B. Leakey, J. F. Evernden, and G. H. Curtis, 1961, Age of Bed I, Olduvai Gorge, Tanganyika: *Nature,* v. 191, p. 478.

45. R. E. F. Leakey, 1970, New hominid remains and early artifacts from northern Kenya: *Nature,* v. 226, p. 223.

46. F. J. Fitch and J. A. Miller, 1970, Radioisotopic age determinations of Lake Rudolf artifact site: *Nature,* v. 226, p. 226.

47. M. D. Leakey, 1970, Early artifacts from the Koobi Fora area: *Nature,* v. 226, p. 228.

48. F. Clarke Howell, 1969, Remains of Hominidae from Pliocene/Pleistocene formations in the lower Omo Basin, Ethiopa: *Nature,* v. 223, p. 1234.

   F. Clarke Howell, 1968, Omo research expedition: *Nature,* v. 219, p. 567.

49. P. V. D. Stern, 1969, *Prehistoric Europe:* Norton, New York.

50. C. V. Haynes, 1969, The earliest Americans: *Science,* v. 166, p. 709.

51. P. J. Munson, 1965, Artifact from deposits of mid-Wisconsin age in Illinois: *Science,* v. 150, p. 1722.

Appendix A

Radiometric Dates on the Phanerozoic Time Scale[a]

Era	General (Period)	General (Epoch)	North America	Europe	Europe (sub)	Age of the Beginning (M.Y.)
Cenozoic	Quaternary	Pleistocene and Recent				1.8
Cenozoic	Tertiary	Pliocene				7
Cenozoic	Tertiary	Miocene		Pontian		
Cenozoic	Tertiary	Miocene		Sarmatian		12
Cenozoic	Tertiary	Miocene		Tortonian		
Cenozoic	Tertiary	Miocene		Helvetian		18.5
Cenozoic	Tertiary	Miocene		Burdigalian		
Cenozoic	Tertiary	Miocene		Aquitanian		26
Cenozoic	Tertiary	Oligocene		Chattian		?
Cenozoic	Tertiary	Oligocene		Rupelian		31.5
Cenozoic	Tertiary	Oligocene		Tongrian		37.5
Cenozoic	Tertiary	Eocene	Jacksonian	Ludian		
Cenozoic	Tertiary	Eocene	Jacksonian	Bartonian		45
Cenozoic	Tertiary	Eocene	Claibornian	Auversian		
Cenozoic	Tertiary	Eocene	Claibornian	Lutetian		49
Cenozoic	Tertiary	Eocene	Wilcoxian	Cuisian		
Cenozoic	Tertiary	Eocene	Wilcoxian	Ypresian		53.5
Cenozoic	Tertiary	Paleocene	Midwayan	Thanetian		58.5
Cenozoic	Tertiary	Paleocene	Midwayan	Montian		65
Mesozoic	Cretaceous	Upper Cretaceous	Laramian	Danian		
Mesozoic	Cretaceous	Upper Cretaceous	Montanan	Maestrichtian		70
Mesozoic	Cretaceous	Upper Cretaceous	Montanan	Senonian	Campanian	76
Mesozoic	Cretaceous	Upper Cretaceous	Montanan	Senonian	Santonian	82
Mesozoic	Cretaceous	Upper Cretaceous	Coloradoan	Senonian	Coniacian	88
Mesozoic	Cretaceous	Upper Cretaceous	Coloradoan	Turonian		94
Mesozoic	Cretaceous	Upper Cretaceous	Dakotan	Cenomanian		100
Mesozoic	Cretaceous	Upper Cretaceous	Washitan	Cenomanian		
Mesozoic	Cretaceous	Lower Cretaceous	Fredericksburgian	Albian		106
Mesozoic	Cretaceous	Lower Cretaceous	Trinitian	Aptian		112
Mesozoic	Cretaceous	Lower Cretaceous		Neocomian	Barremian	118
Mesozoic	Cretaceous	Lower Cretaceous		Neocomian	Hauterivian	124
Mesozoic	Cretaceous	Lower Cretaceous		Neocomian	Valanginian	130
Mesozoic	Cretaceous	Lower Cretaceous		Neocomian	Ryazanian	136
Mesozoic	Jurassic	Upper Jurassic		Purbeckian		141
Mesozoic	Jurassic	Upper Jurassic		Portlandian (Tithonian)		146
Mesozoic	Jurassic	Upper Jurassic		Kimmeridgian		151
Mesozoic	Jurassic	Upper Jurassic		Oxfordian		157
Mesozoic	Jurassic	Upper Jurassic		Callovian		162
Mesozoic	Jurassic	Middle Jurassic		Bathonian		167
Mesozoic	Jurassic	Middle Jurassic		Bajocian		172
Mesozoic	Jurassic	Lower Jurassic		Toarcian		178
Mesozoic	Jurassic	Lower Jurassic		Pliensbachian		183
Mesozoic	Jurassic	Lower Jurassic		Sinemurian		188
Mesozoic	Jurassic	Lower Jurassic		Hettangian		192.5
Mesozoic	Triassic	Upper Triassic		Rhaetian		
Mesozoic	Triassic	Upper Triassic		Norian		
Mesozoic	Triassic	Upper Triassic		Karnian		205
Mesozoic	Triassic	Middle Triassic		Ladinian		
Mesozoic	Triassic	Middle Triassic		Anisian		215
Mesozoic	Triassic	Lower Triassic		Scythian	Olenekian	
Mesozoic	Triassic	Lower Triassic		Scythian	Indulan	225

Era	Period	Subperiod	Series (General)	North America	Europe		Age of the Beginning (M. Y.)
Paleozoic	Permian		Upper Permian	Ochoan	Chideruan (Tartarian)		230
				Guadalupian	Kazanian		240
					Kungurian		256.5
			Lower Permian	Leonardian	Artinskian		266.5
				Wolfcampian	Sakmarian		280
	Carboniferous	Pennsylvanian	Upper Pennsylvanian	Virgilian	Stephanian	Uralian	
				Missourian		Gshelian	292.5
			Middle Pennsylvanian	Desmoinesian	Moscovian (Westphalian)		
				Atokan			312.5
			Lower Pennsylvanian	Morrowan			
				Springeran	Namurian		325
		Mississippian	Upper Mississippian	Chesteran	Viséan		337.5
				Meramecian			
			Lower Mississippian	Osagian	Tournaisian		345
				Kinderhookian	Etroeungtian		
	Devonian		Upper Devonian	Conewangoan	Famennian		353
				Cassadagan			
				Chemungian	Frasnian		359
				Fingerlakesian			
			Middle Devonian	Taghanican	Givetian		
				Tioughniogan			
				Cazenovian	Eifelian		370
			Lower Devonian	Onesquethawan	Coblenzian	Emsian	374
				Deerparkian		Siegenian	390
				Helderbergian	Gedinnian		395
	Silurian		Upper Silurian (Cayugan)	Keyseran	Dowtonian		
				Tonolowayan	Ludlovian		
				Salinan			
			Middle Silurian (Niagaran)	Lockportian			
				Cliftonian	Wenlockian		
				Clintonian	Llandoverian		
			Lower Silurian	Alexandrian			435
	Ordovician		Upper Ordovician (Cincinnatian)	Richmondian	Ashgillian		
				Maysvillian	Caradocian		445
				Edenian			
			Middle Ordovician (Mohawkian)	Trentonian			
				Blackriveran			
				Chazyan	Llandeilian		
					Llanvirian		
			Lower Ordovician	Canadian	Arenigian (Skiddavian)		500
					Tremadocian		
	Cambrian		Upper Cambrian (Croixian)	Trempealeauan	(Lingula Flags)		515
				Franconian			
				Dresbachian			
			Middle Cambrian	Albertan	Menevian		540
					Solvan		
			Lower Cambrian	Waucoban	Comleyan		570

[a] Dates from W. B. Harland, A. G. Smith, and B. Wilcock, Eds., 1964. The Phanerozoic Time-scale: A symposium: Geol. Soc. London, v. 120s.

Appendix B
The Major Fossil Groups

*M*any organisms are entirely soft-bodied and are not generally preserved as fossils. The following brief discussion will be restricted to those groups which have left a significant fossil record.

THE PROTISTS

The single-celled protists are the oldest and most primitive of the fossil groups. Most of these organisms can only be seen with a microscope, but some are almost an inch (several centimeters) in size. Nearly all reproduce asexually by splitting in two.

Modern blue-green algae consist of simple cells joined into long filaments. The filaments are not generally preserved as fossils, but some varieties of blue-green algae precipitate calcium carbonate in concentrically laminated structures called stromatolites (Fig. 9.30).

Diatoms, a more advanced variety of algae, secrete minute siliceous skeletons (Fig. B.1). They appear much later in the fossil record than blue-green algae, and they are present today in both marine and freshwater environments. Locally, they are abundant enough to form a diatomaceous ooze. The consolidation of such an ooze produces a rock called diatomite.

The skeletons of coccoliths are typically very small, discoid in shape, and composed of calcium carbonate (Fig. 12.38). Coccoliths are temperature sensitive and have been used as temperature indicators in oceanic sediments. Their skeletons are important constituents of some marine sediments, especially chalk.

The dinoflagellates are solitary, planktonic marine organisms that may have a shell composed of silica or calcium carbonate (Fig. B.2). The shell is characterized by the presence of an equatorial band girdling it. Dinoflagellates are useful in

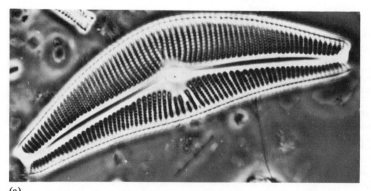

(a)   (b)

Figure B.1   Diatoms in a core from Pickerel Lake, northwestern South Dakota. (a) *Cymbella mexicanum*, X1000, (b) *Stephanodiscus niagarae*, X1000. (From Haworth, Ref. 1.)

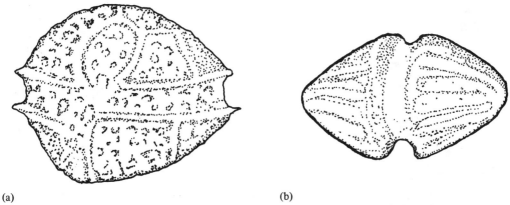

(a)   (b)

Figure B.2 Dinoflagellates. (a) *Gymondinium*, Jurassic to Recent, (b) *Peridinites*, Tertiary. These forms range in size from a few microns to 100 microns. (From Jones, Ref. 2.)

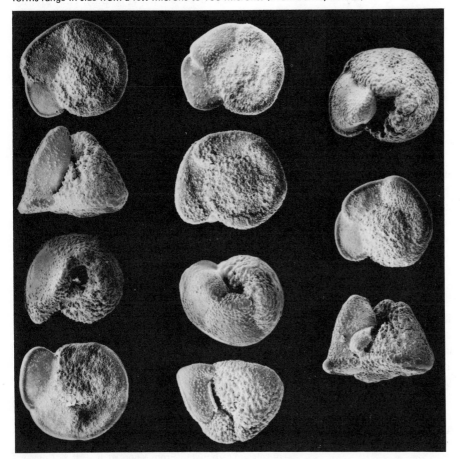

Figure B.3 Scanning electron micrographs of foraminifera from the South Pacific. Top and bottom row, *Globorotalia truncatulinoides*; middle row, *Globorotalia tosaensis*. (From Kennett, Ref. 3.)

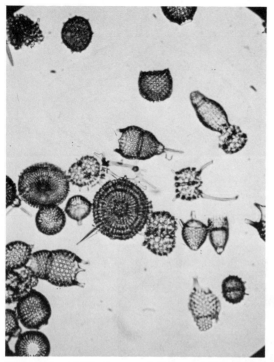

Figure B.4 Middle Eocene radiolarian assemblage from the western tropical Pacific. (Courtesy of Annika Sanfilippo.)

making correlations of marine deposits.

The animal-like protozoans include foraminiferans and radiolarians (Figs. B.3 and B.4). Both are abundant in modern oceans and judging from their fossil records, they were abundant in Paleozoic, Mesozoic, and Cenozoic oceans as well. Most foraminiferal tests are composed of calcite, but some are composed of chitin, an organic cellulose-like substance. Foraminifera are widely used as guide fossils, especially in deposits that lack macrofossils. The radiolarian test is generally composed of silica and is sometimes in the form of concentric shells. Both Foraminifera and Radiolaria are important rock formers.

MULTICELLULAR ANIMALS

Archaeocyathans

The archaeocyathans are an extinct group of invertebrates which had a cone-shaped, double-walled, porous skeleton (Fig. B.5). The affinity of this group is somewhat uncertain. These organisms have been included with the sponges and with the corals, but it is now generally agreed that they should be placed in a separate phylum. They were abundant enough in Early and Middle Cambrian seas to form reeflike deposits. Because of their limited time range, archaeocyathans are very useful as guide fossils.

Sponges

Sponges are sessile (attached) organisms which live mainly in shallow, marine environments both as individuals and in colonies (Fig. B.6). They are the most primitive of the multicellular animals and possess no nervous tissue, circulatory system, or digestive system. The body walls contain numerous small openings through which water flows into a series of canals in the wall of the sponge. After food particles are removed from the water in the

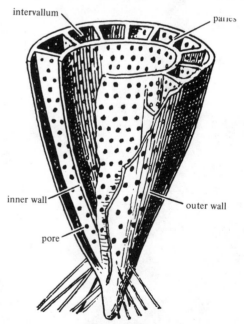

Figure B.5 Diagrammatic sketch of an archaeocyathan. These fossils have porous inner and outer walls composed of calcium carbonate. (From Moore et al., Ref. 4. Copyright © 1952 by McGraw-Hill, by permission of McGraw-Hill Book Co.)

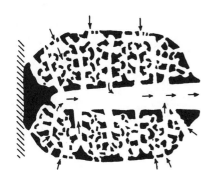

Figure B.6   Diagrammatic cross section of a sponge. Arrows indicate direction of water circulation. (After de Laubenfels, Ref. 5.)

canals, the water is forced into a central cavity and out of an opening in the top of the organism. A network of organic, calcareous, or siliceous spicules imbedded within the tissue walls acts as a skeletal framework. Sponges may be preserved as casts or molds, but commonly only the more resistant spicules are preserved.

Coelenterates

The coelenterates include hydras, jellyfish, corals, and stromatoporoids. This group is characterized by tissues that are not differentiated into organs. Most coelenterates have stinging tentacles that are used to seize and kill small organisms for food.

Jellyfish are free-swimming forms that lack a rigid skeleton. Although they lack hard parts, molds and carbon films of jellyfishlike forms have been preserved as fossils.

Corals have been important reef-building organisms since the early Paleozoic. They occur as individuals (solitary corals) and in colonies that may be several feet across (Figs. B.7 and B.8). The fleshy part of the coral, the *polyp*, possesses a number of armlike tentacles which surround a slitlike mouth (Fig. B.9). Most corals have a hard, calcareous exoskeleton that is the only part of the coral preserved in the fossil record. Sea anemones, which belong to the same class as the stony corals, do not have an exoskeleton.

In many corals, calcareous *septa* radiate inward from the outer walls or *theca* of the

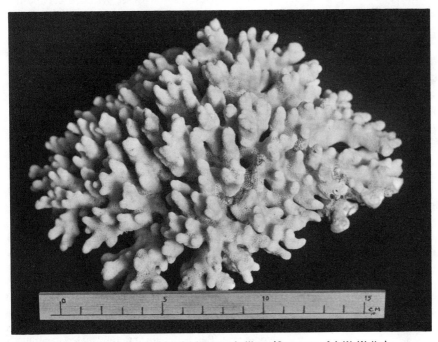

Figure B.7   Modern colonial coral. *Stylopora pistillate.* (Courtesy of J. W. Wells.)

Figure B.8   Modern solitary coral.

exoskeleton (Fig. B.10). There are four primary septa in the Paleozoic corals of the order Rugosa, whereas there are six primary septa in the Meso-zoic and Cenozoic hexacorals of the order Scleractinia. Additional septa are added between the primary septa so that septa occur in multiples of four in the rugose corals and in multiples of six in the hexacorals. Tabulate corals have platforms or *tabulae* within the skeleton that are added during successive growth stages of the coral. Other corals have a series of bubblelike partitions called *dissepiments* within the skeleton.

Bryozoans

Almost all bryozoans are colonial organisms. Some secrete a branchlike calcareous exoskeleton, while others occur as incrustations on other marine forms. The bryozoan polyps bear a superficial resemblance to coral polyps, but are generally smaller and more highly developed (Fig. B.11). Bryozoan polyps have more complex nervous and digestive systems than the corals.

Brachiopods

The brachiopods are anatomically related to the

bryozoans, but their skeletal morphology is totally different. The shells of brachiopods consist of two unequal valves. The valves are symmetrical about a plane perpendicular to the hinge line (Fig. B.12). Living brachiopods are attached to the sea floor by means of a fleshy stalk called the *pedicle* that protrudes through an opening in the pedicle valve of the shell (Fig. B.13). The other valve, known as

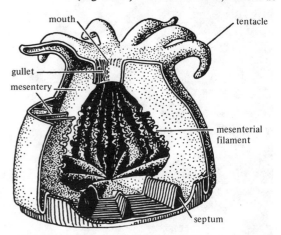

Figure B.9   Diagrammatic sketch of a coral polyp. (From Moore et al., Ref. 3, Copyright © 1952 by McGraw-Hill, Inc. Used by permission of McGraw-Hill Book Co.)

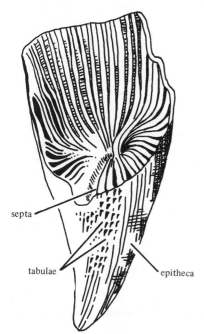

Figure B.10 Diagrammatic sketch of the exoskeleton of a rugose coral. (From D. Hill, Ref. 5.)

the brachial valve, contains the brachidium, an internal support for the animal's tissues. The external ornamentation of the shell may include concentric growth lines and *plications* (ridgelike features) radiating from the beak area.

The two classes of brachiopods, Inarticulata and Articulata, are differentiated on the basis of the manner in which their valves are connected. The valves of inarticulate brachiopods have no definite hinge structures, whereas those of the articulate brachiopods have a definite hinge line with teeth along it. Of the two classes of brachiopods, the inarticulate variety is the more primitive. Their valves are composed of chitin, whereas the valves of articulate brachiopods are generally composed of a double layer of calcite. Teeth of the pedicle valves of articulate brachiopods fit into sockets on the brachial valves. The contraction of muscles attached from the pedicle to the brachial valve opens the valves in a leverlike fashion (Fig. B.13). When the animal dies and the muscles relax, the valves tend to remain closed. Hence, fossil articulate brachiopods are often found intact.

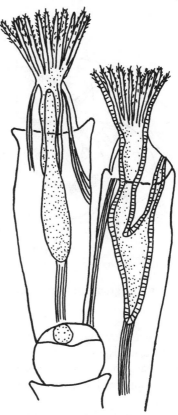

Figure B.11 Diagrammatic sketch of a bryozoan polyp. (From Bassler, Ref. 7.)

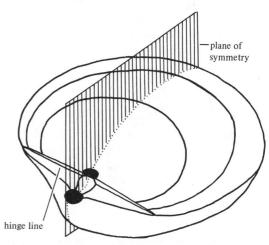

Figure B.12 Diagrammatic sketch of a brachiopod shell, dorsolateral view. (From Williams and Rowell, Ref. 8.)

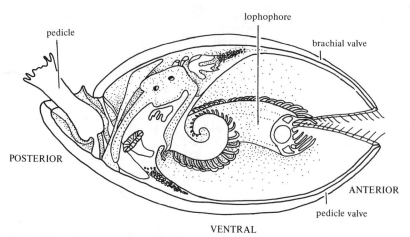

DORSAL

pedicle

lophophore

brachial valve

POSTERIOR

ANTERIOR

pedicle valve

VENTRAL

Figure B.13   Diagrammatic sketch of the internal organs of a brachiopod. (From Williams and Rowell, Ref. 8.)

There are six major groups of articulate brachiopods:

1. The orthids have a relatively straight hinge line and radial plications.
2. Strophomenids have concavo-convex shells, a straight hinge line, and relatively weak radial structures.
3. Most pentamerids have relatively smooth biconvex shells. The interior of the shell is divided near the beak with a series of radiating partitions which are used for the attachment of muscles to the shell.
4. Spiriferids have a spiral coiled internal support for the brachia.
5. Rhynchonellids generally have pointed beaks and a biconvex shell with prominent plications.
6. Most of the Terebratulids have smooth shells with a short hinge line and loop-shaped brachial supports.

Mollusks

The varied classes of mollusks have provided many important index fossils. The most important are the pelecypods (clams, muscles, oysters, and scallops), gastropods (snails), and cephalopods (octopuses and squids). Mollusks possess well-developed nervous and circulatory systems, and they are more advanced than the brachiopods.

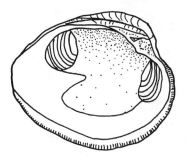

Figure B.14   Diagrammatic sketch of the shell of the pelecypod *Venus mercenaria*. (From Shrock and Twenhofel, Ref. 9, copyright © 1953 by McGraw-Hill, Inc. Used by permission of McGraw-Hill Book Co.)

Pelecypods

Pelecypods superficially resemble brachiopods in that they have two valves (Fig. B.14). However, the pelecypod shells are nearly equal in size and shape, and they are commonly symmetrical about a plane passing between the valves. Furthermore, the internal structure is entirely different from that of brachiopods. The valves of the pelecypod close by muscle contractions and open when the muscle releases. When the organism dies, the two valves of the shells open and are commonly separated. Thus, fossil pelecypods generally consist of single valves. Pelecypod shells have three layers: an outer chitinous layer, a middle layer composed of prismatic calcium carbonate, and an inner layer of laminated aragonite. The inner layer often exhibits a mother-of-pearl luster. Pelecypods are found in both salt and fresh water.

Gastropods

The gastropods are a diverse group both in morphology and in habitat. Most have coiled shells, although certain groups have reduced or eliminated their shells. They are found in marine, freshwater, and terrestrial environments. The terrestrial forms have air-breathing lungs and occupy nearly all terrestrial environments.

Cephalopods

The cephalopods are represented by both coiled and straight marine forms. They may have external shells, as in the nautiloids and ammonoids, an internal skeleton, as in belemnoids and squids, or only restricted skeletal structures, as in octopuses (Fig. B.15). Nautiloid and ammonoid shells have chambers which represent growth stages and which are separated by partitions or *septa*. A tubelike siphuncle connects the gas-filled chambers to the living animal. The cephalopod regulates its buoyancy by adjusting the amount of gas in the chambers. The line of intersection of the septa with the test is the *suture* (Fig. B.16). The

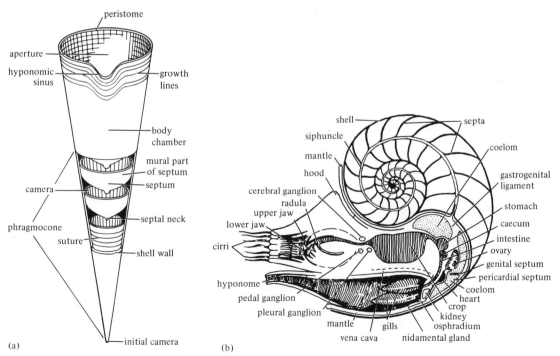

Figure B.15 Diagrammatic sketches of nautiloid cephalopods. (a) Exoskeleton of a straight nautiloid, (b) median longitudinal section of a coiled nautiloid. (From Teichert and Moore, Ref. 10.)

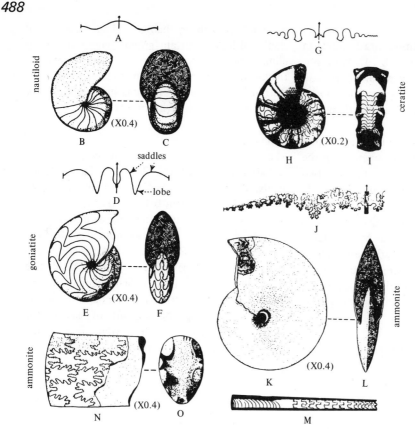

Figure B.16 Sutures of four types of cephalopods. (From Shrock and Twenhofel, Ref. 9, copyright © 1953 by McGraw-Hill, Inc. Used by permission of McGraw-Hill Book Co.)

increasing complexity of cephalopod sutures from late Paleozoic through Mesozoic time has provided a nearly continuous zonation for marine sequences.

Belemnoids are an extinct group of cephalopods which resembled the modern squid. They had a bullet-shaped internal skeleton consisting of a chambered phragmocone and a calcareous guard (Fig. 12.44). The guards are common fossils in Mesozoic rocks.

Annelids and Similar Organisms

Annelids are soft-bodied, segmented worms. Their fossil remains include carbon films on black shale, trails and burrows, and very small toothlike jaws called scolecodonts. Conodonts are similar in size and appearance to scolecodonts, but their biolog-

ical affinity is somewhat uncertain (Fig. B.17). A recent discovery of the carbonized remains of conodont animals (12) indicates that they were bilaterally symmetrical, free swimming, and probably fed on phytoplankton at or near the surface of the water. The toothlike conodonts were located within the digestive tract and may have acted as a filtering system, retaining some food particles and rejecting others. The conodont animals that have been found are about 70 mm long and they do not seem to be directly comparable to any living or fossil animal (12). They may belong to the protochordates, and may have been a direct ancestor to the vertebrates.

Arthropods

The arthropods are a relatively advanced group of

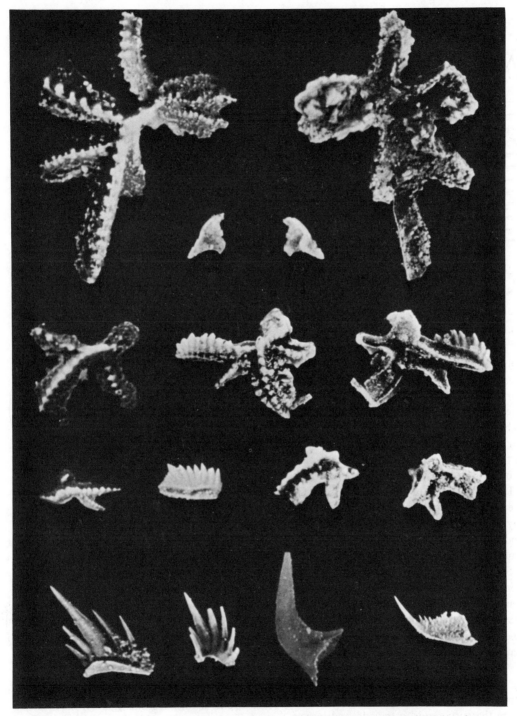

Figure B.17   Ordovician conodonts from western Newfoundland (X28). (From Fahraeus, Ref. 11.)

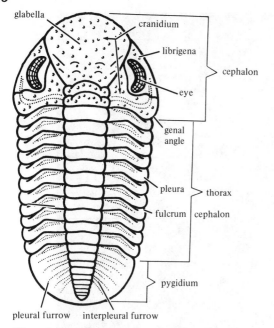

Figure B.18 Diagrammatic sketch of the exoskeleton of a trilobite. (From Harrington, Ref. 13.)

invertebrates that includes insects, spiders, scorpions, centipedes, crustaceans, and trilobites. They are characterized by the presence of a segmented exoskeleton that is composed of calcium carbonate or chitin.

For the purposes of correlation, trilobites are one of the most important fossil groups. Its name refers to the three lobes extending the length of its body. The exoskeleton of the trilobite is divided into the head (cephalon), midsection (thorax), and tail (pygidium) (Fig. B.18). Like many other arthropods, trilobites molted their shells as they grew, and therefore one organism could produce many fossil remains. Frequently, however, the cephalon, thorax, pygidium became separated either after death or after molting and complete trilobites are somewhat uncommon. Trilobites are classified according to shell morphology, such as the relative size of the cephalon and pygidium, the number of body segments, and the presence or absence of eye structures.

Other significant arthropods in the fossil record are the eurypterids, ostracods, and insects.

Eurypterids were scorpionlike forms that apparently lived in brackish water (Figs. 10.46 and 10.47). They were one of the largest of the Paleozoic invertebrates and reached a maximum length of 6 feet (2 meters). Ostracods are bivalved crustaceans whose tests range in size from a few hundredths of a millimeter to a centimeter or two (Fig. B.19). They are useful in correlations since most species have a short time span and they were widely distributed in fresh, brackish, and marine environments. Their tests are common enough and small enough to be retrieved from well cuttings and cores.

Echinoids

The echinoids are an advanced invertebrate group that typically possess a bilateral symmetry or a fivefold radial symmetry. The major repre-

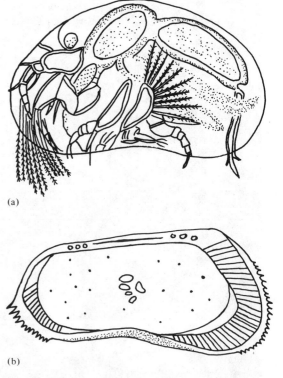

(a)

(b)

Figure B.19 Diagrammatic sketch of the modern ostracode, *Cypris*. (a) Interior anatomy, (b) interior of left valve. (From Jones, Ref. 2.)

sentative groups are the crinoids, asteroids, and echinoids.

Most crinoids, cystoids, and blastoids attach themselves to the sea bottom by rootlike appendages. The rootlike structures and the presence of a "stem" and a tentacle-bearing calyx give these organisms an appearance that is somewhat like that of a plant (Fig. B.20). The tentacles are used to trap food and are commonly arranged in multiples of five. Crinoid stems are composed of a series of disk-shaped segments with a round or star-shaped central canal. When the animal dies, the plates and disks are generally separated. Fragments of the stem are commonly preserved as fossils.

The unattached echinoids include sea urchins and sand dollars. These forms have circular or heart-shaped shells which may be covered with spines. Some spines exhibit distinctive morphological features and are useful as guide fossils.

The asteroids, which include starfish and brittle stars, are mobile benthonic forms. Most starfish have five short arms, but some have ten. Brittle stars typically have five relatively long arms. The undersides of the arms are covered with tube feet which are used for locomotion and opening pelecypod shells.

The Hemichordates

Intermediate between the invertebrates and the vertebrates are the hemichordates. These forms possess a cartilaginous, rodlike supporting structure, the *notochord*. While the notochord is not a backbone, it is believed to represent an early stage in the development of a spinal column.

Graptolites are a group of extinct colonial organisms that were formerly classified with corals and bryozoans, but modern fossil extraction methods have shown that the graptolites are closely related to certain living colonial hemichordates which have a chitinous exoskeleton. Graptolite colonies consist of a stemlike structure, the nema, with a number of cups, zoecia, that house polyplike organisms. The individuals were apparently interconnected and shared biologic functions. The

graptolite colonies consist of one or more banchlike structures called stipes that were often attached to a floating organ or to seaweed.

Vertebrates

Fish. The agnaths or jawless fishes are the most primitive known vertebrate group. They are represented today by the lamprey and hagfish. Most of the early agnaths were covered by a bony armor plate, which is generally all that is preserved of the organism.

The placoderms are an extinct group of fish that possessed simple jaws. This group included huge predatory as well as herbivorous forms.

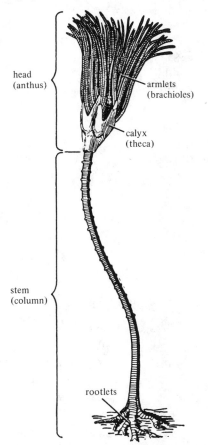

Figure B.20 Reconstruction of the blastoid *Orophocrinus* showing the principal parts. (From Fay, Ref. 14.)

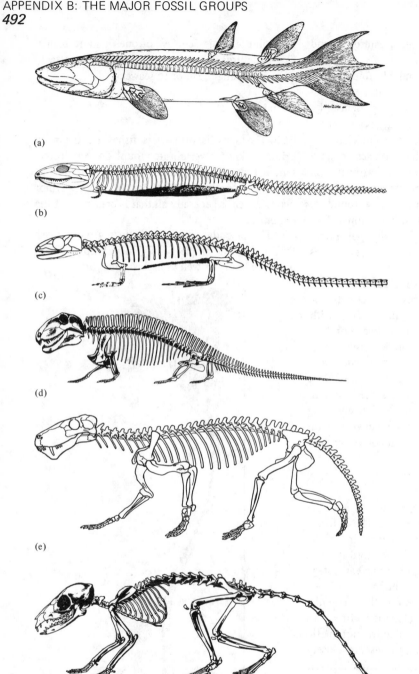

Figure B.21 Series of skeletons illustrating the evolution of placental mammals from crossopterygians: (a) Eusthenopteron, a crossopterygian, (b) *Pholidogaster*, an early labyrinthodont, (c) *Hylonomus*, one of the earliest reptiles, (d) *Sphenacodon*, a pelycosaur, (e) *Lycaenops*, a primitive therapsid, (f) *Tupaia*, a tree shrew. (From Romer, Ref. 15, copyright © 1967 by the American Association for the Advancement of Science.)

Sharks and rays are thought to stem from the placoderms. Sharks have a cartilaginous skeleton and form an important group in the modern marine environment. The bony fish include the ray-finned fish, the lobe-finned fish, and the lungfish. Of these groups, the most prolific are the ray-finned fish which have been found in a wide variety of marine and freshwater environments. This group constitutes the majority of modern fish.

Amphibians. Labyrinthodonts are a primitive group of extinct amphibians that have intricately infolded tooth enamel (Fig. 11.44). Since lobe-finned fish possess similar teeth, it is thought that the amphibians evolved from lobe-finned fish (Fig. B.21). Amphibians are cold-blooded and spend part of their lives in water. Most amphibians are covered with scales which show their close relationship to fish. They lay their eggs in water and the young hatch and develop in water. Young amphibians breathe by means of gills, but later develop lungs.

Reptiles. The reptiles emerged late in the Carboniferous through continued adaptation of the amphibian stock to a terrestrial existence. The evolutionary improvement in this case was the development of the fertilized, shelled egg (the amniote egg) from which the young hatch on dry land. The shell protects the egg from drying out and thus frees the reptiles from dependence on water. Furthermore, the reptilian young are placed out of the grasp of aquatic predators. Like amphibians, reptiles are cold blooded. They have a bony skeleton and are generally covered by scales or bony plates. Reptiles are thought to have evolved from labyrinthodont amphibians.

Birds. Birds are warm blooded animals that have such structures as wings and hollow bones that allow them to fly. They appear reptilian in that they lay eggs, but they resemble mammals in their feeding and care for their young.

Mammals. Mammals are warm-blooded animals, most of which are covered with hair or fur. Almost all mammals give birth to live young which they suckle with milk. However, there are two primitive mammalian forms that lay eggs, the duckbilled platypus and the echidna, an ant-eating mammal. The origin of the mammals is traced to mammal-like reptiles that gave rise to the monotremes, marsupials, and insectivores (Fig. B.21). The placental mammals, which give birth to live young, were derived from the insectivores, a group of primitive mammals. The insectivores were of small size, much like their modern form, the shrew.

MULTICELLULAR PLANTS
The two major groups of multicellular plants are the bryophytes and the tracheophytes. The bryophytes include mosses and liverworts. While they are not a very important group in the fossil record, the mosses are useful as ecological indicators.

The advanced or vascular plants belong to the tracheophytes. The most primitive of these, the psilopsids, were the first terrestrial plants. These have stems, but no true roots, small leaves, if any, and spore cases at the tips of the stems. The psilopsids are presumed to be ancestral to the more advanced forms, the lycopsids (club moss and scale trees), sphenopsids (horse-tails) and pteropsids (ferns). The lycopsids and pteropsids produced the first treelike plants and formed the first forests. These plants also produced large quantities of spores in their reproductive cycles.

Evolution of the higher plants came about through the development of the seed, a biological breakthrough comparable to and nearly contemporaneous with the development of the shelled egg. Important in the reproductive process in the higher plants was the modification of the spore to an entirely male reproductive product, the pollen, formed in a pollen cone. Fertilization of the egg had to occur in the female cone or flower, where the embryonic seed then developed.

The seed-bearing plants, range from the lower Carboniferous, beginning with the seed ferns which evolved from the ferns. The seed ferns, conifers and two later groups, the ginkoes and the cycads are grouped together in the gymnosperms since they share the "naked seed" characteristic.

In the most advanced plants, the angiosperms, the fertilization process was modified by flowers with both pollen and egg and seed-bearing organs. Insects were attracted to nectar-containing flowers where they aided in the pollination process. The application of pollen and spores to environmental studies has been particularly beneficial in tracing forest and grassland composition and determining sequences of climatic change, as for example, in the Pleistocene from glacial to interglacial to glacial climates.

REFERENCES CITED

1. E. Y. Haworth, 1972, Diatom succession in a core from Pickerel Lake, northeastern South Dakota: *Geol. Soc. Amer. Bull.*, v. 83, p. 157.

2. D. J. Jones, 1956, *Introduction to Microfossils*: Harper & Row, New York.

3. J. P. Kennett, 1969, Pliocene-Pleistocene boundary in a South Pacific deep-sea core: *Nature*, v. 225, p. 899.

4. R. C. Moore, C. G. Lalicker, and A. G. Fischer, 1952, *Invertebrate Fossils*: McGraw-Hill, New York.

5. N. D. de Laubenfels, 1955, Poriferat, *in* R. C. Moore, ed., *Treatise on Invertebrate Paleontology, Part E*: Geological Society of America, p. E22.

6. D. Hill, 1956, Rugosa, *in* R. C. Moore, ed., *Treatise on Invertebrate Paleontology, Part F*: Geological Society of America, p. F233.

7. R. S. Bassler, 1953, Bryozoa, *in* R. C. Moore, ed., *Treatise on Invertebrate Paleontology, Part G*: Geological Society of America, p. G1.

8. A. Williams and A. J. Rowell, 1965, Brachiopod anatomy, *in* R. C. Moore, ed., *Treatise on Invertebrate Paleontology, Part H*: Geological Society of America, p. H6.

9. R. R. Schrock and W. H. Twenhofel, 1953, *Principles of Invertebrate Paleontology*: McGraw-Hill, New York.

10. C. Teichert and R. C. Moore, 1964, Mollusca 3, *in*, R. C. Moore, ed., *Treatise on Invertebrate Paleontology, Part K*: Geological Society of America, p. K2.

11. L. E. Fahraeus, 1970, Conodont-based correlations of Lower and Middle Ordovician strata in western Newfoundland: *Geol. Soc. Amer. Bull.*, v. 81, p. 2061.

12. W. G. Melton and H. W. Scott, in press, *Conodont animals from the Bear Gulch Limestone, Montana*: Geological Society of America Special Paper 141.

13. H. J. Harrington, 1959, General description of Trilobita, *in* R. C. Moore, ed., *Treatise on Invertebrate Paleontology, Part O*: Geological Society of America, p. O38.

14. O. Fay, 1967, Introduction, *in* R. C. Moore, ed., *Treatise on Invertebrate Paleontology, Part S*: Geological Society of America, p. S298.

15. A. S. Romer, 1967, Major steps in vertebrate evolution: *Science.* v. 158, p. 1629, Dec. 29.

Index